Illustrated
Dictionary
for
Building Construction

John E. Traister

Published by
THE FAIRMONT PRESS, INC.
700 Indian Trail
Lilburn, GA 30247

Library of Congress Cataloging-in-Publication Data

Traister, John E.
    Illustrated dictionary for building construction / by John E. Traister.
    p. cm.
    ISBN 0-88173-173-0
    1. Building--Dictionaries. I. Title

TH9.T73   1993   690'.03--dc20   92-42335
                                                       CIP

Illustrated Dictionary For Building Construction by John E. Traister.

Published by The Fairmont Press, Inc.
700 Indian Trail
Lilburn, GA 30247

Printed in the United States of America

10 9 8 7 6 5 4 3 2 1

ISBN 0-88173-173-0   FP

ISBN 0-13-066903-2   PH

While every effort is made to provide dependable information, the publisher, authors, and editors cannot be held responsible for any errors or omissions.

Distributed by PTR Prentice-Hall, Inc.
A Simon & Schuster Company
Englewood Cliffs, NJ 07632

Prentice-Hall International (UK) Limited, London
Prentice-Hall of Australia Pty. Limited, Sydney
Prentice-Hall Canada Inc., Toronto
Prentice-Hall Hispanoamericana, S.A., Mexico
Prentice-Hall of India Private Limited, New Delhi
Prentice-Hall of Japan, Inc., Tokyo
Simon & Schuster Asia Pte. Ltd., Singapore
Editora Prentice-Hall do Brasil, Ltda., Rio de Janeiro

Preface

This Illustrated Dictionary for Building Construction has been prepared for the purpose of assisting engineers, architects, contractors, consultants and those involved in construction in securing an understanding of the technical terms with which they come in daily contact.

While many technical books contain glossaries of the terms mentioned in their texts, the authors, in many cases, assume that those interested should be so familiar with the terms described that definitions should not be needed, and they therefore proceed with their descriptions of the application, instead of prefacing their remarks with understandable definitions.

Definitions of many of the terms listed in this book will not be found in any technical text nor in the average dictionary even though they are, almost without exception, terms which are in use on construction projects in all sections of the United States and Canada.

A great effort has been made to keep the definitions simple, yet thorough. The many illustrations further assist the reader in understanding the more than 1000 terms.

No attempt has been made to include all the technical terms in every trade, but the purpose has been to make the scope broad enough to give the workers that knowledge of trade nomenclature which will be extremely useful in the pursuit of his or her vocation.

John Traister
Bentonville, Virginia
1993

A

Aaron's rod: An ornamental molding. It consists of a straight rod with leaves and scroll work emerging from the sides at regular intervals.

abacus: The upper part of the capital of a column, either square or curved. See Fig. A-1.

abate: To remove, as in stone carving, or hammer down, as in metal work, a portion of a surface, in order to produce a figure or pattern in low relief.

abatement: The wastage of wood when lumber is dressed to size.

abbey: A monastery or convent.

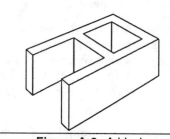

Figure A-2: A-block.

A-block: A concrete masonry unit that is hollow and has one open and one closed end with a web between. See Fig. A-2.

abrade: To scrape off; to wear off or down by friction as in erosion.

abrasion: The process of reducing material by grinding instead of cutting with tools.

abrasion resistance: The ability of a surface to resist erosion-type wear when friction between two objects happens.

abrasive: Grinding material such as sandstone, emery, carborundum, etc. The natural abrasives include the diamond, emery, corundum, sand, crushed garnet and quartz, tripoli, and pumice. The artificial abrasives are in general either silicon carbide or aluminum oxide, and are marketed under many trade names.

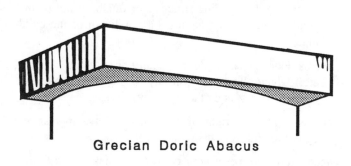

Grecian Doric Abacus

Figure A-1: Abacus.

abrasive paper: Paper or cloth on which flint, garnet, emery, or corundum has been fastened with glue or some other adhesive.

absolute: A term frequently used in the trades to indicate a thing as being perfect or exact.

absorption: The process by which a liquid or gas is drawn into a porous solid material and resulting in a physical/chemical change of the material.

ABS plastic: A strong plastic (acrylonitrile-butadiene-styrene) used for piping.

abstract of title: A condensed history of the title, consisting of a summary of the various links in the chain of title, together with a statement of all liens, charges, or encumbrances affecting a particular property.

abutment: The support of an arch, a beam, or bridge, which sustains the reaction due to the load.

ac (alternating current): 1) A periodic current, the average of which is zero over a period; normally the current reverses after given time intervals and has alternately positive and negative values. 2) The type of electrical current actually produced in a rotating generator (alternator).

acanthus: A Greek conventional leaf ornament used as a decorative feature of carved furniture, and a characteristic of the Corinthian capital. See Fig. A-3.

Figure A-3: Acanthus.

Figure A-4: Typical accent lighting. The recessed lighting fixtures are used to accent the pictures on the wall.

accelerator: 1) A substance that increases the speed of a chemical reaction. 2) Something to increase velocity.

**acceleration clause*:* A clause in a mortgage, land purchase contract or lease stating that, upon default of a payment due, the balance of the obligation should at once become due and payable.

accent lighting: Lighting used to emphasize an object or specific area. See Fig. A-4.

acceptance test: Made to demonstrate the degree of compliance with specified requirements.

**accepted*:* Approval for a specific installation or arrangement of equipment or materials.

accessible: Capable of being removed or exposed without damaging the building structure or finish, or not permanently closed in by the structure or finish of the building.

access door: A small door (usually in a closet, ceiling, etc.) which allows accessibility to provide maintenance to pipes, ducts, and other types of services.

access right: The right of an owner to have ingress and egress to and from his property.

accordion: A type of door or partition that can be folded, or opened and closed, in a manner like the operation of the bellows of the musical instrument.

Figure A-5: Acetylene torch. *Courtesy Uniweld Products, Inc.*

accretion: Addition to the land through natural causes — usually by change in water flow.

accurate: Without error; precise; correct; conforming exactly to a standard.

acetone: A highly flammble solvent which evaporates rapidly; used in lacquers, paint removers, thinners, etc. It is obtained by the destructive distillation of certain wood, acetates, and various organic compounds.

acetylene: A colorless gas with an ether-like odor, when mixed with oxygen is used in welding.

acetylene torch: An instrument used in welding and metal cutting. It is operated by compressed acetylene and oxygen. See Fig. A-5.

acid resistance: The degree to which any given surface will resist damage of an attack by acids.

acknowledgment: A formal declaration made before a notary public or other person empowered to perform the service, by the signatory to the instrument, as to the genuineness of the signature.

acoustical ceiling: A ceiling covered by an acoustical material.

acoustical ceiling system: A structural system, usually incorportating lighting fixtures, air diffusers, etc., to support an acoustical ceiling. See Fig. A-6.

acoustical tile: Special tile for walls and ceilings made of mineral, wood, vegetable fibers, cork, or metal. Its purpose is to control sound volume, while providing cover. See Fig. A-6.

acoustics: The science of sound; the study of the effects of sound upon the ear. The acoustics of a room are said to be "good" or "bad" according to the ease or clearness with which sounds are perceived by the hearers.

acre: A measure of land, 160 square rods (4,840 square yards; 43,560 square feet).

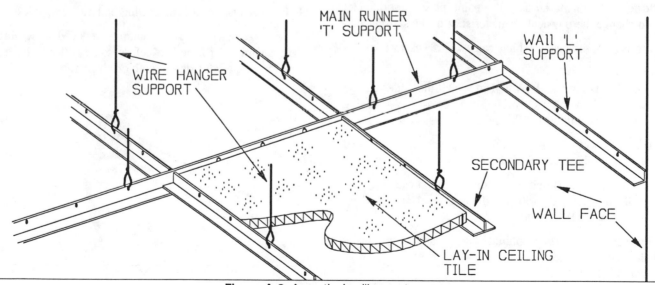

Figure A-6: Acoustical ceiling system.

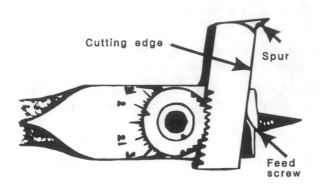

Figure A-7: Adjustable boring tool.

Adam: An English furniture style introduced by Adam brothers (1728-1792), usually decorated with Roman classic ornaments.

addendum: An addition to bidding documents for clarifying, correcting, or adding information to the specifications previously issued.

addition: Any new construction to an existing building which expands it in any way.

additive: Any material added to paint, plaster, mortar, etc. to change the physical characteristics of the substance.

adhesive: A substance which causes surfaces to stick together.

adjustable boring tool: A tool in which the cutter can be set for different jobs, avoiding the necessity of changing both cutter and holder. See Fig. A-7.

adjustable wrench: A wrench having one jaw fixed and the other adjustable. See Fig. A-8.

administrative authority: An organization exercising jurisdiction over the various codes such as National Electrical Code, OSHA, etc.

administrator: Person appointed by court to administer the estate of a deceased person who left no will; that is, who died intestate.

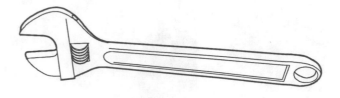

Figure A-8: Adjustable-end wrench.

adobe brick: Large, roughly molded brick made from sun-dried clay.

advance fee: A fee paid in advance of any service rendered in the sale of a property or in obtaining a loan.

adverse possession: The right of an occupant of land to acquire title against the real owner, where possession has been actual, continuous, hostile, visible, and distinct for the statutory period.

adz: A cutting tool with the blade set at right angles to the handle; used for rough-dressing timber. See Fig. A-9.

adz-eye hammer: Usually the claw-type nail hammer in which the eye is extended to give a longer bearing on the handle than is the case with other hammers.

AEIC: Association of Edison Illuminating Companies.

affidavit: A statement of declaration reduced to writing, and sworn or affirmed to before some officer who has authority to administer an oath or affirmation.

Figure A-9: Adz.

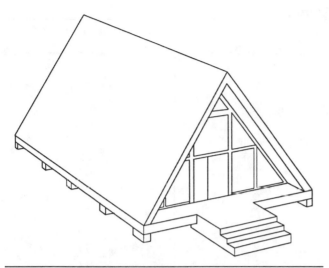

Figure A-10: A-frame building.

A-frame: A roof structure shaped like the letter A. Frequently used for vacation homes and in areas with heavy snow fall. See Fig. A-10.

African mahogany: A remote member of the mahogany family found principally in Africa. The very large tree produces exceptionally fine figured timber of unusual lengths and widths. Used for fine furniture.

agba: A large tree, native to Africa, with lightweight wood and used for plywood, interior millwork, and carpentry.

agent: One who represents another from whom he has derived authority.

agreement of sale: A written agreement whereby the purchaser agrees to buy certain real estate and the seller agrees to sell upon terms and conditions set forth therein.

aggregate: Material mixed with cement and water to produce concrete.

aging: The irreversible change of material properties after exposure to an environment for an interval of time.

agitation: A gentle motion provided to mix several substances together as in making concrete.

agreement: 1) An understanding between two or among several persons which is legally binding. 2) The document on a construction project stating the essential terms of the construction contract.

agricultural lime: A hydrated lime used on yards, gardens, etc. to condition the soil.

AIA: 1) American Institute of Architects. 2) Aircraft Industries Association.

air chamber: Extension pipes used in plumbing supply lines to prevent "waterhammer." See Fig. A-11.

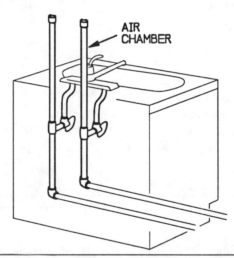

Figure A-11: Domestic cold- and hot-water plumbing pipes showing the placement of air chambers to prevent water hammer.

air changes: The number of times inside air is replaced with outside air. A table giving the recommended air changes in various occupancies is shown in Fig. A-12 on page 10.

air cleaner: Device used for removal of air-borne impurities.

air compressor: A machine that draws air in at atmospheric pressure then compresses it to a higher pressure and delivers it at a rate to operate pneumatic tools or other equipment. See Fig. A-13 on page 10.

air conditioning: The process of treating air to control its temperature, humidity, cleanliness, and distribution within an interior space.

Ventilation Standards

Application	CFM Per Person		CFM per sq. ft. of floor
	Recommended	Minimum	
Apartment	30	25	.33
Bank	10	7½	-
Barber Shops	15	10	-
Beauty Salons	10	7½	-
Cocktail Bars	30	25	-
Corridors, Public Buildings	-	-	.25
Department Stores	7½	5	.05
Drug Stores	10	7½	-
Factories	10	71/2	.10
Garage	-	-	1.0
Hospitals	30	25	.33
Hotel Rooms	30	25	.33
Kitchen, Commercial	-	-	4.0
Kitchen, Residential	-	-	2.0
Meeting Rooms	50	30	1.25
Offices	25	15	.25
Restaurant	15	12	-
Theater	15	10	-

Figure A-12: Table of recommended air changes in various occupancies.

air curtain: High velocity fans placed over outside doors to provide a steady stream of air. This fan provides a curtain of air to prevent heat loss and heat gain through the doorway, and also discourages flying insects from entering the area.

air diffuser: Air distribution outlet designed to direct air-flow into desired patterns. See *diffuser*.

air entrained concrete: Concrete in which a small amount of air is trapped by addition of a special material to produce greater durability.

air-dried: Lumber seasoned by drying in the air as opposed to kiln-dried.

air duct: A duct made from metal, fiberglass, concrete or other substance to take air from one place to another, such as from a furnace or air-conditioning system to areas to be conditioned.

Figure A-13: Air compressor. Courtesy *Sears*.

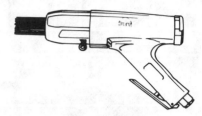

Figure A-14: Portable air hammer. *Courtesy JET Equipment & Tools.*

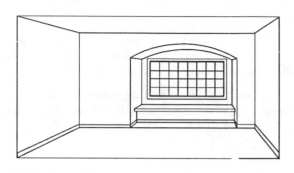

Figure A-15: Typical alcove.

air filter: A device used to remove either solid or gas pollutants from the air.

air (pneumatic) hammer: A portable, air-driven tool into which is set a chisel or hammer. See Fig. A-14.

air line: A tube or hose that supplies compressed air to a pneumatic tool.

air rights: The ownership of the right to use, control or occupy the air space over a designated property.

air terminal: A metal rod and its brace or footing on the roof of a building for lightning protection.

aisle: A passageway, as in a church or assembly room, by which the pews or seats may be reached.

aisleway: A walkway in a factory or the like to permit the flow of inside traffic.

Al-Cu: An abbreviation for aluminum and copper, commonly marked on terminals, lugs and other electrical connectors to indicate that the device is suitable for use with either aluminum conductors or copper conductors.

Al: Aluminum.

alcove: Any large recess in a room, usually separated by an arch. See Fig. A-15.

alette: An ell or wing of a building.

alienation: The transfer of real property by one person to another.

alignment: 1) An adjustment to a line. 2) Lines that establish the position of construction (such as a building) or the shape of an individual element (such as a curved or straight beam).

alive: Energized; having voltage applied.

alkali: A strong chemical base that produces hydroxl ions when dissolved in water.

alkaline soil: Soil containing soluble salts and having a pH value of between 7.3 and 8.5.

Allen head: A screw having a recess in its head shaped like a hexgon. See Fig. A-16.

Allen wrench: A wrench to fit the Allen-head screw. See Fig. A-16.

alley: A narrow passageway between or behind buildings or other properties.

allowable load: Maximum load a structure can support within specified tolerances.

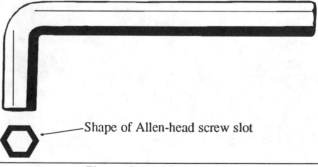

Shape of Allen-head screw slot

Figure A-16: Allen wrench.

alloy: A substance having metallic properties and being composed of elemental metal and one or more chemical elements.

alloy steel: Steel containing one or more alloying elements which have been added to impart particular physical, mechanical, or chemical properties.

alluvion: Also alluvium. Soil deposited by accretion; increase in land on shore or bank of river due to change in flow of stream.

almery: Same as ambry.

altar: In present-day use, commonly applied to the communion table in churches. Originally, a raised platform on which sacrifices or offerings were made to the gods.

alterations: Changes in the original structure or finishes.

alternate bid: A bid submitted by a contractor that deviates from the basic specifications and/or working drawings.

alternating current: An electric current that reverses in direction at rapid, regular intervals, usually 120 times per second for 60 Hz systems.

alternator: A device to produce alternating current. Often used on building sites for temporary power during construction. See Fig. A-17.

Figure A-17: Generac's home standby alternator. *Courtesy Sears.*

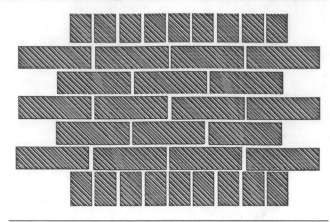

Figure A-18: American bond brick pattern.

ambient temperature: Temperature of fluid (usually air) that surrounds an object on all sides.

ambo: In early Christian churches, a raised pulpit-like platform stand, or desk, where parts of the service were read or chanted.

amboina: An East Indian tree which produces what is commonly considered the most beautiful and most expensive of cabinet woods and veneers.

ambry: A closet near the altar for sacred vessels. A cupboard.

American bond: Brickwork pattern consisting of five courses of stretchers followed by one bonding course of headers. See Fig. A-18.

American standard beam: A type of I-beam; designated by the prefix "S" placed before the size of the member. See Fig. A-19.

American standard channel: A C-shaped structural member of hot-rolled structural steel. See Fig. A-20 beginning on page 14.

American standard pipe threads: Standard pipe threads for commonly used sizes of pipe in the United States.

ammeter: An electric meter used to measure current, calibrated in amperes.

amortization: The liquidation of a financial obligation on an installment basis.

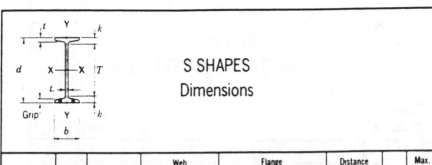

S SHAPES
Dimensions

Designation	Area A	Depth d		Web Thickness t_w		$\frac{t_w}{2}$	Flange Width b_f		Flange Thickness t_f		Distance T	k	Grip	Max. Flge. Fastener
	In.²	In.		In.		In.	In.		In.		In.	In.	In.	In.
S 24x121	35.6	24.50	24½	0.800	13/16	7/16	8.050	8	1.090	1 1/16	20½	2	1 1/8	1
x106	31.2	24.50	24½	0.620	5/8	5/16	7.870	7 7/8	1.090	1 1/16	20½	2	1 1/8	1
S 24x100	29.3	24.00	24	0.745	3/4	3/8	7.245	7 1/4	0.870	7/8	20½	1 3/4	7/8	1
x90	26.5	24.00	24	0.625	5/8	5/16	7.125	7 1/8	0.870	7/8	20½	1 3/4	7/8	1
x80	23.5	24.00	24	0.500	1/2	1/4	7.000	7	0.870	7/8	20½	1 3/4	7/8	1
S 20x96	28.2	20.30	20 1/4	0.800	13/16	7/16	7.200	7 1/4	0.920	15/16	16 3/4	1 3/4	15/16	1
x86	25.3	20.30	20 1/4	0.660	11/16	3/8	7.060	7	0.920	15/16	16 3/4	1 3/4	15/16	1
S 20x75	22.0	20.00	20	0.635	5/8	5/16	6.385	6 3/8	0.795	13/16	16 3/4	1 5/8	13/16	7/8
x66	19.4	20.00	20	0.505	1/2	1/4	6.255	6 1/4	0.795	13/16	16 3/4	1 5/8	13/16	7/8
S 18x70	20.6	18.00	18	0.711	11/16	3/8	6.251	6 1/4	0.691	11/16	15	1 1/2	11/16	7/8
x54.7	16.1	18.00	18	0.461	7/16	1/4	6.001	6	0.691	11/16	15	1 1/2	11/16	7/8
S 15x50	14.7	15.00	15	0.550	9/16	5/16	5.640	5 5/8	0.622	5/8	12 1/4	1 3/8	9/16	3/4
x42.9	12.6	15.00	15	0.411	7/16	1/4	5.501	5 1/2	0.622	5/8	12 1/4	1 3/8	9/16	3/4
S 12x50	14.7	12.00	12	0.687	11/16	3/8	5.477	5 1/2	0.659	11/16	9 1/8	1 7/16	11/16	3/4
x40.8	12.0	12.00	12	0.462	7/16	1/4	5.252	5 1/4	0.659	11/16	9 1/8	1 7/16	5/8	3/4
S 12x35	10.3	12.00	12	0.428	7/16	1/4	5.078	5 1/8	0.544	9/16	9 5/8	13/16	1/2	3/4
x31.8	9.35	12.00	12	0.350	3/8	3/16	5.000	5	0.544	9/16	9 5/8	13/16	1/2	3/4
S 10x35	10.3	10.00	10	0.594	5/8	5/16	4.944	5	0.491	1/2	7 3/4	1 1/8	1/2	3/4
x25.4	7.46	10.00	10	0.311	5/16	3/16	4.661	4 5/8	0.491	1/2	7 3/4	1 1/8	1/2	3/4
S 8x23	6.77	8.00	8	0.441	7/16	1/4	4.171	4 1/8	0.426	7/16	6	1	7/16	3/4
x18.4	5.41	8.00	8	0.271	1/4	1/8	4.001	4	0.426	7/16	6	1	7/16	3/4
S 7x20	5.88	7.00	7	0.450	7/16	1/4	3.860	3 7/8	0.392	3/8	5 1/8	15/16	3/8	5/8
x15.3	4.50	7.00	7	0.252	1/4	1/8	3.662	3 5/8	0.392	3/8	5 1/8	15/16	3/8	5/8
S 6x17.25	5.07	6.00	6	0.465	7/16	1/4	3.565	3 5/8	0.359	3/8	4 1/4	7/8	3/8	5/8
x12.5	3.67	6.00	6	0.232	1/4	1/8	3.332	3 3/8	0.359	3/8	4 1/4	7/8	3/8	—
S 5x14.75	4.34	5.00	5	0.494	1/2	1/4	3.284	3 1/4	0.326	5/16	3 3/8	13/16	5/16	—
x10	2.94	5.00	5	0.214	3/16	1/8	3.004	3	0.326	5/16	3 3/8	13/16	5/16	—
S 4x9.5	2.79	4.00	4	0.326	5/16	3/16	2.796	2 3/4	0.293	5/16	2 1/2	3/4	5/16	—
x7.7	2.26	4.00	4	0.193	3/16	1/8	2.663	2 5/8	0.293	5/16	2 1/2	3/4	5/16	—
S 3x7.5	2.21	3.00	3	0.349	3/8	3/16	2.509	2 1/2	0.260	1/4	1 5/8	11/16	1/4	—
x5.7	1.67	3.00	3	0.170	3/16	1/8	2.330	2 3/8	0.260	1/4	1 5/8	11/16	1/4	—

Figure A-19: Dimensions of S-shape steel beams. *Courtesy American Institute of Steel Construction.*

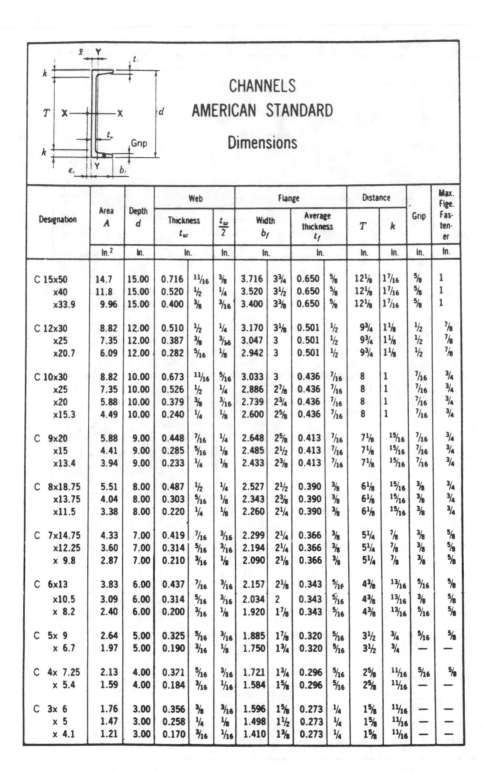

Designation	Area A	Depth d	Web Thickness t_w	Web $\frac{t_w}{2}$	Flange Width b_f		Flange Average thickness t_f		Distance T	Distance k	Grip	Max. Flge. Fastener
	In.^{2}	In.	In.	In.	In.		In.		In.	In.	In.	In.
C 15x50	14.7	15.00	0.716	11/16	3/8	3.716	3¾	0.650	5/8	12⅛	1⁷/₁₆	5/8	1
x40	11.8	15.00	0.520	½	¼	3.520	3½	0.650	5/8	12⅛	1⁷/₁₆	5/8	1
x33.9	9.96	15.00	0.400	3/8	3/16	3.400	3⅜	0.650	5/8	12⅛	1⁷/₁₆	5/8	1
C 12x30	8.82	12.00	0.510	½	¼	3.170	3⅛	0.501	½	9¾	1⅛	½	7/8
x25	7.35	12.00	0.387	3/8	3/16	3.047	3	0.501	½	9¾	1⅛	½	7/8
x20.7	6.09	12.00	0.282	5/16	⅛	2.942	3	0.501	½	9¾	1⅛	½	7/8
C 10x30	8.82	10.00	0.673	11/16	5/16	3.033	3	0.436	7/16	8	1	7/16	¾
x25	7.35	10.00	0.526	½	¼	2.886	2⅞	0.436	7/16	8	1	7/16	¾
x20	5.88	10.00	0.379	3/8	3/16	2.739	2¾	0.436	7/16	8	1	7/16	¾
x15.3	4.49	10.00	0.240	¼	⅛	2.600	2⅝	0.436	7/16	8	1	7/16	¾
C 9x20	5.88	9.00	0.448	7/16	¼	2.648	2⅝	0.413	7/16	7⅛	15/16	7/16	¾
x15	4.41	9.00	0.285	5/16	⅛	2.485	2½	0.413	7/16	7⅛	15/16	7/16	¾
x13.4	3.94	9.00	0.233	¼	⅛	2.433	2⅜	0.413	7/16	7⅛	15/16	7/16	¾
C 8x18.75	5.51	8.00	0.487	½	¼	2.527	2½	0.390	3/8	6⅛	15/16	3/8	¾
x13.75	4.04	8.00	0.303	5/16	⅛	2.343	2⅜	0.390	3/8	6⅛	15/16	3/8	¾
x11.5	3.38	8.00	0.220	¼	⅛	2.260	2¼	0.390	3/8	6⅛	15/16	3/8	¾
C 7x14.75	4.33	7.00	0.419	7/16	3/16	2.299	2¼	0.366	3/8	5¼	7/8	3/8	5/8
x12.25	3.60	7.00	0.314	5/16	3/16	2.194	2¼	0.366	3/8	5¼	7/8	3/8	5/8
x 9.8	2.87	7.00	0.210	3/16	⅛	2.090	2⅛	0.366	3/8	5¼	7/8	3/8	5/8
C 6x13	3.83	6.00	0.437	7/16	3/16	2.157	2⅛	0.343	5/16	4⅜	13/16	5/16	5/8
x10.5	3.09	6.00	0.314	5/16	3/16	2.034	2	0.343	5/16	4⅜	13/16	3/8	5/8
x 8.2	2.40	6.00	0.200	3/16	⅛	1.920	1⅞	0.343	5/16	4⅜	13/16	5/16	5/8
C 5x 9	2.64	5.00	0.325	5/16	3/16	1.885	1⅞	0.320	5/16	3½	¾	5/16	5/8
x 6.7	1.97	5.00	0.190	3/16	⅛	1.750	1¾	0.320	5/16	3½	¾	—	—
C 4x 7.25	2.13	4.00	0.371	5/16	3/16	1.721	1¾	0.296	5/16	2⅝	11/16	5/16	5/8
x 5.4	1.59	4.00	0.184	3/16	1/16	1.584	1⅝	0.296	5/16	2⅝	11/16	—	—
C 3x 6	1.76	3.00	0.356	3/8	3/16	1.596	1⅝	0.273	¼	1⅝	11/16	—	—
x 5	1.47	3.00	0.258	¼	⅛	1.498	1½	0.273	¼	1⅝	11/16	—	—
x 4.1	1.21	3.00	0.170	3/16	1/16	1.410	1⅜	0.273	¼	1⅝	11/16	—	—

Figure A-20: Dimensions of American standard steel channel. *Courtesy American Institute of Steel Construction.*

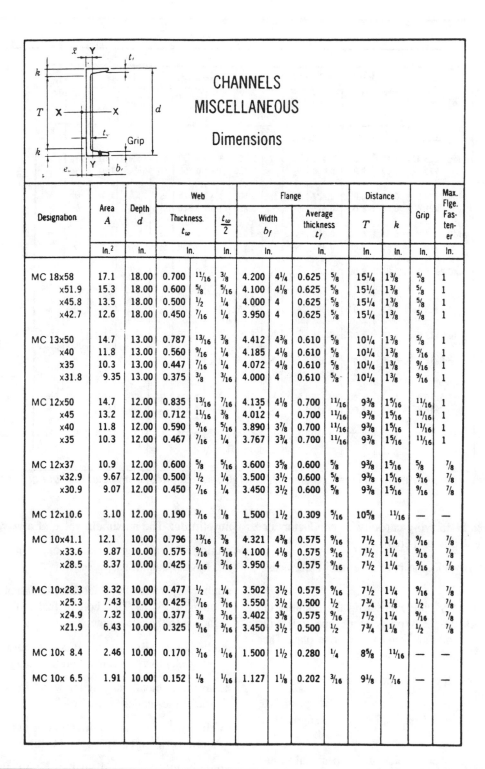

CHANNELS
MISCELLANEOUS
Dimensions

Designation	Area A	Depth d	Web Thickness t_w	$\frac{t_w}{2}$	Flange Width b_f		Average thickness t_f		Distance T	k	Grip	Max. Flge. Fastener
	In.²	In.	In.	In.	In.		In.		In.	In.	In.	In.
MC 18x58	17.1	18.00	0.700	¹¹/₁₆	³/₈	4.200	4¼	0.625	⁵/₈	15¼	1³/₈	⁵/₈	1
x51.9	15.3	18.00	0.600	⁵/₈	⁵/₁₆	4.100	4¹/₈	0.625	⁵/₈	15¼	1³/₈	⁵/₈	1
x45.8	13.5	18.00	0.500	½	¼	4.000	4	0.625	⁵/₈	15¼	1³/₈	⁵/₈	1
x42.7	12.6	18.00	0.450	⁷/₁₆	¼	3.950	4	0.625	⁵/₈	15¼	1³/₈	⁵/₈	1
MC 13x50	14.7	13.00	0.787	¹³/₁₆	³/₈	4.412	4³/₈	0.610	⁵/₈	10¼	1³/₈	⁵/₈	1
x40	11.8	13.00	0.560	⁹/₁₆	¼	4.185	4¹/₈	0.610	⁵/₈	10¼	1³/₈	⁹/₁₆	1
x35	10.3	13.00	0.447	⁷/₁₆	¼	4.072	4¹/₈	0.610	⁵/₈	10¼	1³/₈	⁹/₁₆	1
x31.8	9.35	13.00	0.375	³/₈	³/₁₆	4.000	4	0.610	⁵/₈	10¼	1³/₈	⁹/₁₆	1
MC 12x50	14.7	12.00	0.835	¹³/₁₆	⁷/₁₆	4.135	4¹/₈	0.700	¹¹/₁₆	9³/₈	1⁵/₁₆	¹¹/₁₆	1
x45	13.2	12.00	0.712	¹¹/₁₆	³/₈	4.012	4	0.700	¹¹/₁₆	9³/₈	1⁵/₁₆	¹¹/₁₆	1
x40	11.8	12.00	0.590	⁹/₁₆	⁵/₁₆	3.890	3⁷/₈	0.700	¹¹/₁₆	9³/₈	1⁵/₁₆	¹¹/₁₆	1
x35	10.3	12.00	0.467	⁷/₁₆	¼	3.767	3³/₄	0.700	¹¹/₁₆	9³/₈	1⁵/₁₆	¹¹/₁₆	1
MC 12x37	10.9	12.00	0.600	⁵/₈	⁵/₁₆	3.600	3⁵/₈	0.600	⁵/₈	9³/₈	1⁵/₁₆	⁵/₈	⁷/₈
x32.9	9.67	12.00	0.500	½	¼	3.500	3½	0.600	⁵/₈	9³/₈	1⁵/₁₆	⁹/₁₆	⁷/₈
x30.9	9.07	12.00	0.450	⁷/₁₆	¼	3.450	3½	0.600	⁵/₈	9³/₈	1⁵/₁₆	⁹/₁₆	⁷/₈
MC 12x10.6	3.10	12.00	0.190	³/₁₆	⅛	L.500	1½	0.309	⁵/₁₆	10⁵/₈	¹¹/₁₆	—	—
MC 10x41.1	12.1	10.00	0.796	¹³/₁₆	³/₈	4.321	4³/₈	0.575	⁹/₁₆	7½	1¼	⁹/₁₆	⁷/₈
x33.6	9.87	10.00	0.575	⁹/₁₆	⁵/₁₆	4.100	4¹/₈	0.575	⁹/₁₆	7½	1¼	⁹/₁₆	⁷/₈
x28.5	8.37	10.00	0.425	⁷/₁₆	³/₁₆	3.950	4	0.575	⁹/₁₆	7½	1¼	⁹/₁₆	⁷/₈
MC 10x28.3	8.32	10.00	0.477	½	¼	3.502	3½	0.575	⁹/₁₆	7½	1¼	⁹/₁₆	⁷/₈
x25.3	7.43	10.00	0.425	⁷/₁₆	³/₁₆	3.550	3½	0.500	½	7³/₄	1¹/₈	½	⁷/₈
x24.9	7.32	10.00	0.377	³/₈	³/₁₆	3.402	3³/₈	0.575	⁹/₁₆	7½	1¼	⁹/₁₆	⁷/₈
x21.9	6.43	10.00	0.325	⁵/₁₆	³/₁₆	3.450	3½	0.500	½	7³/₄	1¹/₈	½	⁷/₈
MC 10x 8.4	2.46	10.00	0.170	³/₁₆	¹/₁₆	1.500	1½	0.280	¼	8⁵/₈	¹¹/₁₆	—	—
MC 10x 6.5	1.91	10.00	0.152	⅛	¹/₁₆	1.127	1¹/₈	0.202	³/₁₆	9¹/₈	⁷/₁₆	—	—

Figure A-20: Dimensions of American standard steel channel (Cont.). *Courtesy American Institute of Steel Construction.*

Designation	Area A (In.²)	Depth d (In.)	Web Thickness t_w (In.)		$\frac{t_w}{2}$ (In.)	Flange Width b_f (In.)		Average thickness t_f (In.)		T (In.)	k (In.)	Grip (In.)	Max. Flge. Fastener (In.)
MC 9x25.4	7.47	9.00	0.450	7/16	1/4	3.500	3½	0.550	9/16	6⅝	1³/₁₆	9/16	7/8
x23.9	7.02	9.00	0.400	3/8	3/16	3.450	3½	0.550	9/16	6⅝	1³/₁₆	9/16	7/8
MC 8x22.8	6.70	8.00	0.427	7/16	3/16	3.502	3½	0.525	½	5⅝	1³/₁₆	½	7/8
x21.4	6.28	8.00	0.375	3/8	3/16	3.450	3½	0.525	½	5⅝	1³/₁₆	½	7/8
MC 8x20	5.88	8.00	0.400	3/8	3/16	3.025	3	0.500	½	5¾	1⅛	½	7/8
x18.7	5.50	8.00	0.353	3/8	3/16	2.978	3	0.500	½	5¾	1⅛	½	7/8
MC 8x 8.5	2.50	8.00	0.179	3/16	1/16	1.874	1⅞	0.311	5/16	6½	¾	5/16	5/8
MC 7x22.7	6.67	7.00	0.503	½	1/4	3.603	3⅝	0.500	½	4¾	1⅛	½	7/8
x19.1	5.61	7.00	0.352	3/8	3/16	3.452	3½	0.500	½	4¾	1⅛	½	7/8
MC 7x17.6	5.17	7.00	0.375	3/8	3/16	3.000	3	0.475	½	4⅞	1¹/₁₆	½	3/4
MC 6x18	5.29	6.00	0.379	3/8	3/16	3.504	3½	0.475	½	3⅞	1¹/₁₆	½	7/8
x15.3	4.50	6.00	0.340	5/16	3/16	3.500	3½	0.385	3/8	4¼	7/8	3/8	7/8
MC 6x16.3	4.79	6.00	0.375	3/8	3/16	3.000	3	0.475	½	3⅞	1¹/₁₆	½	3/4
x15.1	4.44	6.00	0.316	5/16	3/16	2.941	3	0.475	½	3⅞	1¹/₁₆	½	3/4
MC 6x12	3.53	6.00	0.310	5/16	1/8	2.497	2½	0.375	3/8	4⅜	13/16	3/8	5/8

Figure A-20: Dimensions of American standard steel channel (Cont.). *Courtesy American Institute of Steel Construction.*

ampacity: The current-carrying capacity of conductors or equipment, expressed in amperes.

amperage: The flow of electric current in a circuit, measured in amperes.

ampere (A): The basic SI unit measuring the quantity of electricity (rate of flow). A pressure of one volt will force one ampere through a resistance of one ohm. The unit of measurement of electric current strength.

ampere-turn: The product of amperes times the number of turns in a coil.

amplification: Procedure of expanding the strength of a signal.

amplifier: 1) A device that enables an input signal to directly control a larger energy flow. 2) The process of increasing the strength of an input.

amplitude: The maximum value of a wave.

analog: Pertaining to data from continuously varying physical quantities.

anchor: 1) Irons of special form to fasten together timbers or masonry. 2) That bolt or fastening device which attaches to the anchorage.

anchor bolts: Bolt which fastens columns, girders, or other members to concrete or masonry. See Fig. A-21.

andiron: A metal support for wood in an open fireplaace. Sometimes called a "firedog."

angle: The difference in direction of two straight lines; the space between two straight lines that do or would meet.

angle bead: A molded strip used in an angle; usually where two walls meet at right angles.

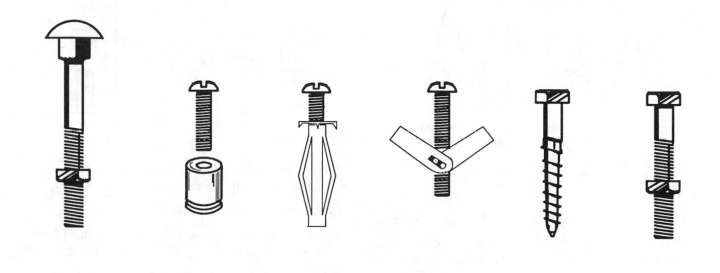

Figure A-21: A variety of anchors used in building construction.

angle bracket: A form of support having two faces generally at right angles to each other. A web is often added to increase strength.

angle dividers: A tool designed primarily for bisecting angles although it can also be used as a try square. See Fig. A-22.

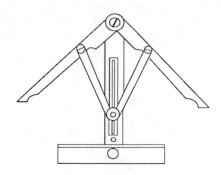

Figure A-22: Angle dividers.

angle, roll over (overhead): The sum of the vertical angles between the conductor and the horizontal on both sides of the traveler; excessive roll over angles can cause premature splice failures.

annealing: The process of preventing or removing objectional stresses in materials by controlled cooling from a heated state; measured by tensile strength.

annealing, bright: Annealing in a protective environment to prevent discoloration of the surface.

annual ring: The circumferential layer of wood seen in a cross section of timber, which represents the yearly growth. See Fig. A-23 on page 18.

annuity: A sum of money or its equivalent that constitutes one of a series of periodic payments.

annulated columns: Columns clustered together by rings or bands; much used in English architecture.

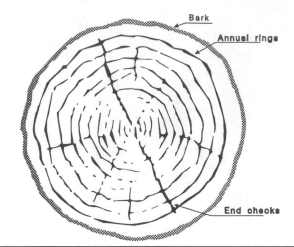

Figure A-23: Annual rings.

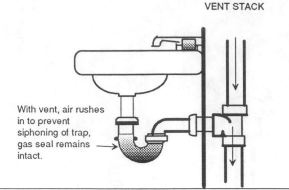

Figure A-25: Antisiphon trap.

annulet: A small square molding used to separate others.

annunciator: A signaling device used in fire-alarm and nurses' call systems. For example, a patient may press a button in a hospital room to alert the nurses' station. See Fig. A-24.

anode: 1) Positive electrode through which current enters a non-metallic conductor such as an electrolytic cell. 2) The negative pole of a storage battery.

ANSI (American National Standards Institute): Organization that publishes nationally recognized standards.

anta: A pilaster opposite another, as on a door jamb.

antenna: A device for transmission or reception of electro-magnetic waves.

antioxidant: Retards or prevents degradation of materials exposed to oxygen (air) or peroxides.

antisiphon trap: Trap in a drainage system designed to preserve a water seal by defeating siphonage. See Fig. A-25.

Operation: When it is necessary to alert the nursing-station supervisor that a patient has called for assistance, a low-voltage switch at the patient's bed can be used to turn on a corridor light outside the room as well as a light in the master-control center at the duty station. The light remains on at the control center until the patient receives attention. Then the light can be turned off by a wall switch in the patient's room.

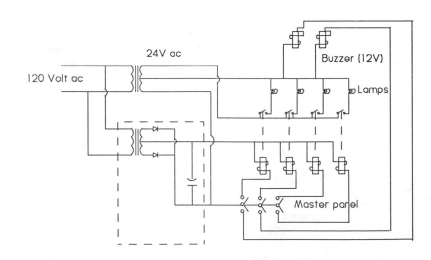

Figure A-24: Diagram of a nurses' call system showing the connection of annunciators.

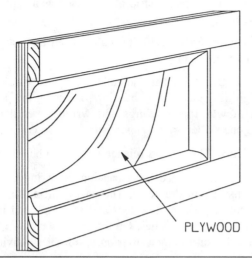

Figure A-25: Sheet of plywood used in combination with molding to simulate wall panels.

aperture seal (nuclear): A seal between containment aperture and the electrical penetration assembly.

appellant: The party who takes an appeal to a higher court.

appellee: The party against whom the appeal is taken to a higher court.

appliance: Equipment designed for a particular purpose which utilizes energy to produce heat, light, mechanical motion, etc., usually complete in itself, generally other than industrial, normally in standard sizes or types. Large appliances include cooking ranges, refrigerators, washers, dryers, dishwashers, etc. Portable appliances include mixers, toasters, coffee makers, etc.

applied molding: Molding placed to give the effect of paneling. See Figure A-25.

appraisal: An estimate of quantity, quality, or value. The process through which conclusions of property value are obtained; also refers to the report setting forth the estimate and conclusion of value.

appraisal by capitalization: An estimate of value by capitalization of productivity and income.

appraisal by comparison: Comparability to the sale prices of other similar properties.

appraisal by summation: Adding together of parts of a property separately appraised to form the whole: for example, value of the land considered as vacant added to the cost of reproduction of the building, less depreciation.

approved: 1) Acceptable to the authority having legal enforcement. 2) Per OSHA: A product that has been tested to standards and found suitable for general application, subject to limitations outlined in the nationally recognized testing lab's listing.

appurtenance: That which belongs to something else; something which passes as an incident to land, such as a right of way.

apron: 1) A plain or molded piece of finish below the stool of a window, put on to cover the rough edge of plastering. 2) That board immediately under the top of a table, which fastens the legs together, gives support to the top, and improves the appearance of the table. 3) A paved area, such as the junction of a driveway with the street or with a garage entrance. See Fig. A-26.

apse: The altar end of a church; a recess.

arabesque: An ornament, painted, inlaid, or carved in low relief, the pattern consisting of plants, fruits, and figures of men and animals interlaced in fantastic devices.

arbor: Detached lattice work; a bower; a nook.

arc: A flow of current across an insulating medium.

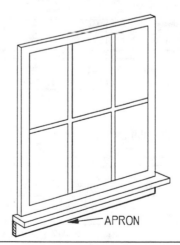

Figure A-26: Window apron.

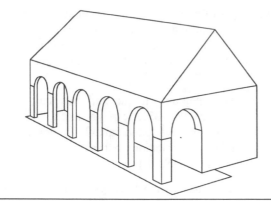

Figure A-27: Arcade.

arcade: A range of arches supported either on columns or on piers, and attached to or detached from the wall. An arched passageway. See Fig. A-27.

arch: A curved or pointed structure supported at the sides or ends only, used to span openings or spaces.

arch bar: A flat bar or strip of iron used as a support for a flat arch.

arch buttress: Same as flying buttress, an arch springing from a buttress or pier.

architect: One skilled in methods of construction and in designing and planning buildings, including the preparation of working drawings and written specifications. Also one skilled in the supervision of their construction.

architecture: The art and science relating to building.

architect's scale: Used for measuing or laying out work on construction drawings. The basic unit at the end of the scale represents one foot and is subdivided into 12 parts to represent inches. On the larger scales, such as 1 inch = 1 foot, the inch division is further subdivided so that the smallest subdivision may represent $\frac{1}{4}$ of an inch. On the smaller scales, such as $\frac{1}{8}$ inch = 1 foot, the basic unit is not divided into as many subdivisions. See Fig. A-28.

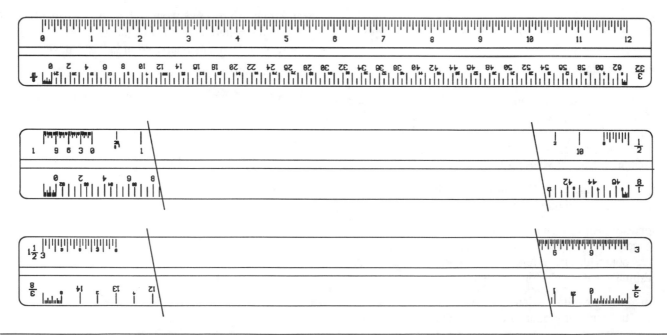

Figure A-28: Architect's scale showing the various graduations available.

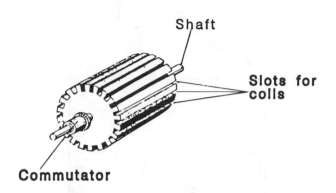

Figure A-29: Typical motor armature.

Figure A-30: Armored cable, often called "BX" cable.

architrave: The lowest member of an entablature; also a door molding.

architrave cornice: An entablature of two members only, an architrave and a cornice, the frieze being omitted.

archivolt: An ornamented or adorned band or frame running over the face of the arch stones of an arch, and bearing upon imposts; also a collection of members forming the inner contour of an arch.

archway: The space or passage under an arch.

area: An open space or court; an uncovered space.

areaway: Open space below the ground level immediately outside a building. It is enclosed by substantial walls.

armature: 1) Rotating machine: the member in which alternating voltage is generated. 2) Electromagnet: the member which is moved by magnetic force. See Fig. A-29.

armor: Mechanical protector for cables; usually a helical winding of metal tape, formed so that each convolution locks mechanically upon the previous one (interlocked armor); may be a formed metal tube or a helical wrap of wires. See Fig. A-30.

arpen: French measurement term, being ⅞ of one acre.

arrester: Wire screen secured to the top of an incinerator to confine sparks and other products of burning.

artisan: One who works with his hands and manufactures articles in metal, wood, etc.

asbestos shingles: A fireproof roof covering made in the form of shingles; asbestos is the principal part of its composition.

ash: A light-colored, coarse-grained wood used frequently for hammer handles and generally in work requiring flexibility combined with moderate strength.

ashlar: Squared and dressed stones used for facing a masonry wall; short upright wood pieces extending from the attic floor to the rafters forming a dwarf wall.

askarel: A synthetic insulating oil which is nonflammable but very toxic—being replaced by silicone oils.

ASME: American Society of Mechanical Engineers.

aspen: A common tree in many parts of the U.S. The wood is of little commercial importance except for making paper pulp in which it ranks next to spruce and hemlock.

assessed valuation: Assessment of real estate by a unit of government for taxation purposes.

assessment: A charge against real estate made by a unit of government to cover the proportionate cost of an improvement, such as a street or sewer.

assignee: The person to whom an agreement or contract is assigned.

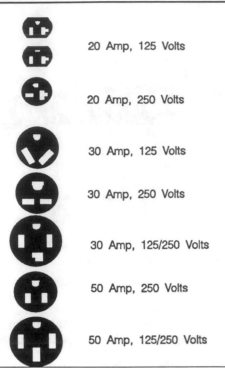

20 Amp, 125 Volts

20 Amp, 250 Volts

30 Amp, 125 Volts

30 Amp, 250 Volts

30 Amp, 125/250 Volts

50 Amp, 250 Volts

50 Amp, 125/250 Volts

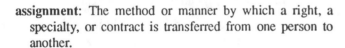

Figure A-31: Receptacle configurations.

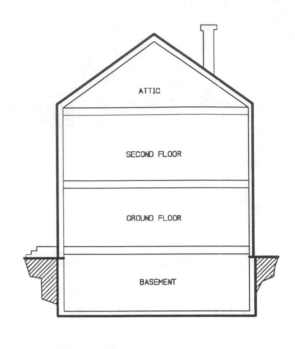

Figure A-32: Parts of a building, showing attic.

assignment: The method or manner by which a right, a specialty, or contract is transferred from one person to another.

associate broker: A person who has qualified as a real estate broker, but works for a broker named in the associate broker's license.

ASTM (American Society for Testing and Materials): A group writing standards for testing materials, and specifications for materials.

astragal: A small semicircular molding, either plain or ornamented, frequently used to cover the joint between doors.

asymmetrical: Not identical on both sides of a central line; unsymmetrical.

atrium: A large hall or lobby with galleries on three or more sides at each floor level.

attachment plug or cap: The male connector for electrical cords that fits into a receptacle. See Fig. A-31.

attic: That space between the roof and the ceiling of the upper story. In classical structures, that part extending above the level of the cornice. See Fig. A-32.

audible: Capable of being heard by humans.

auditable data: Technical information that is documented and organized so as to be readily understandable and traceable to independently verify inferences of conclusions based on these records.

auger: A wood-boring tool of large size with attached handle at right angles to the tool line. Several types are made for different purposes.

augur bit: An auger without a handle, to be used in a brace. See Fig. A-33.

automatic: Operating by own mechanism when actuated by some impersonal influence; nonmanual; self-acting.

automatic transfer equipment: A device to transfer a load from one power source to another, usually from normal to emergency source and back. See Fig. A-34.

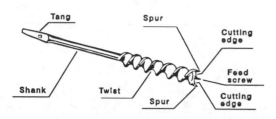

Figure A-33: Auger bit.

autotransformer: Any transformer where primary and secondary connections are made to a single cell. See Fig. A-35.

auxiliary: A device or equipment which aids the main device or equipment.

avodire: A fine cabinet wood native to the west coast of Africa. Color dull white to golden cream with beautiful grain effects.

avoirdupois weight: A system of measuring. See table in appendices.

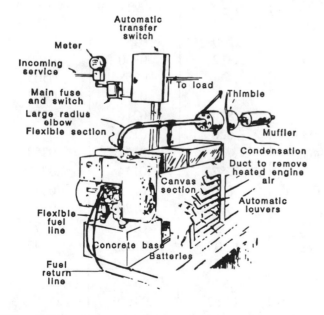

Figure A-34: Practical application of automatic transfer switch.

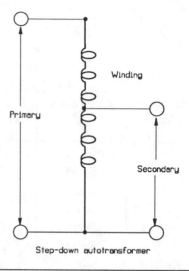

Figure A-35: Basic parts of an autotransformer.

avulsion: Removal of land from one owner to another when a stream suddenly changes its channel.

AWG (American Wire Gage): The standard for measuring wires in America. See table in Fig. A-37 on page 24.

awl: A small pointed tool for making holes for nails or screws. See *scratch awl*.

awning: A rooflike shelter of canvas, wood, metal, or other material that extends over windows, doorways, and other openings to provide protection from the sun.

awning window: A window that is hinged at the top to swing outward. See Fig. A-36.

Figure A-36: Awning window.

Size AWG, Kcmil	Area Cir. Mils	Concentric Lay Standard Conductors		Bare Conductors		DC Resistance Ohms/M Ft. At 25° F.	
		No. Wires	Diam. Each Wire inches	Diam. Inches	Area Sq. Inches	Copper	Aluminum
18	1620	Solid	.0403	.0403	.0013	6.51	10.7
16	2580	Solid	.0508	.0508	.0020	4.10	6.72
14	4110	Solid	.0641	.0641	.0032	2.57	4.22
12	6530	Solid	.0808	.0808	.0051	1.62	2.66
10	10380	Solid	.1019	.1019	.0081	1.018	1.67
8	16510	Solid	.1285	.1285	.0130	.6404	1.05
6	26240	7	.0612	.184	.027	.410	.674
4	41740	7	.0772	.232	.042	.259	.424
3	52620	7	.0867	.260	.053	.205	.336
2	66360	7	.0974	.292	.067	.162	.266
1	83690	19	.0664	.332	.087	.129	.211
0	105600	19	.0745	.372	.109	.102	.168
00	133100	19	.0837	.418	.137	.0811	.133
000	167800	19	.0940	.470	.173	.0642	.105
0000	211600	19	.1055	.528	.219	.0509	.0836
250	250000	37	.0822	.575	.260	.0431	.0708
300	300000	37	.0900	.630	.312	.0360	.0590
350	350000	37	.0973	.681	.364	.0308	.0505
400	400000	37	.1040	.728	.416	.0270	.0442
500	500000	37	.1162	.813	.519	.0216	.0354
600	600000	61	.0992	.893	.626	.0180	.0295
700	700000	61	.1071	.964	.730	.0154	.0253
750	750000	61	.1109	.998	.782	.0144	.0236
800	800000	61	.1145	1.030	.833	.0135	.0221
900	900000	61	.1215	1.090	.933	.0120	.0197
1000	1000000	61	.1280	1.150	1.039	.0108	.0177
1250	1250000	91	.1172	1.289	1.305	.00863	.0142
1500	1500000	91	.1284	1.410	1.561	.00719	.0118
1750	1750000	127	.1174	1.526	1.829	.00616	.0101
2000	2000000	127	.1255	1.630	2.087	.00539	.00885

Figure A-38: Dimensions and specifications of American Wire Gauge sizes.

B

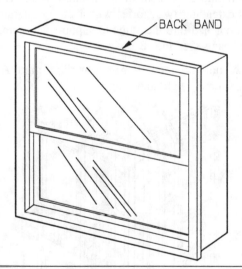

Figure B-1: Window details showing back band.

back band: The outside member of a window or door casing. See Fig. B-1.

backfill: The gravel or earth replaced in the space around a building wall after foundations are in place.

back filling: Fill of broken stone or other coarse material outside a foundation or basement wall to provide drainage.

Figure B-2: Backhoe.

backhoe: A machine that digs narrow, deep trenches for foundations, draintile, cable, etc. See Fig. B-2.

back pressure: Pressure in the low side of a refrigerating system; also called suction pressure or low side pressure.

backing of a joist or rafter: Owing to the variation in widths of joists, it is necessary, in order to obtain even floors or roofs, to block up the narrower pieces until all the upper surfaces are at the same level.

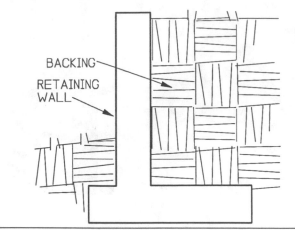

Figure B-3: Retaining wall with backing.

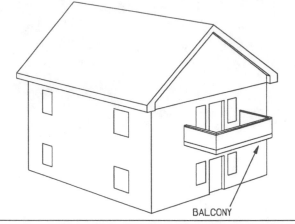

Figure B-5: Building with balcony.

backing of a wall: The rough inner face of a wall. The fill deposited behind a retaining wall. See Fig. B-3.

backing-up: The using of bricks of a cheaper grade for the inner face of a wall.

backsaw: Any saw whose blade is stiffened with a metallic back. Tenon and dovetail saws are backsaws. See Fig. B-4.

badger: A wide rabbet plane having a skew mouth.

balcony: A platform or gallery projecting from the wall of a building, enclosed by a balustrade or parapet. See Fig. B-5.

balk: A squared beam or timber.

ballast: A device designed to stabilize current flow.

balloon framing: System of small house framing; two by fours extending two stories with inch by quarter ledger strips notched into the studs to support the second-story floor beams. See Fig. B-6.

balsa: A common, second-growth tree in Central America. It is extremely light and tough.

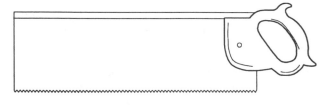

Figure B-4: Backsaw.

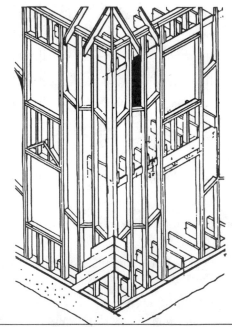

Figure B-6: Balloon framing.

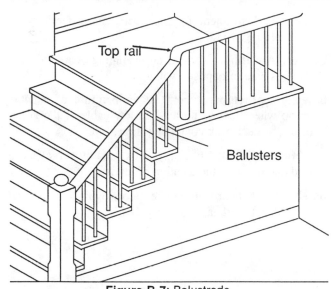

Figure B-7: Balustrade.

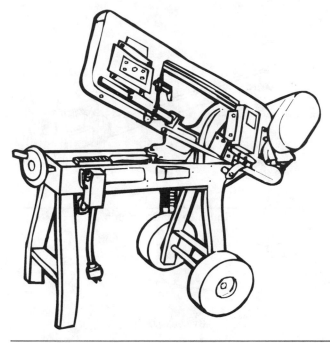

Figure B-8: Band saw.

baluster: Upright supports of a balustrade or handrail of an open stair; usually made of wood. Also, a small pillar or column; one of the units of a balustrade. A splat with the outline of a baluster. See Fig. B-7.

balustrade: A railing made up of balusters, top rail, and sometime bottom rail. See Fig. B-7.

band saw: An endless saw running on revolving pulleys, used for cutting work in wood. See Fig. B-8.

banister: A baluster.

bank: An installed grouping of a number of units of the same type of electrical equipment; such as "a bank of transformers" or "a bank of capacitors" or a "meter bank," etc. See Fig. B-9.

bar clamp: A clamp consisting of a long bar and two clamping jaws, used by woodworkers for clamping large work. See Fig. B-10 on page 28.

bare (conductor): Not insulated; nor coated.

barefaced tenon: A tenon shouldered on one side only.

bargeboard: Same as verge board. A board suspended from the verge of a gable. Ornamented board covering the roof boards and projecting over the slope of the roof. See Fig. B-11 on page 28.

bargain and sale deed: Deed that conveys the property for a consideration, but without any warranties.

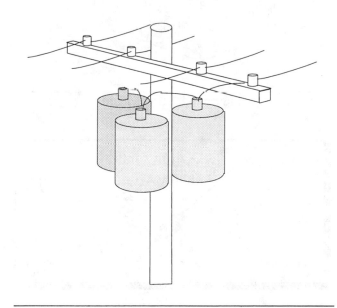

Figure B-9: Bank of transformers.

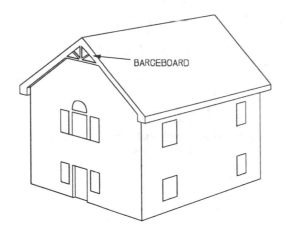

Figure B-10: Bar clamp.

barometer: Instrument for measuring atmospheric pressure. See Fig. B-12.

barrier: A partition; such as an insulating board to separate bus bars of opposite voltages.

base ambient temperature: The temperature of a cable group when there is no load on any cable of the group or on the duct bank containing the group.

base and meridian: Imaginary lines used by surveyors to find and describe the location of lands.

baseboard: A board along the floor, against walls and partitions, to hid gaps. See Fig. B-13.

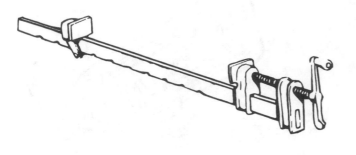

Figure B-11: Bargeboard.

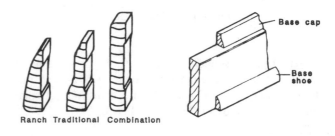

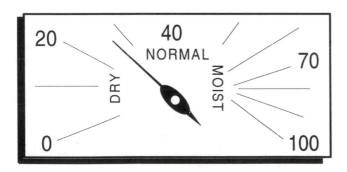

Figure B-12: Barometer.

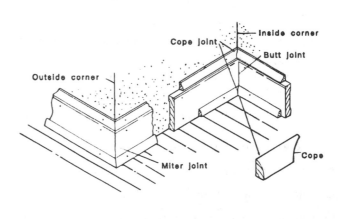

Figure B-13: Application of baseboards and base moldings.

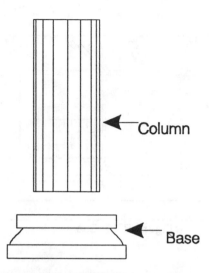

Figure B-14: Base of a column.

base line: A definite line from which measurements are taken in the layout of work.

base load: The minimum load over a period of time.

basement: The finished portion of a building below the main floor.

basement floor: The lowest floor level in a building.

base molding: The molding immediately above the plinth of a wall, pillar, or pedestal. See Fig. B-13.

base of a column: That part between the shaft and the pedestal. See B-14.

base shoe: Molding used next to the floor on baseboard. Sometimes called carpet strip.

base trim: A board of molding used for finishing off the base of a piece of work, as on a baseboard.

batt: Insulation in the form of a blanket, rather than loose filling.

batten: A strip of wood placed across a surface of one or more boards to prevent warping, to strengthen, etc. See Fig. B-15.

batter: Slope of the exposed face of a retaining wall.

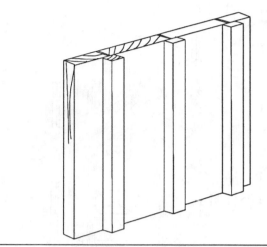

Figure B-15: Board and batten siding.

batter board: One of a pair of horizontal bards nailed to posts set at the corners of an excavation, used to indicate the desired level. They are also used as fastenings for stretched strings to indicate outlines of foundation walls. See Fig. B-16.

battery: A device that changes chemical to electrical energy, used to store electricity.

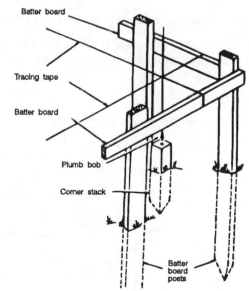

Figure B-16: Application of batter boards.

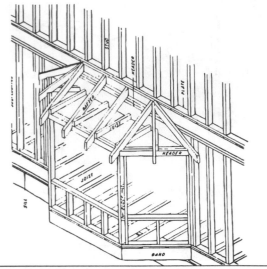

Figure B-17: Framing for a bay window. *Courtesy U.S. Dept. of Agriculture.*

bay window: Any window space projecting outward from the walls of a building, either square or polygonal in plan. See Fig. B-17.

bead: Narrow projecting molding with a rounded surface; or in plastering, a metal strip embedded in plaster at the projecting corner of a wall.

bead plane: A special form of plane for cutting beads.

beakhead: A sort of drip mold used on the extreme lower edge of the lowest member of a cornice. See Fig. B-18.

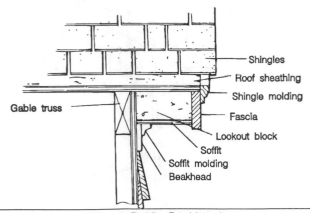

Figure B-18: Beakhead.

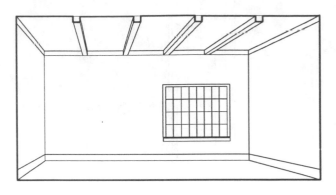

Figure B-19: Beam ceiling.

beam: A piece of timber or other material, or a built-up structure whose length is greater than its width or depth and whose strains are those due to leverage, tension, or compression. A horizontal member of wood, reinforced concrete, steel, or other material used to span the space between posts, columns, girders, or over an opening in a wall.

beam ceiling: A ceiling in which beams, either false or true, usually in a horizontal position, are exposed to view. See Fig. B-19.

bearing: That portion of a beam, truss, etc., which rests on the supports.

bearing partition: A partition that carries the floor joists and other partitions above it.

bearing plate: Steel plate placed under one end of a beam or truss for load distribution.

bearing wall: Wall supporting a load other than its own weight.

bedding: A layer of mortar into which brick or stone is set.

bed molding: Molding used as a finish underneath an overhang; the finish molding at the joining of eaves and outside. See Fig. B-20.

bed plate: A metal plate used as a rest or support of some structural part.

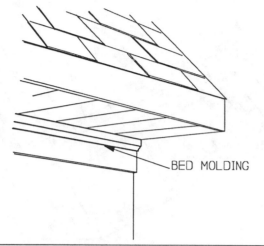

Figure B-20: Application of bed molding.

bed: Place or material in which stone or brick is laid; horizontal surface of positioned stone; lower surface of brick.

bedding: A layer of material to act as a cushion or interconnection between two elements of a device, such as the jute layer between the sheath and wire armor in a submarine cable; sometimes incorrectly used to refer to extruded insulation shields.

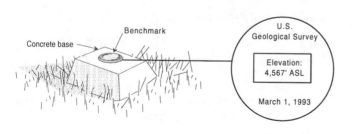

Figure B-21: Bench mark.

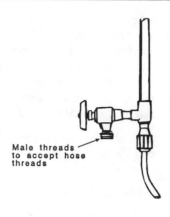

Figure B-22: Hose bib.

beech: A large tree whose wood is hard, strong and tough but not durable.

bench mark: Point of reference from which measurements are made. Fig. B-21.

bench plane: Plane kept on the bench, being in constant use, as jack, trying, and smoothing planes.

bent gouge: A gouge curved in the direction of its length, used by patternmakers for hollowing out the concave portions of core boxes.

berm: A raised area of earth such as earth pushed against a wall.

bevel: A sloping edge; a tool similar to a square. Also a tool used for testing the accuracy of work cut to an angle or bevel.

bib: A water faucet to which a hose may be attached; also called hose bib or sill cock. See Fig. B-22.

bid: A proposal to furnish supplies or equipment, to carry out or to perform certain work for a specified sum.

bilateral contract: Both parties expressly enter into mutual engagements (reciprocal).

bimetal strip: Temperature regulating or indicating device that works on the principle that two dissimilar metals with unequal expansion rates, welded together, will bend as temperature changes.

binder: 1) An agreement to cover a down payment for the purchase of real estate as evidence of good faith on the part of the purchaser; 2) in insurance; a temporary agreement given to one having an insurable interest, and who desires insurance subject to the same conditions which will apply if, as, and when a policy is issued. 3) Material used to hold assemblies together.

bird's-eye maple: The very beautiful effect obtained with boards or veneers cut from maple burl.

bit brace: A device for holding bits, so constructed that good leverage is had for the turning thereof. *See brace.*

black bean: An Australian tree which produces wood similar to teak. It is popular for fine interiors.

black birch: A strong hard wood. Used for interior trim and as a substitute for mahogany and cherry in furniture construction.

black gum: Also known as sour gum. Wood not hard but cross grained and hard to split. Used for wooden kitchenware, baskets and crates.

black spruce: Found in the northern states and Canada; lightweight; reddish in color; easy to work, though tough in fiber.

black walnut: Heavy, hard, porous wood of a brownish color; durable. Used for small cabinetwork, gunstocks, furniture and interior decorating.

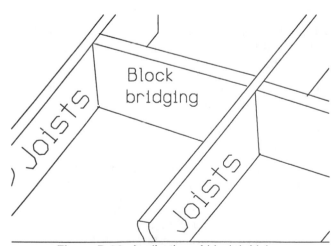

Figure B-23: Application of block bridging.

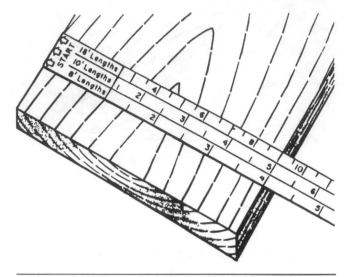

Figure B-24: Principles of determining amount of board feet in a board with a board rule.

blanket mortgage: A single mortgage that covers more than one piece of real estate.

bleeding: Seeping of resin or gum from lumber. This term is also used in referring to the process of drawing air from water pipes.

blind-nailing: Nailing in such a way that the nailheads are not visible on the face of the work. Blind nailing is usually done at the tongue of matched boards.

block bridging: Solid wood members nailed between joists to stiffen a floor. See Fig. B-23.

blueprint: See definition of drawing.

board feet: The unit of measure for lumber. One foot square and one inch or less in thickness. See Appendices.

board measure: The term in which quantities of lumber are designated and prices determined.

board rule: A graduated scale used in checking lumber for quantity. See Fig. B-24.

board: Thin timber as distinguished from planks or strips. Long pieces less than 2 in. thick and more than 4 in. wide are called boards. See Fig. B-25.

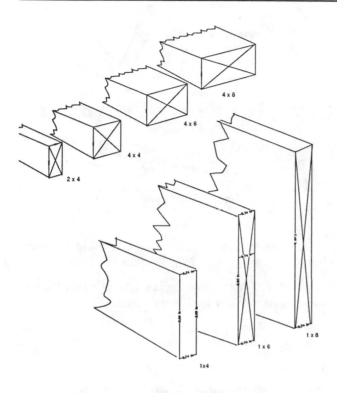

Figure B-25: Common types of boards.

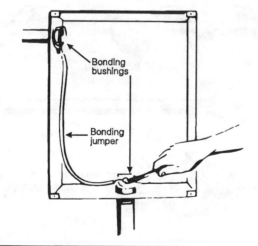

Figure B-26: Bonding bushing and bonding jumper used on electrical panelboard.

boiler: Closed container in which a liquid may be heated and vaporized.

boiling point: Temperature at which a liquid boils or generates bubbles of vapor when heated.

bolster: A short horizontal wood or steel beam on top of a column to support and decrease the span of beams or girders.

bona fide: In good faith, without fraud.

bond: 1) Any obligation under seal. A real estate bond is a written obligation, usually issued on security of a mortgage or a trust deed. 2)A mechanical connection between metallic parts of an electrical system, such as between a neutral wire and a meter enclosure or service conduit to the enclosure for the service equipment with a bonding locknut or bushing; the junction of welded parts; the adhesive for abrasive grains in grinding wheels.

bonding bushing: A special conduit bushing equipped with a conductor terminal to take a bonding jumper; also has a screw or other sharp device to bite into the enclosure wall to bond the conduit to the enclosure without a jumper when there are no concentric knockouts left in the wall of the enclosure. See Fig. B-26.

bonding jumper: A bare or insulated conductor used to assure the required electrical conductivity between metal parts required to be electrically connected. Frequently used from a bonding bushing to the service equipment enclosure to provide a path around concentric knockouts in an enclosure wall, also used to bond one raceway to another. See Fig. B-26.

bonding locknut: A threaded locknut for use on the end of a conduit terminal, but a locknut equipped with a screw through its lip. When the locknut is installed, the screw is tightened so its end bites into the wall of the enclosure close to the edge of the knockout.

boring tool: Brace and bit and similar tools for wood or metal.

Boston hip roof: Formed by laying a double row of shingles or slate lengthwise along the hip; joints must be securely made to obtain a watertight job.

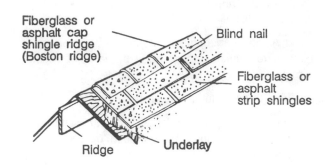

Figure B-27: Boston ridge shingling.

Boston ridge: A method of applying shingles at the ridge or hips of a roof as a finish. See Fig. B-27.

bow: Any projecting part of a building in the form of an arc or of a polygon. See Fig. B-28.

bow saw: A saw with a thin narrow blade held in tension by the leverage obtained through the twisting of a cord, or by means of rods and turnbuckle. See Fig. B-29.

brace: An inclined piece of timber used as a support stiffener.

brace bit: Ordinary bit used for wood boring, and having square, tapered shanks to fit in the socket of a common brace. See Fig. B-30.

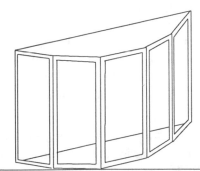

Figure B-28: Bow window is one type of building bow.

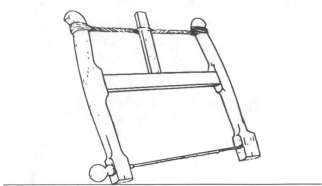

Figure B-29: Bow saw.

braced framing: Construction technique using posts and cross-bracing for greater rigidity. See Fig. 31.

brace frame: Framework of a building in which the corner posts are braced to sills and plates.

brace jaws: Those parts of a bit brace which clamp about the tapered shank of a bit.

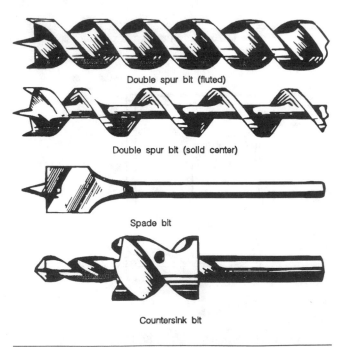

Figure B-30: Types of wood bits used on building construction projects.

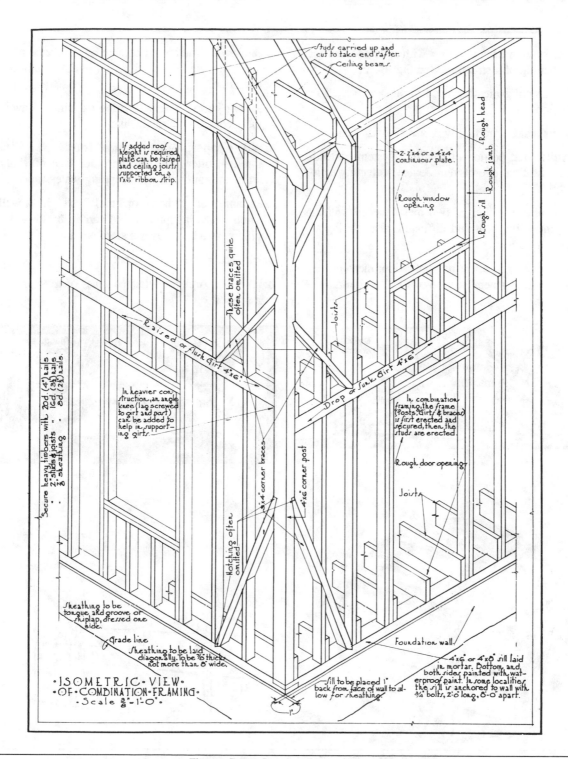

Studs carried up and cut to take end rafter.

Ceiling beams.

If added roof height is required, plate can be raised and ceiling joists supported on a 1"x6" ribbon strip.

2-2"x4" or a 4"x4" continuous plate.

Rough head

Rough jamb

Rough sill

Rough window opening

These braces quite often omitted.

Raised or flush Girt 4"x6".

Joists.

Drop or Sunk Girt 4"x6".

In heavier construction, an angle knee (lag screwed to girt and post) can be added to help in supporting girts.

In combination framing, the frame (Posts, Girts & braces) is first erected and secured, then the studs are erected.

Rough door opening.

4"x4" corner braces

4"x6" corner post

Joists

Secure heavy timbers with 20d (4") nails.
2"slabs & joists ... 16d (3½") nails.
    " 3" sheathing ... 8d (2½") nails.

Notching often omitted.

Sheathing to be tongue and groove or shiplap, dressed one side.

Grade line

Sheathing to be laid diagonally. To be ⅞ thick, not more than 8" wide.

·ISOMETRIC·VIEW·
·OF·COMBINATION·FRAMING·
·Scale ⅜"-1'-0"·

Sill to be placed 1" back from face of wall to allow for sheathing.

Foundation wall

4"x6" or 4"x8" sill laid in mortar. Bottom and both sides painted with waterproof paint. In some localities the sill is anchored to wall with ¾" bolts, 2'-0" long, 8'-0" apart.

Figure B-31: Details of braced framing.

bracing: The staying or supporting with rods and ties for the strengthening of a structure. See Fig. B-32.

bracket: A projecting member used to support a shelf or ornament. See Fig. B-33.

brad: A long slender wire nail with a very small head.

branch circuit: That portion of a wiring system extending beyond the final overcurrent device protecting a circuit.

braze: The joining together of two metal pieces, without melting them, using heat and diffusion of a jointing alloy of capillary thickness.

break: A term very general in its application, meaning any projection from the general surface of a wall or building.

breakdown: The abrupt change of resistance from high to low, allowing current flow: an initial rolling or drawing operation.

breaking joints: The staggering of joints to avoid having them come in a straight line.

breather paper: A paper that lets water vapor pass through, often used on the outer face of walls to stop wind and rain while not trapping water vapor.

breastsummer: A beam or lintel flush with a wall or partition which it supports, and carried by the side walls or pillars, as a beam over a store window.

brick facing: See brick veneer.

brick pier: A plain, detached mass of masonry serving as a support.

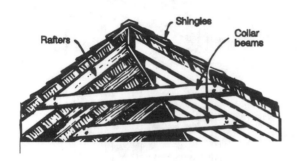

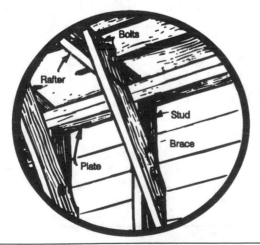

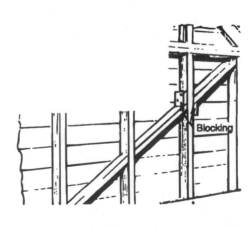

Figure B-32: Methods used to brace existing buildings.

Figure B-33: Types of brackets in common use.

brick veneer: A facing of brick applied to a frame or other structure.

bridging: A method of bracing joists or studding by the use of short strips or braces. See Fig. B-34.

british thermal unit (Btu): Quantity of heat required to raise the temperature of 1 pound of water 1 degree Fahrenheit.

brittle point: The highest temperature at which a chilled strip of polymer will crack when it is held at one end and impacted at the other end.

broker: One employed by another, for a fee, to carry on any of the activities listed in the license law definition of the word.

brush: A conductor between the stationary and rotating parts of a machine, usually of carbon.

bubinga: A large tree native to Africa. Wood is hard and heavy and is used for furniture and paneling.

buck: Rough wood door frame placed on a wall partition to which the door moldings are attached; completely fabricated steel door frame set in a wall or partition to receive the door.

buff: To lightly abrade.

builder's tape: A long measuring tape of steel or fabric contained in a circular case.

building code: Regulating the construction of buildings within a municipality by ordinance or law.

building line: The line of the outside face of a building wall. Also the line on a lot beyond which the law forbids that a building be erected.

building paper: Heavy paper used between sheathing and siding, or as an undercovering on roofs as a protection against weather.

building stone: A general term applied to stone used in building construction. B-35.

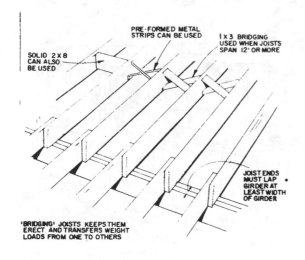

Figure B-34: Types of bridging used for house framing.

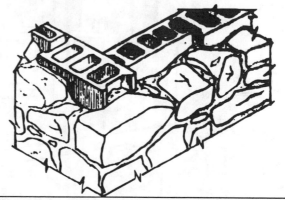

Figure B-35: Building stone used to veneer a concrete block wall.

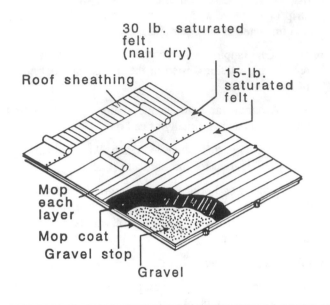

30 lb. saturated felt (nail dry)

Roof sheathing

15-lb. saturated felt

Mop each layer

Mop coat

Gravel stop

Gravel

Figure B-36: Principles of built-up roofs.

built-up roof: A roofing material applied in sealed, waterproof layers, where there is only a slight slope to the roof. See Fig. B-36.

bull-cutters: A larger, long-handled tool for cutting the larger sizes of wire and cable, up to MCM sizes.

bull pine: Common throughout western U.S. It is of medium strength but very resinous. Used for both interior and exterior work in building construction.

buna: A synthetic rubber insulation of styrenebutadiene; was known as GR-S, now as SBR.

bundle of legal rights: Establishes real estate ownership, consists of right to sell, to mortgage, to lease, to will, to regain possession at end of a lease (reversion); to build and remove improvements; to control use within the law.

bungalow: A one-story house with verandas. Sometimes the attic is finished as a second story. See Fig. B-37.

burner: Device in which combustion of fuel takes place.

Figure B-37: Typical bungalow.

bus: The conductor(s) serving as a common connection for two or more circuits. The neutral bus in a circuit breaker panel, for example, contains all neutral connections.

busbars: The conductive bars used as the main current supplying elements of panelboards or switchboards; also the conductive bars duct; an assembly of bus bars within an enclosure which is designed for ease of installation, have no fixed electrical characteristics, and allowing power to be taken off conveniently, usually without circuit interruption.

business chance broker: One who negotiates the sale of a mercantile business for another for a fee.

butane: Liquid hydrocarbon commonly used as fuel for heating purposes.

butt: Term usually applied to a hinge other than a strap hinge.

butt hinge: A hinge secured to the edge of a door and the face of the jamb which butts against the edge of the door when it is shut, as distinguished from strap hinge. See Fig. B-38.

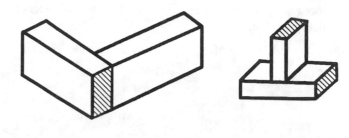

Figure B-39: Straight butt joints.

butt joint: Where the ends of two pieces of timber come together without over lapping. See Fig. B-39.

butternut: A medium-sized tree native to eastern U.S. Wood is soft and porous, it is used for furniture and interior finishing.

buttress, flying: A detached buttress or pier of masonry at some distance from a wall and connected thereto by an arch or a portion of an arch. See *flying buttress.*

buttress: Projecting structure built against a wall to give it greater strength. See Fig. B-40.

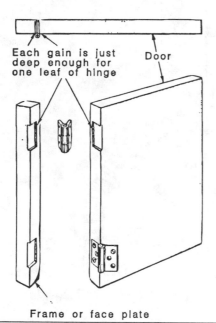

Each gain is just deep enough for one leaf of hinge

Door

Frame or face plate

Figure B-38: Application of butt hinges.

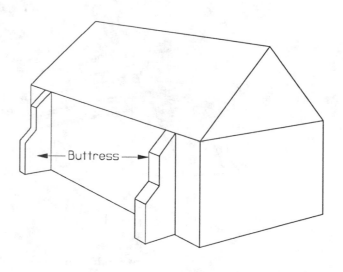

Buttress

Figure B-40: Buttress.

buttress saw: A saw or saw blade that has every other tooth missing and is used on work where there is a tendency to clog.

buzz saw: A name often applied to a portable circular saw.

BX cable: Electrical cable wrapped in rubber with a flexible steel outer covering. Also see *armored cable*. See Fig. B-41.

bypass: Passage at one side of or around a regular passage.

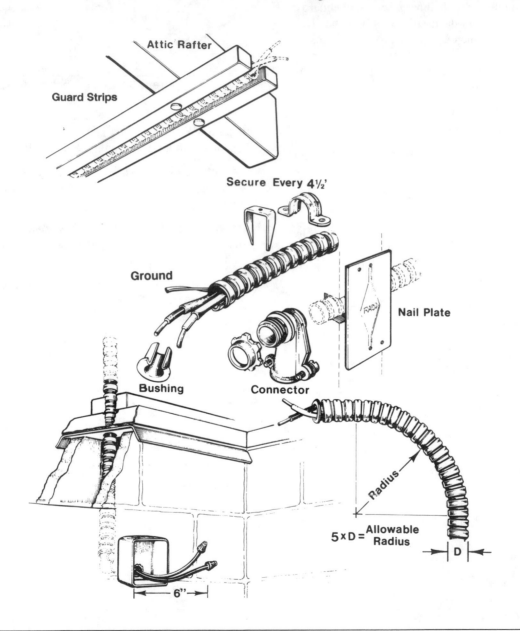

Figure B-41: National Electrical Code rules governing the installation of BX cable.

C

cabbaging press: A press for compressing loose sheet-metal scrap into a convenient form for handling and remelting.

cabin hook: A small hook and eye used on doors in cabinetwork. See Fig. C-1.

cabinet: A cupboard containing shelves, drawers, etc., enclosed by doors and used as a repository for various articles.

cabinet burnisher: A piece of hardened steel often oval shape in section, inserted in a wooden handle; used to turn the edge of scrapers.

cabinet latch: Name given to a wide variety of catches depending on their use, ranging from the type of latch used on refrigerator doors to the horizontal spring-and-bolt type operated by turning a knob as on kitchen cabinets, etc. See Fig. C-2.

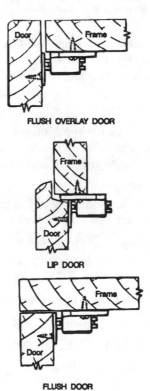

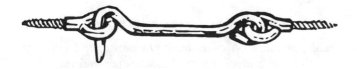

Figure C-1: Cabin hook and eye.

Door
Frame
FLUSH OVERLAY DOOR
Frame
Door
LIP DOOR
Frame
Door
FLUSH DOOR

Figure C-2: Method of mounting magnetic cabinet latches.

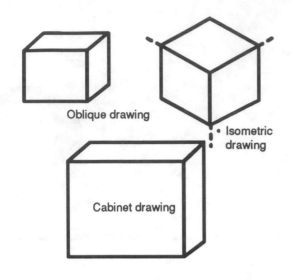

Figure C-3: Cabinet projection compared with isometric and oblique drawings.

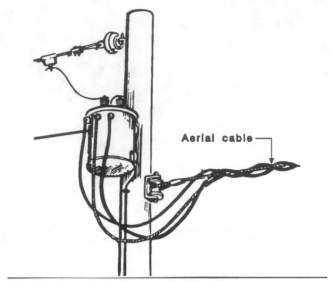

Figure C-4: Aerial cable.

cabinet projection: A system of drawing in which one face of the object is drawn as parallel to the observer and the faces perpendicular thereto are drawn at an angle usually 45 degrees, the slant edges being drawn to half scale. See Fig. C-3.

cabinet scraper: A flat piece of steel having an edge of such shape that when drawn over a wood surface irregularities can be removed and smoothness can be obtained.

cable: An insulated electrical conductor or group of conductors, protected by a waterproof coat.

cable, aerial: An assembly of one or more conductors and a supporting messenger. Sometimes referred to as "messenger cable." See Fig. C-4.

cable, armored: A cable having armor (see armor).

cable, belted: A multiconductor cable having a layer of insulation over the assembled insulated conductors.

cable, bore-hole: The term given vertical-riser cables in mines.

cable clamp: A device used to clamp around a cable to transmit mechanical strain to all elements of the cable.

cable, coaxial: A cable used for high frequency, consisting of two cylindrical conductors with a common axis separated by a dielectric; normally the outer conductor is operated at ground potential for shielding.

cable, control: Used to supply voltage (usually ON or OFF).

cable, duplex: A twisted pair of cables.

cable, festoon: A cable draped in accordion fashion from sliding or rolling hangers, usually used to feed moving equipment such as bridge cranes.

cable, hand: A mining cable used to connect equipment to a reel truck.

cable, parkway: Designed for direct burial with heavy mechanical protection of jute, lead, and steel wires.

cable, portable: Used to transmit power to mobile equipment.

cable, power: Used to supply current (power).

cable, pressure: A cable having a pressurized fluid (gas or oil) as part of the insulation; paper and oil are the most common fluids.

cable, ribbon: A flat multiconductor cable.

cable, service drop: The cable from the utility line to the customer's property.

cable, signal: Used to transmit data.

cable, spacer: An aerial distribution cable made of covered conductors held by insulated spacers; designed for wooded areas.

cable, spread room: A room adjacent to a control room to facilitate routing of cables in trays away from the control panels.

cable, tray: A multiconductor having a nonmetallic jacket, designed for use in cable trays; (not to be confused with type TC cable for which the jacket must also be flame retardant).

cable tray: A rigid structure to support cables; a type of raceway; normally having the appearance of a ladder and open at the top to facilitate changes. See Fig. C-5.

cable, triplexed d: Helical assembly of 3 insulated conductors and sometimes a bare grounding conductor.

cable, unit: A cable having pairs of cables stranded into groups (units) of a given quantity, then these groups form the core.

cable, vertical riser: Cables utilized in circuits of considerable elevation change; usually incorporate additional components for tensile strength.

cabling: Flutes of columns partly occupied by solid convex masses; or a column having convex projecting parts in place of concaved fluting.

CAD: 1) Computed-aided drafting; 2) Computer-aided design.

caisson: Sunken panel in a ceiling, contributing to a pattern.

caliber: Internal diameter or bore of pipe.

calibrate: Compare with a standard.

caliper: A tool for measuring the diameter of circular work. Both inside and outside calipers are available. See Fig. C-6.

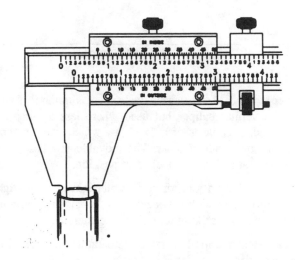

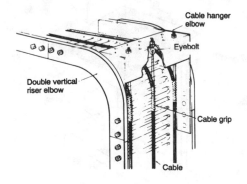

Figure C-5: Typical cable tray containing cable-tray cable.

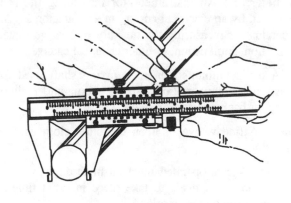

Figure C-6: Using calipers for measuring cylindrical work.

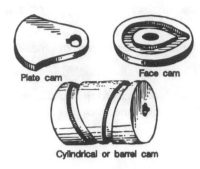

Figure C-7: Typical cams.

caliper square: A measuring instrument similar in shape to the vernier caliper but used where less accuracy is required. It consists of a fixed jaw which is an integral part of the graduated bar. The movable jaw has screw adjustment and can be locked in position.

calking: Making a joint or seam watertight or steam-tight by filling it in with rust cement or by closing the joint by means of a calking tool. Also see caulking.

calorie: Heat required to raise temperature of 1 gram of water 1 degree centigrade.

calorimeter: 1.) An instrument for measuring the heat generated by an electrical current in a conductor. 2.) An apparatus for measuring the quantity of heat generated by friction, combustion, or other chemical change.

cam: A device mounted on a revolving shaft used for transposing rotary motion into an alternating, reciprocating, or back-and-forth motion. See Fig. C-7.

cambric: A fine weave of linen or cotton cloth used as insulation.

cam drive: A cam-operated mechanism through which a certain motion is made to take place in exact time or relation to some other motion.

cam vise: A vise whose opening and closing depend on cam action.

candela (cd): The basic SI unit for luminous intensity: the candela is defined as the luminous intensity of $\frac{1}{600,000}$ of a square meter of a blackbody at the temperature of freezing platinum.

candlefoot: A unit of illumination. The light given by a British standard candle at a distance of one foot. See footcandle.

candle power: The illuminating power of a standard sperm candle used as a measure for other illuminants.

canopy: An ornamental roof-like structure projecting from a wall, or supported on pillars. See Fig. C-8.

canopy switch: A small switch usually located in the canopy of an electric fixture.

cant: To tilt, to set up on a slant; also a molding formed of plain surfaces and angles rather than curves.

cantilever: A beam fixed at one end and loaded at the other, or loaded uniformly.

canting strip: In frame buildings same as water table. See *water table*.

cap: A coping of a wall; a cornice over a door. Also the lintel over a door or window frame; a top piece. See Fig. C-9.

capacitance: The measure of the ability to store electrical charges, and to oppose changes of voltage across a condenser.

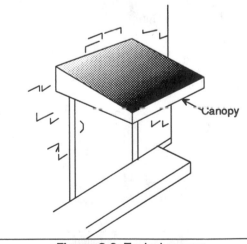

Figure C-8: Typical canopy.

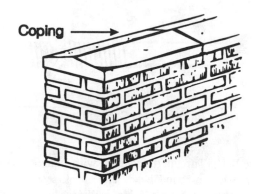

Figure C-9: Application of capping; coping a wall.

capacitive reactance: The measure of resistance to the flow of an alternating current through a resistor.

cape chisel: A narrow-blade chisel for cutting channels or keyways.

Cape Cod: A style of cottage developed mainly on Cape Cod, Massachusetts during the 18th and 19th centuries. A Cape Cod house style is traditionally a one-story cottage covered by a gable roof having a rise equal to the span, and without dormers. Each room usually contained a fireplace and a large central chimney was used to carry off the smoke. A partial basement is almost always found in early designs which was used as a root cellar. Other characteristics of early Cape Code homes, besides the large central chimney and roof design, include narrow clapboard siding, small-paned windows, and a paneled entrance door. The basic concept of the Cape Cod style is simplicity — providing a basic shelter without frills. See Fig. C-10.

capacitor: An apparatus consisting of two conducting surfaces separated by an insulating material. It stores energy, blocks the flow of direct current, and permits the flow of alternating current to a degree depending on the capacitance and frequency.

capillary action: The traveling of liquid along a small interstice due to surface tension.

capital: The upper part of a pier, pilaster, or column. See Fig. C-11.

Figure C-10: Typical Cape Code house style.

capping: A topping or placing of a cap at the upper end of a piece of work; Stone used for the top or crowning of a structure. See Fig. C-9.

capstan: A rotating drum used to pull cables or ropes by friction; the cables are wrapped around the drum.

carbon-arc welding: A welding process in which heat is generated by an electric arc between the metal workpiece and a carbon electrode.

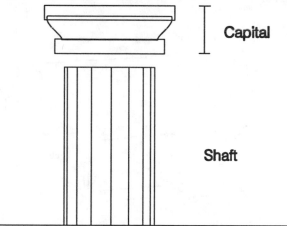

Figure C-11: Capital or upper part of a column.

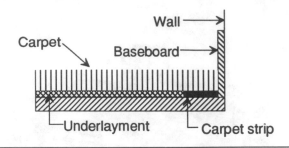

Figure C-12: Carpet strip.

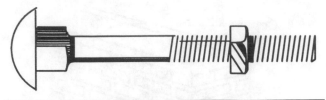

Figure C-14: Carriage bolt.

carbon black: A black pigment produced by the incomplete burning of natural gas or oil; used as a filler.

carbon dioxide: Compound of carbon and oxygen that is sometimes used as a refrigerant.

carbonizing: The reduction of a substance to carbon by subjecting it to intense heat in a closed vessel.

carcase: Also carcass. The frame, as of a house.

carpentry: The art of cutting and assembling timber as in buildings, boats, etc.

carpet strip: A strip attached to the floor beneath a door. See Fig. C-12.

carport: An open structure, usually attached to the house itself, to shelter vehicles. See Fig. C-13.

carriage: The timber or steel joist which supports the steps of a wooden stair.

carriage bolt: Threaded fastener with square shank directly below a round head. The shank secures the bolt and prevents it from turning while torque is being applied to the bolt's nut. See Fig. C-14.

cartouche: An ornamental tablet or scroll bearing an inscription.

cartridge fuse: A fuse enclosed in an insulating tube to confine the arc when the fuse blows. See *Fuse.*

cascade: The output of one device connected to the input of another.

casement frame and sash: A frame of wood or metal enclosing part or all of a sash, which can be opened by means of hinges affixed to the vertical edge.

casement window: A window hinged on its vertical edge to permit opening inward or outward. More weather-tight if opened outward. See Fig. C-15.

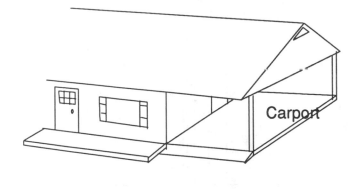

Figure C-13: Typical carport.

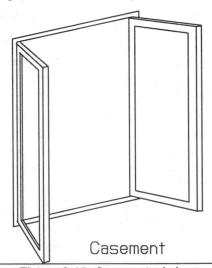

Figure C-15: Casement window.

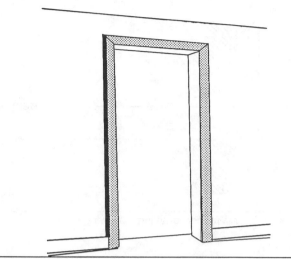

Figure C-16: Casing around door.

casing: The framework about a window or door. See Fig. C-16.

casing nails: Nails made of mild steel for use in casework, window and door frames, cornices, corner boards, and similar wood construction. See Fig. C-17.

caulk: To fill or close a joint with a seal to make it watertight and airtight. The material used to seal a joint.

caulking: Making a joint or seam watertight or steamtight by filling it in with rust cement or by closing the joint by means of a caulking tool. Oakum is frequently caulked into the seams of wooden vessels. See *calking*.

Casing Nails

SIZE	LENGTH	GAUGE	DIAMETER HEAD GAUGE	APPROX. NO. TO POUND
4d	1½ inch	No. 14	11	450
6d	2 inch	No. 12½	9½	240
8d	2½ inch	No. 11½	8½	145
10d	3 inch	No. 10½	7½	94

Figure C-17: Specifications of casing nails; used mostly for internal trim.

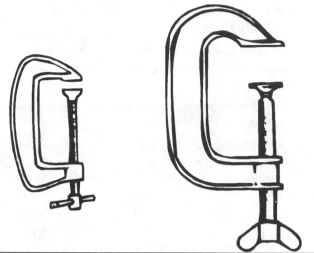

Figure C-18: C clamps used on building construction projects.

caveat emptor: "Let the purchaser beware"; the buyer is duty-bound to examine the property he is purchasing and he assumes conditions which are readily ascertainable upon view.

cavetto: A quarter-round, concave molding.

cavity wall: Wall built of solid masonry units arranged to provide airspace within the wall.

C-C: Center to center.

C clamp: A tool, shaped like the letter C, for holding portions of work together, both in wood and metal. See Fig. C-18.

ceiling joists: Support joists over a ceiling. They also serve as floor joists if there is a floor above. See Fig. C-19.

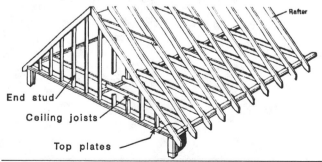

Figure C-19: Ceiling joists.

cell: A single element of an electric battery, either primary or secondary.

cellar: The room or rooms below the main portion of a building, usually containing the heating plant and accessories.

Celsius: German word for centigrade; the metric system's temperature scale. See Fig. C-20.

cement: Generally used in reference to Portland cement, stucco, natural cements, etc. Mortar, plaster of Paris, or any substance which causes bodies to adhere to one another.

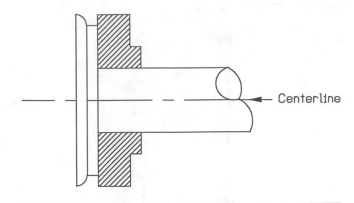

Figure C-21: Working drawings showing how centerlines appear.

cement mortar: A mixture of cement with sand and water used as a bonding agent between bricks or stones.

cementing trowel: A tool similar to the plasterer's trowel but often of heavier gauge stock.

center gage: A flat gage used for setting a tool for cutting threads.

center head: A device attached to a scale or blade for use in locating the center of some round object, such as the center point on the end of a shaft preliminary to centering.

center line: A broken line usually dot and dash indicating the center of the object and a very convenient line from which to lay off measurements. See Fig. C-21.

centerpiece: Ornament placed in the center of a ceiling.

centering: The false work over which an arch is formed. In concrete work the centering is known as the frames.

centrifugal: Proceeding from the center.

certificate of no defense: An instrument, executed by the mortgagor, upon the sale of the mortgage, to the assignee, as to the validity of the full mortgage debt.

Centigrade scale: Temperature scale used in metric system. Freezing point of water is 0 degrees; boiling point is 100 degrees. See *Celsius*.

centimeter: Metric measure equalling 0.394," or 2.54 centimeters per inch.

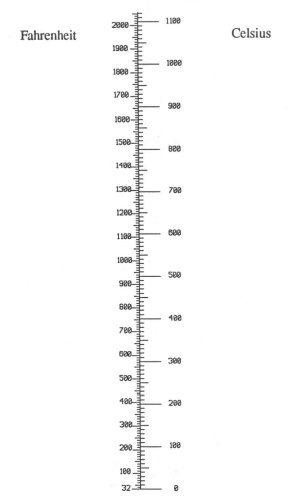

Figure C-20: Celsius and Fahrenheit scales.

centrifugal switch: A switch used on the single-phase, split-phase motor to open the starting winding after the motor has almost reached synchronous speed.

cesspool: Lined or covered pit in the ground that receives domestic wastes from the drainage system. Retains organic matter and solids; lets liquids seep into the ground through its porous bottom and sides.

chain: Unit of land measurement — 66 feet.

chain tongs: Used for holding pipe from turning or to turn pipe. It is a heavy bar with sharp teeth at one end, which are held against the pipe by a chain wrapped around the pipe and attached to the bar.

chain transmission: A means of transmitting power, useful when the distance between driver and driven shafts is too great for gearing and not sufficient for belting.

chair rail: Molding along the walls of a room, positioned at chair-back height, to protect the wall finish. See Fig. C-22.

chalking string: A string covered with chalk that, when stretched between two points and snapped, marks a straight line between the points.

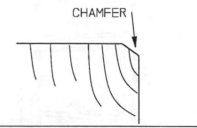

Figure C-23: Chamfered corner.

chamfer: A beveled edge or cut-off corner. See Fig. C-23.

charge: The quantity of positive or negative ions in or on an object; the quantity of electricity residing on an electrostatically charged body. Unit: coulomb.

charged cell: A storage cell which has had direct current passed through it until the positive plate has changed chemically from $PbSO_4$ to $PbSO_2$ and the negative plate has changed from $PbSO_4$ to Pb.

charging current: Direct current applied to a storage battery to create a chemical action to charge the battery. Its direction is always the reverse of the discharge current.

charging rate: The rate of flow, in amperes, of an electric current flowing through a storage battery while it is being charged.

chase: Recess in inner face of masonry wall providing space for pipes and/or ducts. See Fig. C-24.

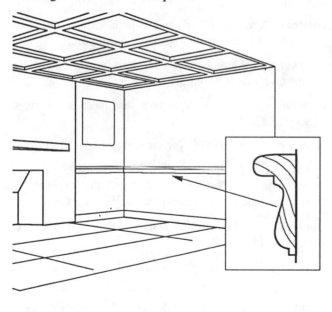

Figure C-22: Room showing chair rail molding extending along the wall.

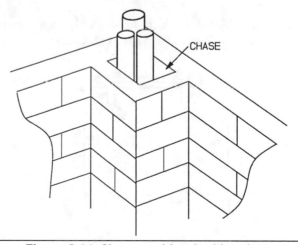

Figure C-24: Chase used for plumbing pipes.

Figure C-25: Check rails on a double-hung window.

chattel: Personal property, such as household goods or removable fixtures.

checking: Fissures that can appear with age in exterior paint coatings. Such fissures, at first superficial, may in time penetrate entirely through the coating.

check rails: Also called meeting rails. The upper rail of the lower sash and the lower rail of the upper sash of a double-hung window. Meeting rails re made sufficiently thicker than the rest of the sash frame to close the opening between the two sashes. Check rails are usually beveled to insure a tight fit between the two sashes. See Fig. C-25.

check valve: A valve which automatically closes to prevent the back flow of water. See Fig. C-26.

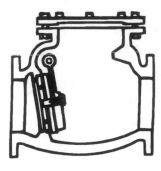

Figure C-26: Cross-section of check valve.

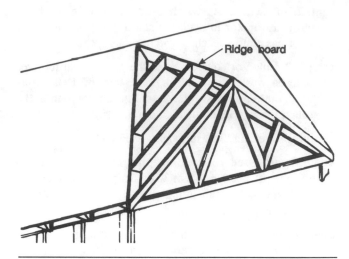

Figure C-27: Rafters in a gable rood meeting at the roof ridge.

chemical resistance test: Checking performance of materials immersed in different chemicals; loss of strength and dimensional change are measured.

cherry: A hard, close-grained, red or brown-colored wood.

chestnut: A medium hard, coarse-grained wood, used for building trim and for inexpensive furniture.

chestnut oak: Medium size tree, dark brown with lighter sapwood; heavy, strong, and close grained.

chevron: Rafters in a gable roof that meet at the ridge. See Fig. C-27.

chimney: A vertical flue for drawing off the products of combustion from a stove, furnace, etc.

chimney breast: That surface of a wall that projects into a room at the place where a chimney passes through it.

chimney cap: The finishing course at the top of the chimney. See Fig. C-28.

chimney effect: Tendency of air or gas to rise when heated.

chimney lining: The tile flues placed within a chimney.

chisel: Tool of great variety whose cutting principle is that of the wedge. See Fig. C-29.

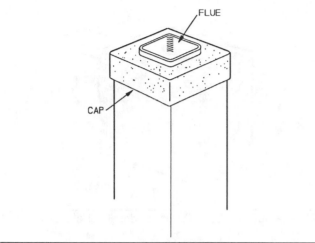

Figure C-28: Chimney cap.

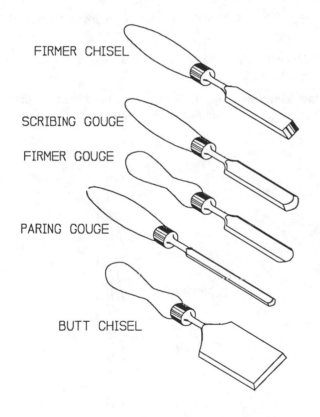

Figure C-29: Types of chisels used on building construction woodworking projects.

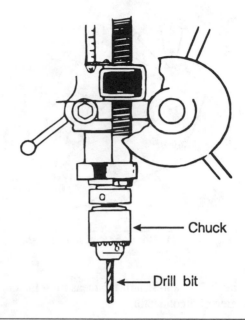

Figure C-30: Chuck used on electric drill.

chlorination: Application of chlorine to water or treated sewage to disinfect or accomplish other biological or chemical results.

chuck: A device for holding a rotating tool or work during an operation. There are many different kinds of chucks for various purposes. See Fig. C-30.

cincture: A ring or fillet at the top and bottom of a column serving to divide the shaft from its capital and its base.

circassian walnut: A fawn-colored walnut with dark streaks.

circuit: A closed path through which current flows from a generator, through various components, and back to the generator.

circuit breaker: A resettable fuse-like device designed to protect a circuit against overloading. See Fig. C-31 on page 52.

circuit foot: One foot of circuit; i.e., if one has a 3-conductor circuit, then each lineal foot of circuit would have 3 circuit feet.

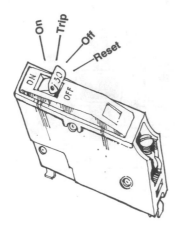

Figure C-31: Common circuit breakers.

circular mil: The non-SI unit for measuring the cross-sectional area of a conductor.

circular saw: A widely-used portable power saw. See Fig. C-32.

cistern: A reservoir for the storage of rain water. Usually made of poured concrete or plastered concrete blocks. Some older cisterns are dug in the ground and plastered directly against the earth. See Fig. C-33.

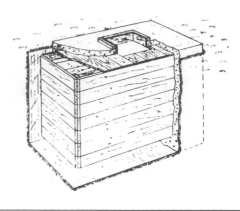

Figure C-33: Construction details of a cistern.

CL: Center line; sometimes written C_L.

clamp: A tool for holding portions of work together, both in wood and metal.

clapboard: A lapping weather-board used for siding. See Fig. C-34.

claw hammer: Carpenter's hammer with a head for driving nails and a split claw for removing nails. See Fig. C-35.

Figure C-32: Circular saw. *Courtesy Sears.*

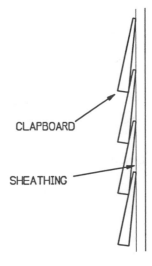

CLAPBOARD

SHEATHING

Figure C-34: Clapboard installation.

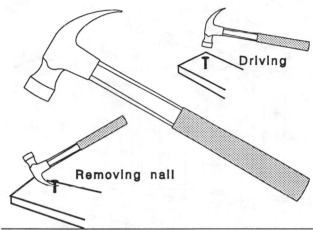

Figure C-35: Claw hammer and its uses.

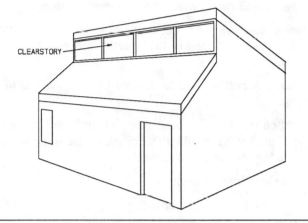

Figure C-37: Clearstory.

cleanout: A capped Y fitting installed in plumbing pipe that allows access to the interior of the pipe for cleaning. See Fig. C-36.

clearstory: A break in a roof system where two roofs meet at different locations. See Fig. C-37.

clear wood: Wood that has no knots.

cleat: A strip of wood or metal fastened to other material or nailed against a wall usually for the purpose of fastening something to it.

clinch: The process of bending over the end of a nail. See Fig. C-38.

close nipple: Twice as long as standard pipe thread, with no shoulder between the two sets of threads. See Fig. C-39.

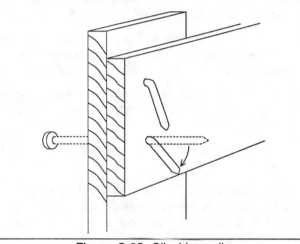

Figure C-38: Clinching nails.

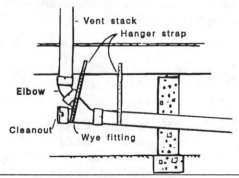

Figure C-36: Plumbing cleanout.

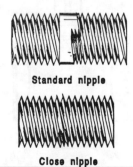

Figure C-39: Comparison of standard and close nipple.

closet: Small room or compartment used for storage.

closing statement: An accounting of funds in a real estate sale made by a broker to the seller and buyer, respectively.

closure: A portion of a brick used to close the end of a course.

cloud on the title: An outstanding claim or encumbrance which, if valid, would affect or impair the owner's title; a judgement, or dower interest.

clustered: Grouped together.

coaxial cable: A cable consisting of two conductors concentric with and insulated from each other.

cock: A plug type of valve which has an opening to permit passage of liquids or gases. A quarter turn opens or closes the valve. See Fig. C-40.

code: Any set of rules devised for the purpose of securing uniformity in work and for the maintaining of proper standards is usually called a code; e.g., a building code.

code installation: An installation that conforms to the local code and/or the national code for safe and efficient installation.

coefficient of expansion: The change in dimension due to change in temperature.

coefficient of friction: The ratio of the tangential force needed to start or maintain uniform relative motion between two contacting surfaces to the perpendicular force holding them in contact.

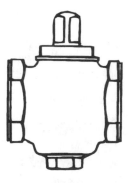

Figure C-40: Cock valve.

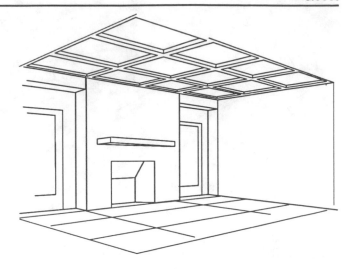

Figure C-41: Coffered ceiling.

coffer: A deeply recessed panel, in a ceiling or dome. See Fig. C-41.

cognovit note: Note authorizing confession of judgement.

coil: A wire or cable wound in a series of closed loops. Successive turns of insulated wire which create a magnetic field when an electric current passes through them.

cold-rolled steel: Steel made by either the open-hearth or the Bessemer process. The carbon content runs from 0.12 to 0.20 percent. This steel is marketed with a bright, smooth surface and is made quite accurate to size so that for many purposes no machining is necessary. It may be casehardened, but it will not temper.

collar beam: A beam above the lower ends of the rafters and attached to them. See Fig. C-42.

collector: The part of a transistor that collects electrons.

collet: A clamping ring or holding device; in the shop the term is freely applied to sockets for tapered-shank drill and to reducing sleeves and bushings of various types.

colonial: Relating to the style of architecture in vogue from early settlements in America up to the establishment of the U.S. Government; although the term is frequently applied to buildings erected as late as 1840. See Fig. C-43 for several examples.

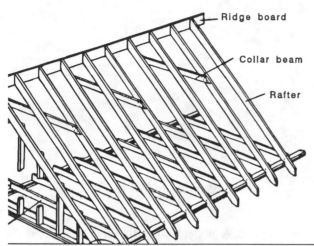

Figure C-42: Application of collar beams.

Figure C-44: Combination pliers.

colonnade: A row of columns.

color code: Identifying conductors by the use of color.

color of title: That which appears to be good title, but as a matter of fact, is not good title.

column: A vertical shaft or pillar receiving pressure in the direction of its longitudinal axis.

combination pliers: Those adjustable for size of opening by means of a slip joint; having the outer grip scored and the inner grip notched for grasping round objects. See Fig. C-44.

combination square: A square containing bevel protractor, level and center head in addition to a movable square head. See Figs. C-45.

Figure C-43: Colonial style houses.

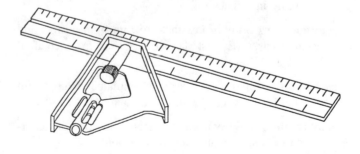

Figure C-45: Combination square.

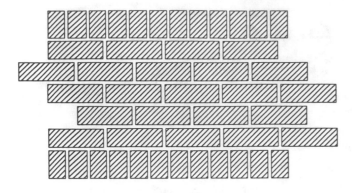

Chapter C-46: Common bond brick pattern.

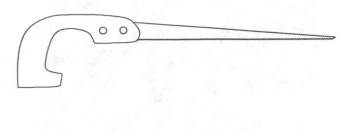

Figure C-47: Compass saw.

combustion: The chemical union of a combustible substance with oxygen, resulting in the production of heat.

comfort zone: Area on psychometric chart that shows conditions of temperature, humidity, and sometimes air movement in which most people are comfortable.

commission: Sum due a real estate broker for services in that capacity; the administrative and enforcement tribunal of real estate license laws.

common bond: Similar to a running bond, but with every fifth, sixth, or seventh course a header course, either full or Flemish, the former being all headers, the latter with headers and stretchers alternately. See Fig. C-46.

common brick: Brick such as is used for rough work or for filling in or backing.

common law: Body of law that grew up from custom and decided cases (English law) rather than from codified law (Roman law).

community property: Property accumulated through joint efforts of husband and wife living together.

commutator: Device used on electric motors or generators to maintain a unidirectional current.

compact: To increase the bearing capacity of soil, or similar material, by tamping to eliminate any void spaces.

compass plane: A plane with adjustable sole, used for smoothing concave or convex surfaces.

compass saw: A type of handsaw with small tapering blade, used for cutting in a small circle or the like. See Fig. C-47.

compound arch: A type of arch made up of a number of concentric archways, successively placed within and behind each other.

compo board: A trade name for a building board made of narrow strips of wood glued together to make a large sheet, both sides being faced with a heavy paper.

compound wound: A generator or motor having a part of a series-field winding wound on top of a part of a shunt-field winding on each of the main pole pieces.

compressibility: A density ratio determined under finite testing conditions.

compression lug or splice: Installed by compressing the connector onto the strand, hopefully into a cold weld.

compressor: The pump of a refrigerating mechanism that draws a vacuum or low pressure on the cooling side of a refrigerant cycle and squeezes or compresses the gas into the high pressure or condensing side of the cycle.

computer: An electronic apparatus 1) For rapidly solving complex and involved problems, usually mathematical or logical. 2) For storing large amounts of data. See Fig. C-48.

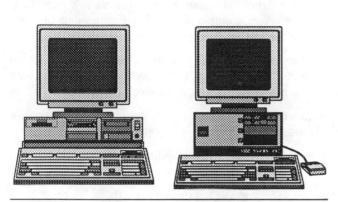

Figure C-48: Computers.

concealed: Rendered inaccessible by the structure or finish of the building. Wires in concealed raceways are considered concealed, even though they may become accessible by withdrawing them.

concentrated load: A load localized on a beam, girder, or structure.

concentricity: The measurement of the center of the conductor with respect to the center of the insulation.

concrete: Cement, sand, and gravel, with proportions varying to suit conditions.

concrete blocks: Molded blocks, usually with hollow spaces used in construction of walls. See Fig. C-49.

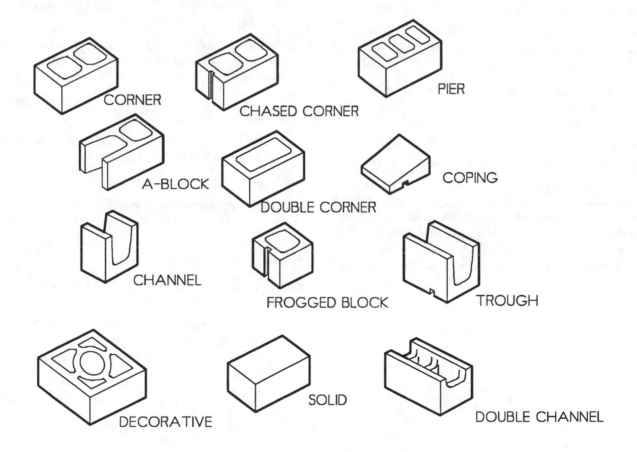

Figure C-49: Types of concrete blocks used on building construction projects.

Figure C-50: Concrete mixer.

concrete mixer: Apparatus for mixing concrete on the job site; either gas or electric powered. See Fig. C-50.

concrete truck: Truck with concrete drum that agitates and delivers ready-mixed concrete to the job site. See Fig. C-51.

condemnation: Taking private property for public use, with compensation to the owner, under the right of eminent domain.

condenser: The accumulator of electrical energy, as in a capacitor. A vessel in which the condensation of gases is effected.

condensation: Beads or films of water, or frost in cold weather, that accumulate on the inside of the exterior covering of a building when warm, moisture-laden air from the interior reaches a point where the temperature no longer permits the air to sustain as vapor the moisture it holds.

Figure C-51: Concrete truck.

condominium: Individual ownership units in a multi-family structure, combined with joint ownership of common areas of the building and ground.

conductance: The ability of material to carry an electric current. The opposite of resistance; that is, the ease with which a conductor carries an electric current.

conduction: The flow of an electric current through a conducting body, such as a copper wire.

conductivity: The relative ability of materials to carry an electrical current.

conductor: Any substance that allows energy flow through it with the transfer being made by physical contact but excluding net mass flow.

conductor, bare: Having no covering or insulation whatsoever.

conductor, covered: A conductor having one or more layers of nonconducting materials that are not recognized as insulation under the National Electric Code.

conductor, insulated: A conductor covered with material recognized as insulation.

conductor load: The mechanical loads on an aerial conductor—wind, weight, ice, etc.

conductor, plain: Consisting of only one metal.

conductor, segmental: Having sections isolated, one from the other and connected in parallel; used to reduce ac resistance.

conductor, solid: A single wire.

conductor, stranded: Assembly of several wires, usually twisted or braided.

conductor stress control: The conducting layer applied to make the conductor a smooth surface in intimate contact with the insulation; formerly called extruded strand shield (ESS).

conduit: A tubular raceway such as electrical metallic tubing (EMT), rigid metal conduit, rigid nonmetallic conduit, etc.

conduit fill: Amount of cross-sectional area used in a raceway.

conduit, rigid metal: Conduit made of Schedule 40 pipe, normally 10 foot lengths.

Condulet: A trade name for conduit fitting.

cone pulley: A stepped pulley, one having two or more faces of different diameters. Used in pairs, the large end of one being opposite the small end of the other, so that a shifting of the belt will give a change of speed.

connector: A device used to physically and electrically connect two or more conductors.

connector, pressure: A connector applied using pressure to form a cold weld between the conductor and the connector.

connector, reducing: Used to join two different size conductors.

console: A supporting bracket usually ornamented by a reverse scroll.

construction, frame: A type of construction in which the structural parts are wood or depend upon a wood frame for support. In codes, if masonry veneer is applied to the exterior walls, the structure is still classified as frame construction.

continuous beam: A beam that rests on more than two supports. See Fig. C-52.

continuous vent: A continuation of a vertical, or approximately vertical, waste pipe above the connection at which liquid wastes enter the waste pipe. The extension may or may not continue in a vertical direction.

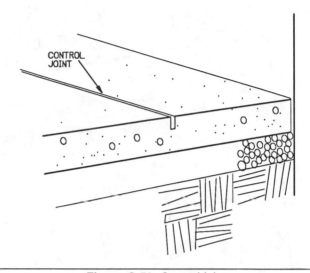

Figure C-53: Control joint.

control: Automatic or manual device used to stop, start, and/or regulate flow of gas, liquid, and/or electricity.

control joint: A joint that penetrates only partially through a concrete slab or wall so that if cracking occurs it will be a straight line at that joint. See Fig. C-53.

control, temperature: A thermostatic device that automatically stops and starts a motor, the operation of which is based on temperature changes.

controller: A device or group of devices that serves to govern in some predetermined manner the electric power delivered to the apparatus to which it is connected.

contour: The profile or section of a molding.

contour lines: Lines on a site or plot plan to indicate the elevation of the land; all points have the same elevation. See Fig. C-54 on page 60.

contractor: One who agrees to supply materials and to perform work for a specified sum.

constructive eviction: Breach of a covenant of warranty or quiet enjoyment; for example, the inability of a purchaser or lessee to obtain possession by reason of a paramount outstanding title.

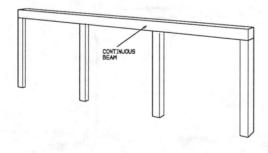

Figure C-52: Application of continuous beam.

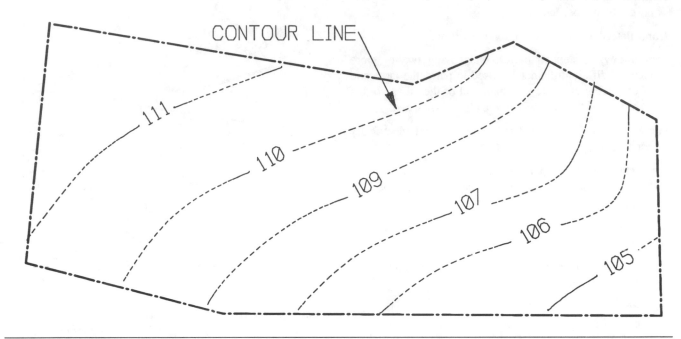

Figure C-54: Plot plan with contour lines.

constructive notice: Contract for sale.

convection: The transfer of heat to a fluid by conduction as the fluid moves past the heat source.

convenience outlet: A point on the wiring system at which current is taken to supply portable 120-volt appliances such as TV sets, toasters, etc.

conversion: Change in character, form, or function. See conversion table in appendices.

conveyance: The means or medium by which title to real estate is transferred.

cooling tower: Device that cools water by water evaporation in air. Water is cooled to the wet bulb temperature of air.

coped joint: A joint between molded pieces in which a portion of one member is cut out to receive the molded part of the other member.

coping: The cap or top course of a wall, frequently projecting.

coping machine: Used for cutting away the flanges and corners of beams and bending the ends.

coping saw: Consists of a narrow blade carried on pins set in a steel bow frame. Used for cutting curves, coping out molding, etc. See Fig. C-55.

corbel: A bracket or support; a stepping out of courses in a wall to form a ledge. See Fig. C-56.

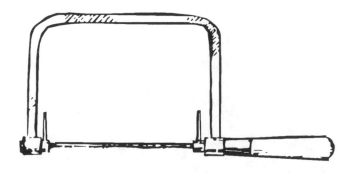

Figure C-55: Coping saw.

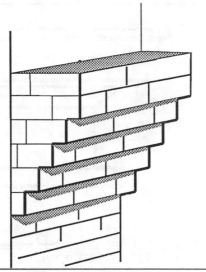

Figure C-56: Corbeled wall.

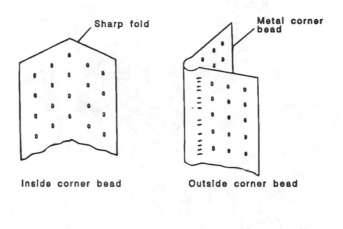

Inside corner bead   Outside corner bead

Figure C-58: Corner bead.

corbeling: Projecting courses of brick stepped out from the face of the wall forming a bracket for the wall above.

corbel out: To build out one or more courses of masonry from the face of a wall to form a support for timbers.

core: The portion of a foundry mold that shapes the interior of a hollow casing.

Corinthian order: One of the five classic orders of architecture in which the conventionalized acanthus leaf is freely used as a decoration of the capital. See Fig. C-57.

cornerbead: A metal bead built into plastered corners to prevent accidental breaking of the plaster. See Fig. C-58.

corner board: A board used as trim for the external corner of a house or other frame structure, against which the ends of siding are butted.

corner brace: A diagonal brace placed at the corner of a frame structure to stiffen and strengthen the wall. See Fig. C-59.

Figure C-57: Corinthian capital.

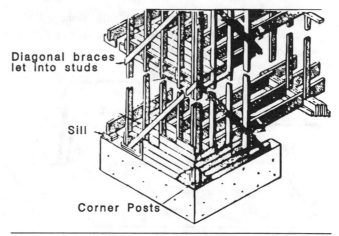

Figure C-59: Application of corner braces.

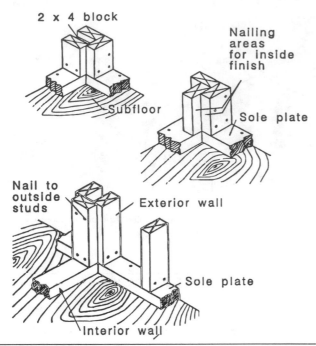

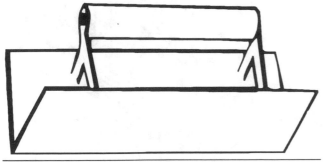

Figure C-60: Types of corner studs.

corner bit brace: A bit brace designed for use in difficult positions where the regular bit brace could not be operated.

corner stud: Framing studs arranged so that paneling, sheetrock, or similar wall covering can be attached in the corner. See Fig. C-60.

corner trowel: A plasterer's trowel with V-shaped blade for working on corners, made in both inside and outside patterns. See Fig. C-61.

Figure C-61: Corner trowel.

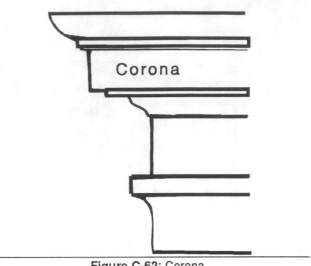

Figure C-62: Corona.,

cornice: The projection at the top of a wall.

cornice return: The underside of the cornice at the corner of the roof where the walls meet the gable end roofline. The cornice return serves as trim rather than as a structural element, providing a transition from the horizontal eave line to the sloped roofline of the gable.

corona: The portion of the cornice which projects over the bed molding to throw off the water. See Fig. C-62.

corridor: A passageway in a building, usually having rooms opening upon it.

corrosion: The deterioration of a substance (usually a metal) because of a reaction with its environment.

corrugated: Sheet metal formed into parallel ridges and depressions, alternately concave and convex.

corundum: An extremely hard aluminum oxide used as an abrasive.

cosine: The cosine of an angle is the quotient of the adjacent side divided by the hypotenuse.

cotangent: The cotangent of an angle is the quotient of the adjacent side divided by the opposite side.

cotter pin: Usually a form of split pin that is inserted into a hole near the end of a bolt to prevent a nut from working loose. See Fig. C-63.

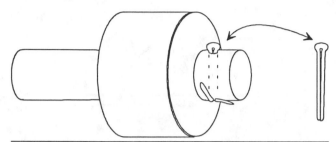

Figure C-63: Cotter pin.

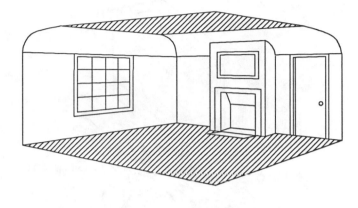

Figure C-65: Cove ceiling.

cottonwood: Also known as Carolina poplar; a large tree 50 to 75 feet in height and up to 6 feet in diameter. The wood is soft, warps easily, is hard to split. Used mainly for paper pulp, boxes, crates, etc.

counterbore: To enlarge a hole through part of its length by boring. For example, the enlarging of the beginning of a drilled hole so that the head of a screw will be flush with the surrounding surface.

counter-bracing or cross bracing: Diagonal bracing used in a truss or girder for giving additional support to the beam and relieving it of transverse stress.

counterflashing: Sheet metal strip in the form of an inverted L built into a wall to overlap the flashing and make the roof watertight.

coupling: A fitting with inside threads only, used for connecting two pieces of pipe. See Fig. C-64.

course: To arrange in a row or course. A row of bricks when laid in a wall is called a "course."

court: An open space surrounded partly or entirely by a building.

Straight coupling

Reducing coupling

Figure C-64: Sample pipe couplings.

cove ceiling: A ceiling that springs from the walls with a curve. See Fig. C-65.

cove molding: A concave molding.

covenant: An agreement between two or more persons, by deed, whereby one of the parties promises the performance or nonperformance of certain acts, or that a given state of things does or does not exist.

craft: Skill in the execution of manual work. An occupation; a trade.

cramp: Iron rod with ends bent to a right angle; used to hold blocks of stone together.

crawling: A defect appearing during application of paint or lacquer in which the film breaks, separates, or raises, due to application over a slick or glassy surface, or to surface tensions caused by heavy coatings, or use of an elastic film over a surface that is hard and brittle.

crawl space: Shallow space between the first tier of beams and the ground (no basement).

creosote: A light-colored, oily distillate of coal tar. Used as a disinfectant and as a wood preservative.

creosoting: The injecting of creosote into timber which is to be exposed to weather, in order to increase its durability.

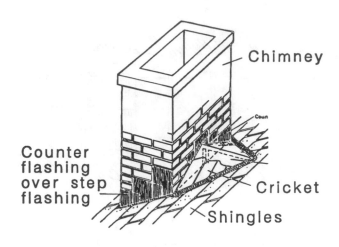

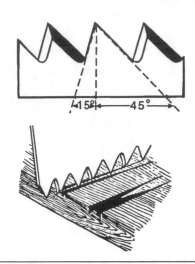

Figure C-67: Characteristics of crosscut saw teeth.

Figure C-66: Chimney flashing details, showing cricket.

cresting: An ornamental finish in the wall or ridge of a building.

cricket: Small false roof to throw off water from behind an obstacle. See Fig. C-66.

crimp: A crease formed in sheet metal for fastening purposes or to make the material less flexible.

cripple rafter: A rafter which extends from a hip to a valley rafter.

critical temperature: Temperature at which vapor and liquid have the same properties.

cross: A pipe fitting with four branches arranged in pairs, each pair on one axis and the axes of right angles.

crossbelt: A belt changed to run from the top of one pulley to the bottom of another to produce a reversal of direction.

cross connection: Physical connection between portable water supply and any non-portable water source.

crosscut saw: A saw made for cutting wood across the grain. See Fig. C-67.

crosshatching: The representation of different kinds of material by means of lines, usually drawing obliquely on a part that has been sectioned on a working drawing. See Fig. C-68.

crossover: A fitting shaped like the letter U with the ends turned out. It is used to pass the flow of one pipe past another when the pipes are in the same plane.

cross section: The resulting view if an object were sliced into two parts and one part removed.

cross tap: An electrical joint similar to a cross joint except it is accomplished by tapping two wires off the main conductor instead of using a single cross wire. See Fig. C-69.

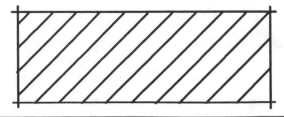

Figure C-68: Example of crosshatching on a working drawing.

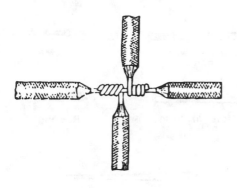

Figure C-69: Cross tap.

crowbar: A heavy pinch bar of round iron flattened to a chisel-like point at one end, the other end curved similar to a claw hammer; used as a lever in removing nails and separating nailed boards. See Fig. C-70.

crown: The uppermost member of a cornice.

crown molding: A molding with a double-curved face. The upper member of a closed cornice placed immediately below the roof proper. See Fig. C-71.

crown pulley: A pulley whose diameter is greater at the middle than at the edges of its face. This crown tends to prevent the belt from running off the pulley provided the belt is not slipping.

cryogen: A substance that becomes a superconductor at extremely cold temperatures.

cubage: Front or width of building multiplied by depth of building and by the height, figured from basement floor to the outer surfaces of walls and roof.

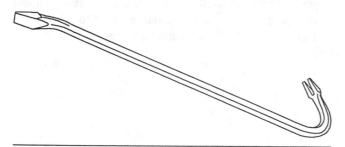

Figure C-70: Crowbar.

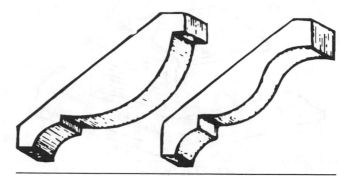

Figure C-71: Crown molding.

Cuban mahogany: A cabinet wood. The hardest and darkest mahogany of commercial importance.

cul de sac: A passage way with one outlet; a blind alley.

culvert: An artificially covered channel for the passage of water, as under a road.

cut-in: The connection of electrical service to a building, from the power company line to the service equipment, e.g., "the building was cut-in" or "the power company cut-in the service."

cut-in-card: The certificate of approval issued by the electrical inspection authority to the electrical contractor, to be given to the power company as evidence that the building electrical system is safe for connection or "cut-in" by the power company.

cup joint: A lead pipe joint in which one end of the pipe is opened enough to receive the tapered end of the adjacent pipe.

cupola: A dome; generally any small structure above the roof of a building. See Fig. C-72 on page 66.

cup shake: A division or opening between two concentric layers of timber.

curb roof: See Mansard roof.

current (I): The time rate of flow of electric charges; unit: ampere.

current-carrying capacity: The current in amperes a conductor can carry continuously under the conditions of use without exceeding its temperature rating.

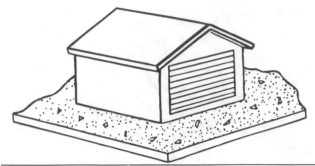

Figure C-72: Typical cupola.

current, charging: The current needed to bring the cable up to voltage; determined by capacitance of the cable; after withdrawal of voltage, the charging current returns to the circuit; the charging current will be 90° out of phase with the voltage.

current density: The current per unit cross-sectional area.

current-induced: Current in a conductor due to the application of a time-varying electromagnetic field.

current, leakage: That small amount of current which flows through insulation whenever a voltage is present and heats the insulation because of the insulation's resistance; the leakage current is in phase with the voltage, and is a power loss.

current limiting: A characteristic of short-circuit protective devices, such as fuses, by which the device operates so fast on high short circuit currents that less than a quarter wave of the alternating cycle is permitted to flow before the circuit is opened, thereby limiting the thermal and magnetic energy to a certain maximum value, regardless of the current available.

curtain wall: A thin wall supported independently of the wall below, by the structural steel or concrete frame of the building.

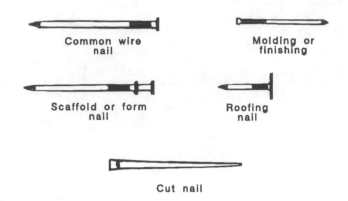

Figure C-73: Cut nail shown in comparison with others.

curtesy: The right which a husband has in his wife's estate at her death.

curtilage: Area of land occupied by a building and its yard and outbuildings, actually enclosed or considered enclosed.

cut nails: Machine-cut iron nails as distinguished from wire nails. See Fig. C-73.

cycle: 1) An interval of space or time in which one set of events or phenomena is completed. 2) A set of operations that are repeated regularly in the same sequence. 3) When a system in a given state goes through a number of different processes and finally returns to its initial state.

cyma: A commonly used molding having a reverse curve.

cypress: Grows in southern U.S.; one of the most durable of woods; somewhat like cedar, adapted to both inside and outside work in the building trades.

D

dado: 1) A plain, flat, often ornamented surface at the base of a wall, as of a room. 2) A type of joint cut at a right angle to the grain of the wood. See Fig. D-1.

dais: A raised platform at the end of a room, or a portion of the floor raised a step above the rest of the floor. See Fig. D-2.

damper: Vanes for controlling air flow. See Fig. D-3.

damping: The dissipation of energy with time or distance.

damping coil: A coil mounted near a galvanometer to produce a damping effect; i.e., to bring the needle quickly to a point of rest after deflection.

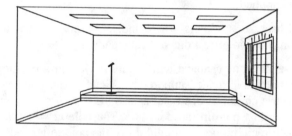

Figure D-2: Application of a dais.

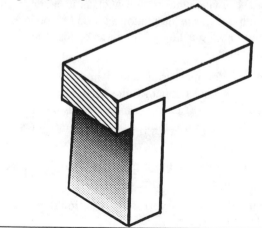

Figure D-1: Dado joint.

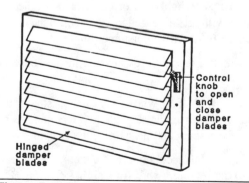

Control knob to open and close damper blades

Hinged damper blades

Figure D-3: Dampers used in HVAC ductwork.

Daniell cell: A closed-circuit type of primary cell.

darby: A two-handed, flat tool used by plasterers, especially for working on both walls and ceilings.

D'Arsonval galvanometer: A very sensitive periodic or dead-beat galvanometer in which the indicating coil is suspended in the field of a powerful horseshoe magnet.

data: Generally refers to tabulated statistical information concerning a piece of work or project.

datum: Point of reference.

datum line: Any base or fundamental line from which dimensions are taken or graphic calculations made.

DBE (Design Basis Event) (nuclear power): Postulated abnormal events used to establish the performance requirements of the structures, systems and components.

d.c.: Direct current.

dead: 1) Not having electrical charge. 2) Not having voltage applied.

dead beat: Instruments where indicators come promptly to a position of rest due to heavy damping.

dead-end: A mechanical terminating device on a building or pole to provide support at the end of an overhead electric circuit. A dead-end is also the term used to refer to the last pole in the pole line. The pole at which the electric circuiting is brought down the pole to go underground or to the building served.

deadening: The soundproofing of floors and walls by the use of insulating materials made for that purpose.

dead-front: A switchboard or panel or other electrical apparatus without "live" energized terminals or parts exposed on front where personnel might make contact.

dead level: An emphatic expression in the sense of absolutely level.

dead load: 1) A load whose pressure is steady and constant. 2) The weight, expressed in pounds per square foot, of elements that are part of the structure.

deadman: Reinforced concrete anchor set in earth and tied to the retaining wall for stability. See Fig. D-4.

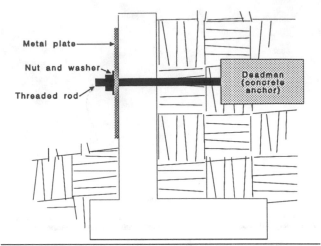

Figure D-4: Application of a deadman in a retaining wall.

deadman's switch: A switch necessitating a positive action by the operator to keep the system or equipment running or energized.

deadman timber: A large buried timber used as an anchor as for anchoring a retaining wall.

deal: A board or plank, or the wood of which the board or plank is made.

debug: To examine or test a procedure, routine, or equipment for the purpose of detecting and correcting errors especially during start-up.

decay: 1) The transmutation of a nucleus to a stable energy condition. 2) Disintegration of wood or other substance through the action of fungi, as opposed to insect damage.

decibel: A unit for measuring sound intensity, named in honor of Alexander Graham Bell. When sound or noise is created it gives off energy which is measured in decibels; i.e., the noise of an airplane engine measures 120 decibels.

deciduous: Pertaining to those trees that shed their leaves at specific seasons.

decimal: A method of expressing fractional parts by tenths, hundredths, etc.

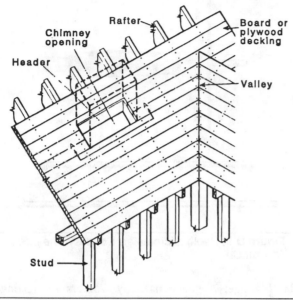

Figure D-5: Roof deck.

decimal equivalent: The value of a fraction expressed as a decimal. See Appendices.

deck: An outside platform such as a roof surface on which finish roofing is applied. See Fig. D-5:

deck paint: An enamel with a high degree of resistance to mechanical wear, designed for use on such surfaces as porch floors.

decomposition: The act, process or result of decomposing by natural decay or by chemical action.

deeping: Cutting out to a depth; placing comparatively far below the surrounding surface.

defeater: A means to deactivate a safety interlock system.

defense in depth (nuclear power): A basic design philosophy to keep nuclear power plants safe during normal operations and the worst imagined accidents.

deflection: Deviation of the central axis of a beam from normal when the beam is loaded.

deflection plate: The part of a certain type of electron tube that provides an electrical field to produce deflection of an electron beam.

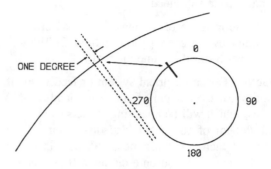

Figure D-6: Degrees in a circle.

deformed shank nail: A nail with ridges on the shank to provide better withdrawal resistance.

degree: A 360th part of the circumference of a circle. See Fig. D-6.

dehumidifying: The lessening of the moisture content of the air.

delta connection: The connection of the circuits in a three-phase system in which the terminal connections are triangular like the Greek letter delta. See Fig. D-7.

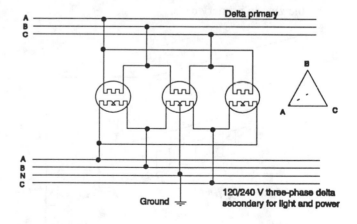

Figure D-7: Delta transformer connection.

demagnetization: The process of removing magnetism from a magnetized substance.

demand: 1) The measure of the maximum load of a utility's customer over a short period of time. 2) The load integrated over a specified time interval.

demand factor: For an electrical system or feeder circuit, this is a ratio of the amount of connected load (in kVA or amperes) which will be operating at the same time to the total amount of connected load on the circuit. An 80% demand factor, for instance, indicates that only 80% of the connected load on a circuit will ever be operating at the same time. Conductor capacity can be based on that amount of load.

demarcation line: A limit or line fixed as a boundary when performing masonry work.

demonstration (nuclear power): A course of reasoning showing that a certain result is a consequence of assumed premises: an explanation or illustration.

density: Closeness of texture or consistency.

dentil: A rectangular supporting block such as that commonly used in the bed mold of a Corinthian entablature. See Fig. D-8.

depolarization: The process of preserving the activity of a voltaic cell by preventing polarization.

depolarizer: An oxidizing substance used for fixing the hydrogen derived from the decomposition of the acid by the zinc in primary cells.

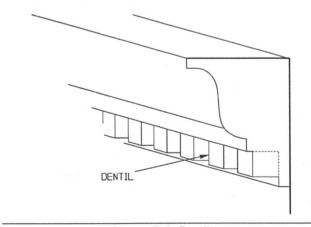

Figure D-8: Dentil.

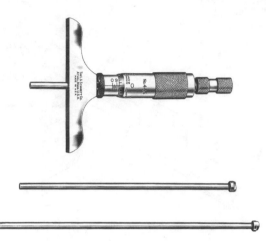

Figure D-9: Depth micrometer. *Courtesy The L.S. Starrett Co.*

depth gauge: A gauge used by workers for testing the depth of holes and recessed portions of work. A depth micrometer is shown in Fig. D-9.

derating: The intentional reduction of stress/strength ratio in the application of a material; usually for the purpose of reducing the occurrence of a stress-related failure.

derating factor: A factor used to reduce ampacity when the cable is used in environments other than the standard, or when more than the standard number of conductors are installed in a raceway.

design: To plan, originate, or show by drawings, some figure or object which can be built or reproduced; that is, an architect's design of a building; an engineer's design of the electrical and mechanical systems for a building; an artist's design of materials used for decoration.

detection: The process of obtaining the separation of the modulation component from the received signal.

deterioration: The state or process of growing worse.

determine: To find the value of, mathematically, by exact measurement.

detonation: Rapid combustion replacing normal combustion disclosed by loss of power, engine overheating, etc.; also, an explosion.

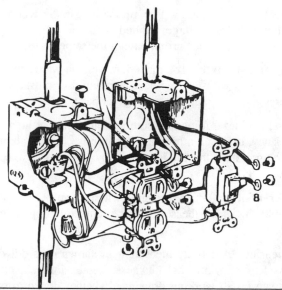

Figure D-10: A wall switch and receptacle help carry electric current but do not utilize any energy in the process.

detonator: A cap in which a spark from an electrically heated wire sets off an explosive charge with sufficient violence to explode a greater charge of dynamite.

develop: To draw a pattern of. To unfold gradually; make known in detail.

development: A drawing representing a pattern or layout.

deviation: Departure from the exact or from a set standard.

device: An item intended to carry, or help carry, but not utilize electrical energy. A wall switch, for example, helps carry current to a lighting fixture, but does not actually use any current in its operation. See Fig. D-10.

dew point: The temperature at which vapor starts to condense (liquify) from a gas-vapor mixture at constant pressure.

diagonal: A straight, oblique line, such as would divide a rectangular figure into equal parts. The braces, struts, and ties of a lattice girder are its diagonals.

diagonal bond: A brick pattern whereas the bricks are laid obliquely in a wall; used frequently in colonial houses.

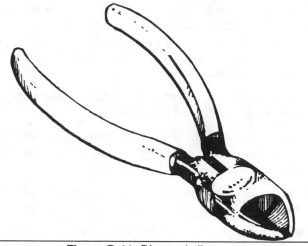

Figure D-11: Diagonal pliers.

diagonal pliers: Pliers with side-cutting edges at an angle for cutting electrical wires; diagonal pliers are also frequently used to cut the insulation from around the conductors, since the angle keeps the cutting edges from damaging the conductors. See Fig. D-11.

diagram: An outline drawing, or graphic representation.

dial: A graduated plate, usually circular or oval, on which a reading is indicated by a needle or pointer.

diameter: A straight line passing through the center of a circle, its ends terminating in the circumference. Diameter = circumference $\times \pi$.

diamond-point tool: A tool having two cutting edges inclined to each other and meeting at an acute angle. One face of the tool is flat, the other is beveled from the cutting edges. See Fig. D-12.

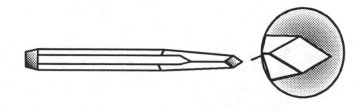

Figure D-12: Diamond-point chisel.

die: 1) Wire: a metal device having a conical hole which is used to reduce the diameter of wire which is drawn (pulled) through the die or series of dies. 2) Extruder: the fixed part of the mold. 3) An internal screw used for cutting an outside thread.

dielectric: An insulator or a term referring to the insulation between the plates of a capacitor.

dielectric absorption: The storage of charges within an insulation; evidenced by the decrease of current flow after the application of dc voltage.

dielectric constant: The ratio of the conductivity of a dielectric for electrostatic lines of force to that of air.

dielectric dispersion: The change in relative capacitance due to change in frequency.

dielectric heating: The heating of an insulating material by ac induced internal losses; normally frequencies above 10 mHz are used.

dielectric loss: The time rate at which electrical energy is transformed into heat in a dielectric when it is subjected to a changing electric field.

dielectric phase angle: The phase angle between the sinusoidal ac voltage applied to a dielectric and the component of the current having the same period.

dielectric strength: The maximum voltage which an insulation can withstand without breaking down; usually expressed as a gradient—vpm (volts per mil).

diesel engine: An internal-combustion engine of high efficiency, in which a high temperature is obtained through compression; the oil is ignited by the heat developed. See Fig. D-13.

differential motor: A motor with a compound-wound field, in which the series and shunt coils oppose each other.

dimension: A definite measurement shown on a drawing, as length, width, or thickness. Dimensions should be given on all working drawings, whether drawn to scale or not, but not on assembly drawings, except "over-all" dimensions.

dimensioning: Indicating on a drawing the sizes of various parts. See Fig. D-14.

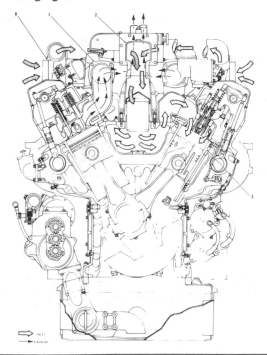

Figure D-13: Typical diesel engine.

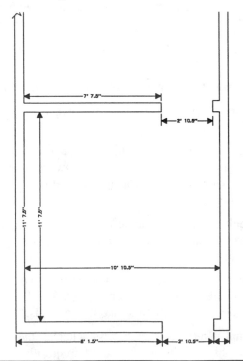

Figure D-14: Dimensioning of working drawings.

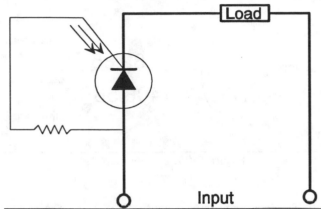

Figure D-15: Diode used in an electronic circuit.

dimension line: A line on a drawing which indicates to what part or line the dimension has reference.

diode: A device having two electrodes, the cathode and the plate or anode—used as a rectifier and detector. See Fig. D-15.

direct current (dc): Electricity which flows only in one direction; produced by a battery and dc generators.

direct drive: A compact arrangement of driving a generator by direct connection with the prime mover or the load, avoiding the use of shafts and belts. See Fig. D-16.

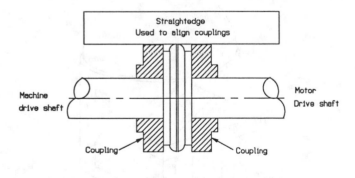

Figure D-16: Apparatus connected directly to an electric motor, rather than using belts or pulleys.

direction of lay: The lateral direction, designated as left-hand or right-hand, in which the elements of a cable run over the top of the cable as they recede from an observer looking along the axis of the cable.

direction of magnetic flux: The direction is always from the north to the south pole in a magnetic field.

disconnect: A switch for disconnecting an electrical circuit or load (motor, transformer, panel) from the conductors which supply power to it, e. g., "he pulled the motor disconnect," meaning he opened the disconnect switch to the motor. See Fig. D-17.

disconnecting means: A device, a group of devices, or other means whereby the conductors of a circuit can be disconnected from their supply source.

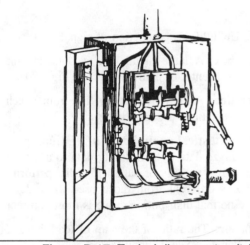

Figure D-17: Typical disconnect switch.

dispersion: Holding fine particles in suspension throughout a second substance.

displacement current: An expression for the effective current flow across a capacitor.

dissipation factor: Energy lost when voltage is applied across an insulation because of capacitance; the cotangent of the phase angle between voltage and current in a reactive component; because the shift is so great, we use the complement (angle) of the angle ⌀ which is used for power factor; dissipation factor = tan = cot ⌀; is quite sensitive to contamination and deterioration of insulation; also known as power factor (of dielectrics).

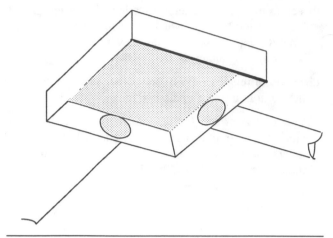

Figure D-18: Typical distribution box.

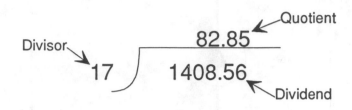

Figure D-20: Division depicted.

dividend: A number or quantity divided or to be divided into equal parts. See Fig. D-20.

dividers: Compasses for measuring or setting off distances.

DOAL: Diameter Overall.

DOC: Diameter Over Conductor; note that for cables having a stress control, the diameter over the stress control layer becomes DOC.

documents (nuclear power): Pertaining to Class 1E equipment and systems.

DOI: Diameter Over Insulation.

DOJ: Diameter Over Jacket.

dome: The vaulted roof of a rotunda; a cupola. See Figs. D-21 and D-22.

distortion: Unfaithful reproduction of signals.

distribution box: A large metal box used in conduit installation as a center of distribution. See Fig. D-18.

distribution line: An exterior supply line from which individual installations are supplied.

distribution, statistical analysis: A statistical method used to analyze data by correlating data to a theoretical curve to a) Test validity of data. b) Predict performance at conditions different from those used to produce the data; The normal distribution curve is most common.

diversity factor: The ratio of the sum of load demands to a system demand. See demand factor.

divide: To cut or part into two or more pieces.

divided light: A window having several small panes of glass. See Fig. D-19.

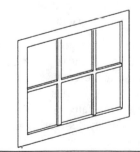

Figure D-19: Window with divided lights.

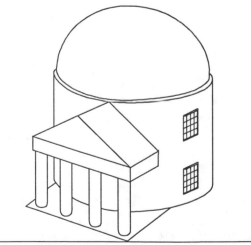

Figure D-21: Building dome.

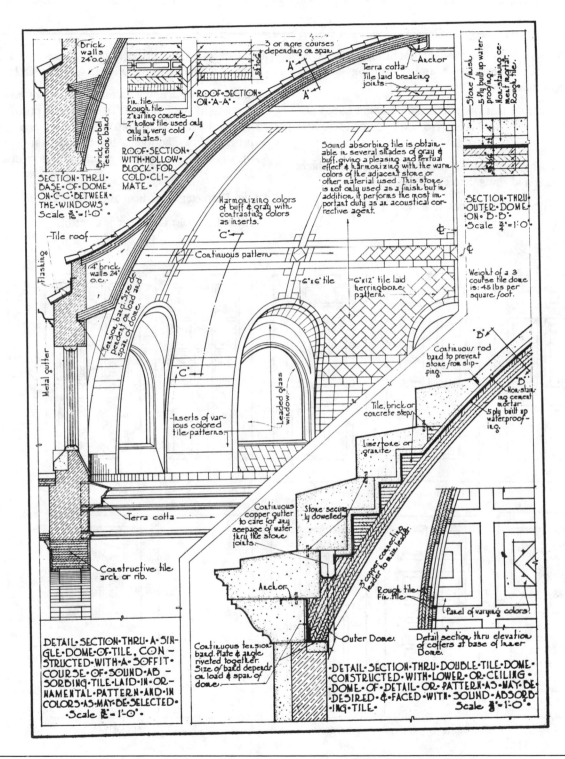

Figure D-22: Cross-section and construction details of a building dome.

75

donkey: A motor-driven power machine, on legs or wheels, used for threading and/or cutting conduit. Sometimes called "mule."

door: An assembly that fits into a door frame to keep out the elements and intruders; also used for privacy. See Fig. 23.

door frame: The surrounding case into and out of which a door opens and shuts. See Fig. D-24.

doorhead: The upper portion of a door frame. See Fig. D-24.

dormer: A roofed projection from a sloping roof, into which a dormer window is set. See Fig. D-25.

doorstop: A device, which may or may not be attached to a door, to hold it open in a desired position; usually attached near the bottom of the door and operated by pressure from the foot. Also, the strip on the inside face of a door frame against which a door closes. See Fig. D-26.

DOS: Diameter Over Insulation Shield.

DOSC: Diameter Over Stress Control.

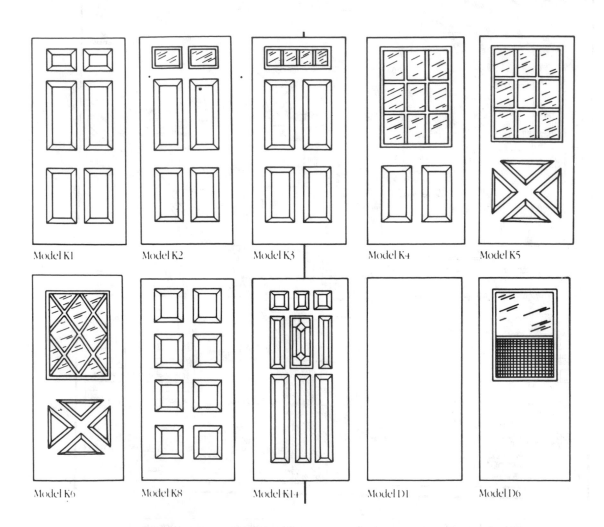

Model K1   Model K2   Model K3   Model K4   Model K5

Model K6   Model K8   Model K14   Model D1   Model D6

Figure D-23: Types of doors in common use. *Courtesy The Stanley Works.*

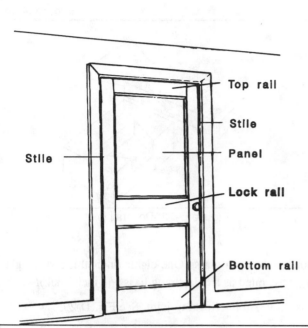

Figure D-24: Parts of a door.

Top rail
Stile
Panel
Lock rail
Bottom rail
Stile

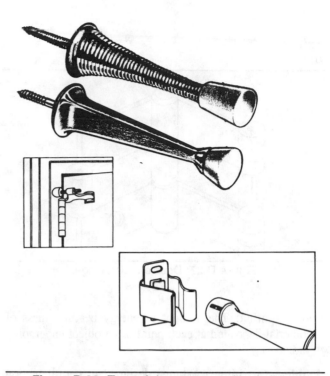

Figure D-26: Type of doorstops used on building projects. *Courtesy The Stanley Works.*

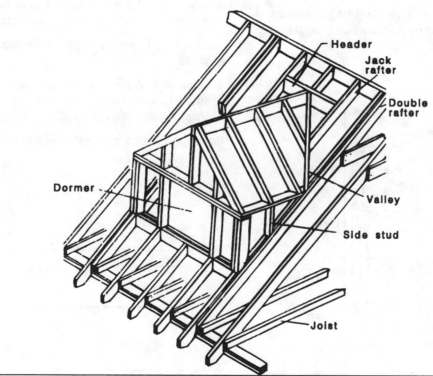

Header
Jack rafter
Double rafter
Valley
Side stud
Joist
Dormer

Figure D-25: Details of dormer construction.

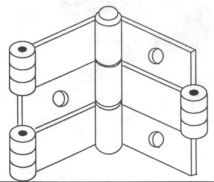

Figure D-27: Double-acting hinge.

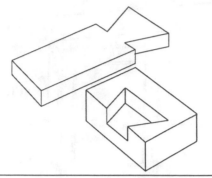

Figure D-29: Dovetail joint.

dose, radiation: The amount of energy per unit mass of material deposited at each point of an object undergoing radiation.

double-acting hinge: A hinge designed to permit motion in either direction as on folding draught screens, swinging doors, etc. See Fig. D-27

double-hung window: A window consisting of upper and lower sash, both carried by sash cord and weights. See Fig. D-28.

double-strength glass: One-eighth inch thick sheet glass (glass rated as single strength is $\frac{1}{10}$ inch thick).

dovetail: An interlocking joint. See Fig. D-29.

dovetail cutter: Inner and outer dovetails are milled with this tool, and edges of work conveniently beveled.

dovetail-halved joint: A halved joint in which both cuts are narrowed at the heel as in a dovetail.

dovetailing: Fastening together by means of dovetailed joints.

dovetail saw: A saw similar to a backsaw but somewhat smaller, with finer teeth and usually with different shape handle. See Fig. D-30.

dowel: A pin of wood used in making permanent joints.

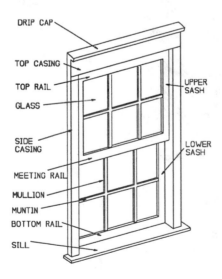

Figure D-28: Double-hung window and its related parts.

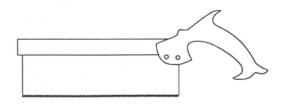

Figure D-30: Dovetail saw.

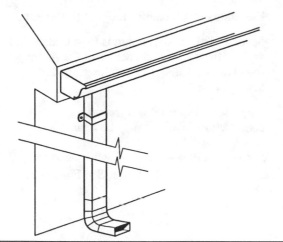

Figure D-31: Gutter and downspout.

dowel screw: A wood screw threaded at both ends.

downspout: A pipe, usually of metal, for carrying rainwater from roof gutters. See Fig. D-31.

draft indicator: An instrument used to indicate or measure chimney draft or combustion gas movement.

draft stop or fire stop: Obstructions placed in air passages to prevent the passage of flames up or across a building. See *fire stop*.

drain: Any pipe that carries waste water or water-borne wastes in a building drainage system.

drainage system: All the piping that carries sewage, rainwater, or other liquid wastes to the point of disposal or sewer.

draintile: Hollow tile used for draining wet places. See Fig. D-32.

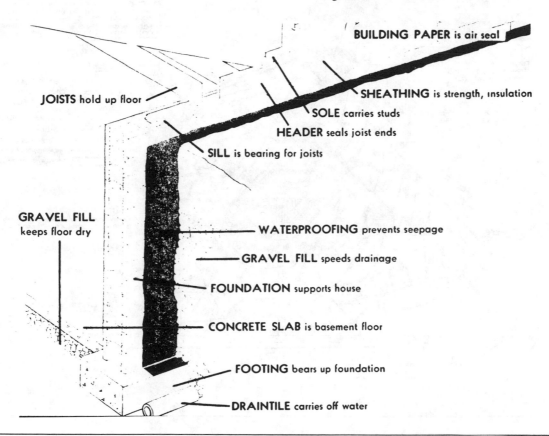

Figure D-32: Cut-away view of a building foundation showing the placement of drain tile.

drain wire: A bonding wire laid parallel to and touching shields.

drawing: Reducing wire diameter by pulling through dies.

drawing, architectural: Working drawings consisting of plans, elevations, details, and other information necessary for the construction of buildings. Architectural drawings usually include:

- A plot plan indicating the location of the building on the property.

- Floor plans showing the walls and partitions for each floor or level.

- Elevations of all exterior faces of the building.

- Several vertical cross sections to indicate clearly the various floor levels and details of the footings, foundations, walls, floors, ceilings, and roof construction.

- Large-scale detail drawings showing such construction details as may be required.

Written specification also usually accompany a set of construction working drawings. See Figs. D-33 through D-37.

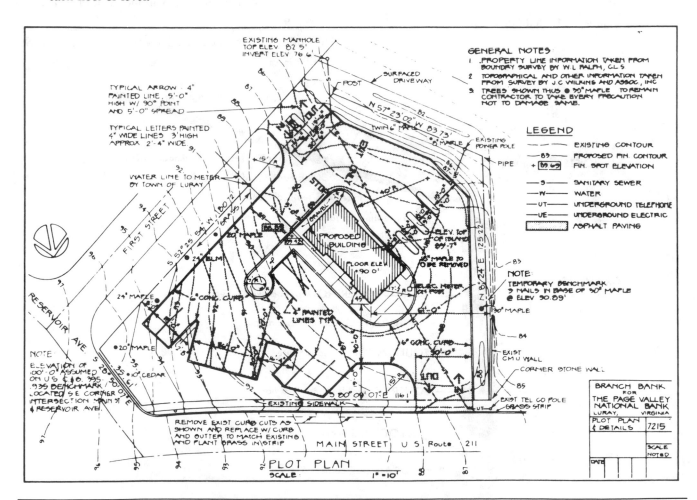

Figure D-33: Plot plan of a building-construction project.

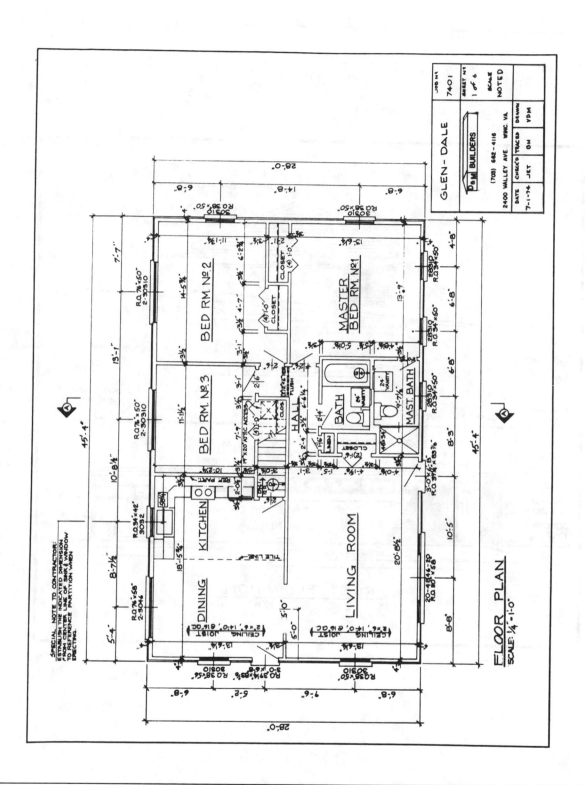

Figure D-34: Floor plan of a building construction project.

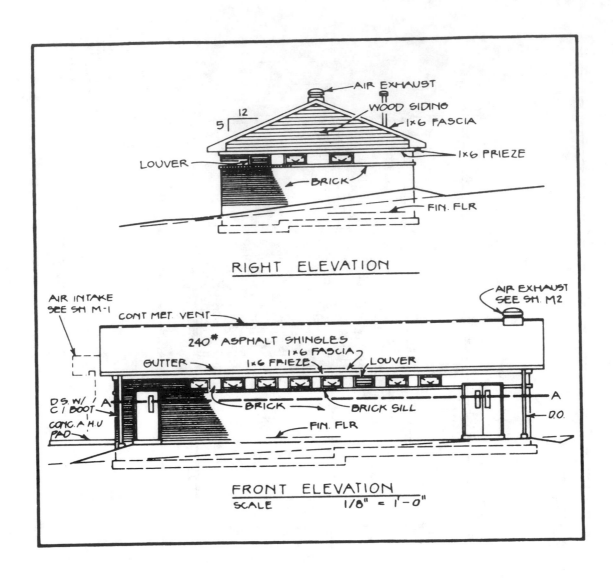

Figure D-35: Elevations of a building construction project.

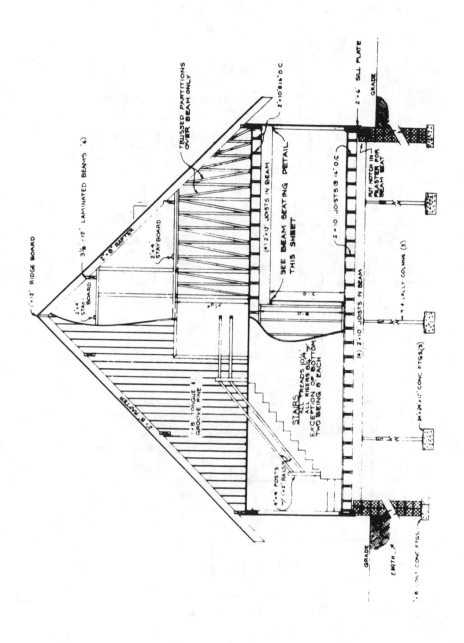

Figure D-36: Cross-section of a building construction project.

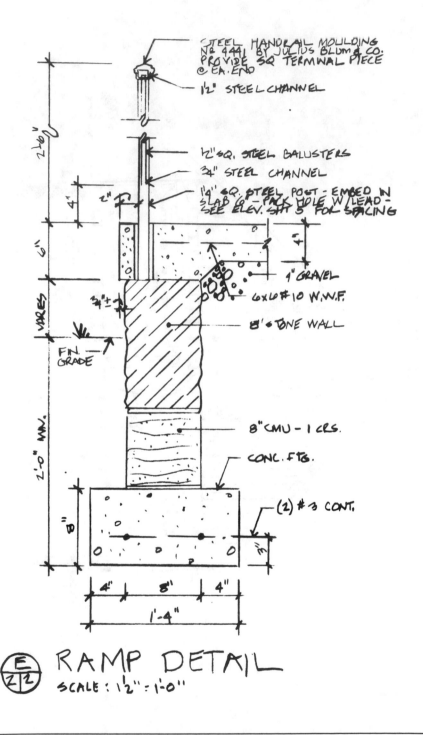

STEEL HANDRAIL MOULDING
No 4441 BY JULIUS BLUM & CO.
PROVIDE SQ. TERMINAL PIECE
@ EA. END

1½" STEEL CHANNEL

½" SQ. STEEL BALUSTERS

¾" STEEL CHANNEL

1½" SQ. STEEL POST - EMBED IN
SLAB 6" - PACK HOLE W/ LEAD -
SEE ELEV. SHT 5 FOR SPACING

1" GRAVEL

6x6#10 W.W.F.

8" STONE WALL

8" CMU - 1 CRS.

CONC. FTG.

(2) #3 CONT.

FIN. GRADE

2'-6"

4"

6"

VARIES

1'-0" MIN.

8"

4"

8"

4"

1'-4"

E 2/2 RAMP DETAIL
SCALE : 1½" = 1'-0"

Figure D-37: Construction details of a building construction project.

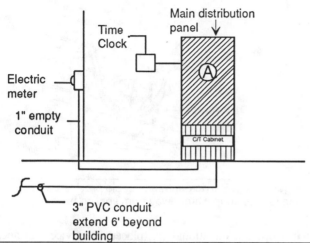

Figure D-38: Block diagram.

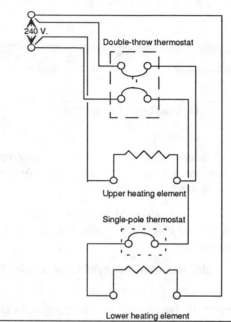

Figure D-40: Wiring diagram.

drawing, block diagram: A simplified drawing of a system showing major items as simplified blocks; normally used to show how the system works and the relationship between major items. Fig. D-38.

drawing, electrical: Consists of lines, symbols, dimensions, and notations to accurately convey an engineer's design to workmen who install the electrical system on the job. A means of conveying a large amount of exact, detailed information in an abbreviated language.

drawing, line schematic (diagram): Shows how a circuit works. Fig. D-39.

drawing, plot or layout: Shows the "floor plan." Fig. D-33 on page 78.

drawing, wiring diagram: Shows how electrical devices are interconnected. Fig. D-40.

drawknife: A two-handled wood-cutting tool having a long, narrow blade, the handles being at right angles to the blade. See Fig. D-41.

dry well: Underground excavation used for leaching of other than sewage into the ground.

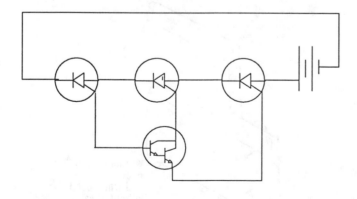

Figure D-39: Schematic diagram.

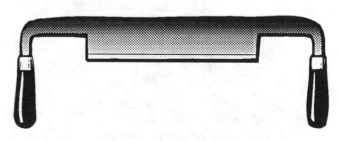

Figure D-41: Drawknife.

Figure D-42: Typical drill bit.

drift pin: A round tapered pin driven into rivet holes when they are not in perfect alignment. Also used in assembly of mechanical objects.

drift punch: A tool designed for removing drift pins or aligning rivet or bolt holes in adjacent parts so that they will coincide.

drill: A circular tool used for machining a hole. See Fig. D-42.

drip: Projecting horizontal course sloped outward to throw water away from a building.

drip cap: A molding placed on the exterior top side of a door or window frame to cause water to run off beyond the outside of the frame. See Fig. D-43.

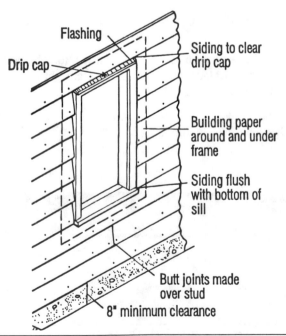

Figure D-43: Drip cap above window.

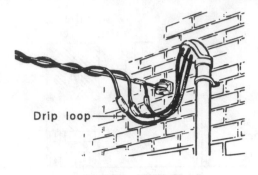

Figure D-44: Service-entrance cable provided with drip loop to keep water away from service head.

drip loop: An intentional sag placed in service en-trance conductors where they connect to the utility service drop conductors on overhead services; the drip loop will conduct any rain water to a point lower than the service head, to prevent moisture being forced into the service conductors by hydrostatic pressure and then running through the service head into the service conduit or cable. Fig. D-44.

drip mold: A molding designed to prevent rain water from running down the face of a wall.

dripstone: A label molding used over a window for throwing off rain; also called "weather molding."

drop siding: A type of weatherboard used on the exterior surface of frame buildings. See Fig. D-45.

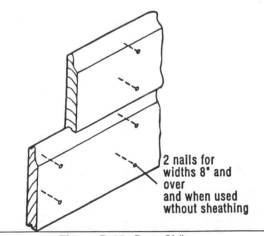

Figure D-45: Drop Siding.

drop window: A window that can be lowered into a pocket below the sill.

drum: The part of a cable reel on which the cable is wound.

drum armature: A generator or motor armature having its coils wound longitudinally or parallel to its axis.

dry: Not normally subjected to moisture.

dry battery: An electric battery made up of a number of dry voltaic cells. The term is often wrongly applied to a single dry cell.

dry bulb: An instrument with a sensitive element that measures ambient (moving) air temperature.

dry cell: A primary cell which does away with the liquid electrolyte so that it may be used in any position.

dry kiln: A chamber in which the seasoning of wood is hastened artificially.

dry well: A hole in the ground lined with stone in such a manner that liquid effluent or other sanitary wastes will leach into the surrounding soil.

dry wood: Timber from which the sap has been removed by seasoning.

drying oil: An oil which becomes dry and solid on exposure to the air. Linseed oil is a very good example.

drywall: Interior wall construction consisting of plasterboard, wood paneling, or plywood nailed directly to the studs without application of plaster. See Fig. D-46.

dual extrusion: Extruding two materials simultaneously using two extruders feeding a common cross head.

duct: A tube or channel through which air is conveyed or moved.

duct bank: Several underground conduits grouped together. See Fig. D-47 on page 88.

ductility: The ability of a material to deform plastically before fracturing.

dumbbell: A die-cut specimen of uniform thickness used for testing tensile and elongation of materials.

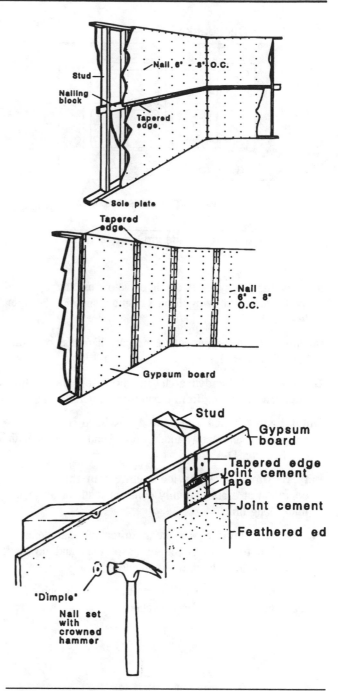

Figure D-46: Principles of installing drywall as a wall finishing material. *Courtesy U.S. Dept. of Agriculture.*

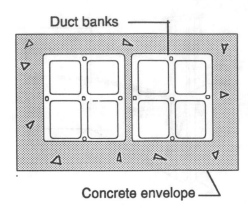

Duct banks

Concrete envelope

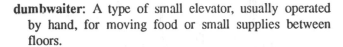

Figure D-47: Underground duct banks.

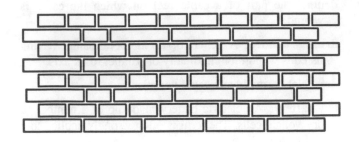

Figure D-48: Dutch bond brick pattern.

dumbwaiter: A type of small elevator, usually operated by hand, for moving food or small supplies between floors.

durometer: An instrument to measure hardness of a rubber-like material.

Dutch arch: A bonded arch, flat both top and bottom, with bricks sloping from a common center.

Dutch bond (English cross bond): A bond in which the courses are alternately made up of headers and stretchers. See Fig. D-48.

duty, continuous: A service requirement that demands operation at a substantially constant load for an indefinitely long time.

duty, intermittent: A service requirement that demands operation for alternate intervals of load and no load, load and rest, or load, no load, and rest.

duty, periodic: A type of intermittent duty in which the load conditions regularly reoccur.

duty, short-time: A requirement of service that demands operations at loads and for intervals of time, both which may be subject to wide variation.

dwarf partition: Partition that ends short of the ceiling.

dwell: A planned delay in a timed control program.

dwelling: A term of rather broad use, meaning a house or residence; the National Electrical Code recognizes a house as a single-family dwelling and an apartment as a multi-family dwelling.

dynamic: A state in which one or more quantities exhibit appreciable change within an arbitrarily short time interval.

dynamo: An electrical machine for converting mechanical energy into electrical energy. An electrical generator, especially for producing direct current.

dynamometer: A device for measuring power output or power input of a mechanism.

E

earnest money: Down payment made by a purchaser of real estate as evidence of good faith.

earth work: Excavation and piling of earth in connection with an engineering operation.

easement: 1) The curved part of a handrail or baseboard. Also, in stair construction, an ease-off or triangular piece to match the inside string and the wall base where these join at the bottom of the stair. 2)The right, liberty, advantage or privilege which one individual has in lands of another (a right of way).

eastern hemlock: A hemlock of Eastern North America.

eaves: The projecting edges of a roof. See Fig. E-1.

eaves trough: A channel or trough at the eaves of a roof for carrying off rain water; wood gutter. See Fig. E-2.

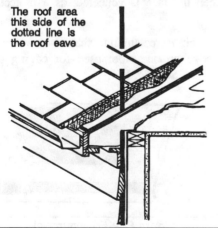

The roof area this side of the dotted line is the roof eave

Figure E-1: Eave of a roof.

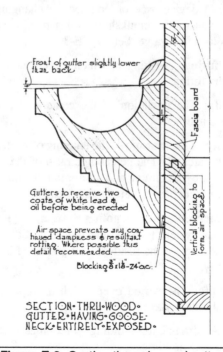

Front of gutter slightly lower than back

Fascia board

Gutters to receive two coats of white lead & oil before being erected

Air space prevents any continued dampness & resultant rotting. Where possible this detail recommended.

Vertical blocking to form air space.

Blocking ⅝ x1⅛-24"o.c.

SECTION·THRU·WOOD· GUTTER·HAVING·GOOSE- NECK·ENTIRELY·EXPOSED·

Figure E-2: Section through wood gutter.

Figure E-3: Eccentric.

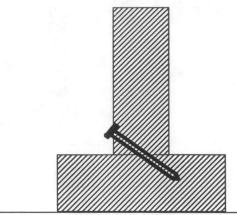

Figure E-4: Edge nailing.

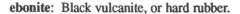

ebonite: Black vulcanite, or hard rubber.

ebonize: To stain or finish in imitation of ebony.

ebony: A hard, heavy, very dark wood used in cabinet-work.

ebulition: The bubbling of a liquid; boiling.

EC: Electrical Conductor of Aluminum.

eccentric: A device used on engines for changing the rotary motion of the crankshaft into a reciprocating motion on the slide valve. See Fig. E-3.

eccentric fitting: A fitting in which the center line is offset in the fitting.

eccentricity: The deviation of the centers of two circles from one another.

ecdentricity: 1) A measure of the entering of an item within a circular area. 2) The percentage ratio of the difference between the maximum and minimum thickness to the minimum thickness of an annular area.

ECCS (Emergency Core Cooling System): (n u c l e a r power): A system to flood the fueled portion of the reactor and remove the residual heat produced by radioactive decay.

economic life: The period over which a property may be profitably utilized.

echinus: An ornamental molding, sometimes spoken of as egg-and-dart molding.

economy wall: A brick wall four inches thick blanketed with back mortaring, strengthened at intervals with vertical pilasters, having brick corbelling for the support of floors and roof, providing a four-inch outside reveal for doors and windows, and with every window and door frame bricked in.

eddy currents: Circulating currents induced in conducting materials by varying magnetic fields; usually considered undesirable because they represent loss of energy and cause heating.

edgenailing: Nailing into the edge of a board. See Fig. E-4.

edging: The small, solid squares set in on the edge of a top when the face is veneered, as a protection to the veneer.

edging trowel: A rectangular shaped trowel with turned down edge on one side; used for edging pavements, curbing, etc. See Fig. E-5.

Figure E-5: Edging trowel.

Figure E-6: Edifice.

edifice: A building. A term usually applied to architecture distinguished for its size or grandeur. See Fig. E-6.

edison base: The standard screw base used for ordinary lamps. See Fig. E-7.

EEI: Edison Electric Institute.

effective temperature: Overall effect on a person of air temperature, humidity, and air movement.

effervescence: The escape of bubbles of gas from a liquid otherwise than by boiling.

efficiency: The ratio of the output to the input.

efficiency apartment: An apartment consisting typically of a combination living room and bedroom area, bathroom, and a small kitchen alcove.

efflorescence: The formation of a whitish powder on the face of a brick or stone wall.

effluent: The liquid discharged from a septic tank or sewage disposal plant.

ejectment: A form of action to regain possession of real property, with damages for the unlawful retention.

elasticity: That property of recovering original size and shape after deformation.

elastomer: A material which, at room temperature, stretches under low stress to at least twice its length and snaps back to the original length upon release of stress.

elbow: A short conduit that is bent.

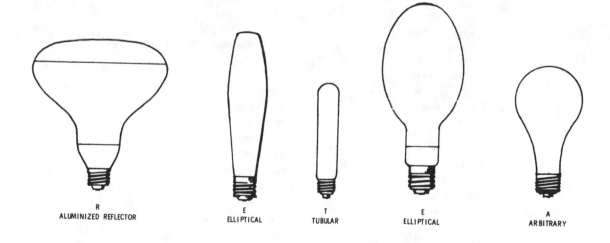

R
ALUMINIZED REFLECTOR

E
ELLIPTICAL

T
TUBULAR

E
ELLIPTICAL

A
ARBITRARY

Figure E-7: Types of Edison-base lamps.

electrical engineer: One well versed in the theory and practice of the development and utilization of electrical energy.

electrical symbols: Graphical symbols used on electrical drawings. In preparing electrical drawings, most engineers and designers used symbols adopted by the United States of America Standards Institute (USASI). However, many designers frequently modify these standard symbols to suit their own needs. For this reason, most drawings will have a symbol list or legend. Electrical drawing symbols found on building construc-

tion drawings appear in the appendices. Symbols used by one consulting engineering firm appear in Fig. E-8.

electric defrosting: Use of electric resistance heating coils to melt ice and frost off evaporators during defrosting.

electric heating: House heating system in which heat from electrical resistance units is used to heat rooms. See Fig. E-9.

electric power: The rate of doing work by electricity; measured in watts or kilowatts.

⊢——b	Fluorescent strip
▭a	Fluorescent fixture
▢c	Incandescent fixture, recessed
○b	Incandescent fixture, surface or pendant
⊢○a	Incandescent fixture, wall-mounted
Ⓔ	Letter "E" inside fixtures indicates connection to emergency lighting circuit
	Note: on fixture symbol, letter outside denotes switch control.
⊗	Exit light, surface or pendant
⊢⊗	Exit light, wall-mounted
△A	Indicates fixture type
⊖	Receptacle, duplex-grounded
⊖wp	Receptacle, weatherproof
⊖s	Combination switch and receptacle
⊙	Receptacle, floor type
⊢◑ 3P-30A	Receptacle, polarized (poles and amperes indicated)
s	Switch, single-pole
s3,4	Switch, three-way, four-way
sP	Switch and pilot light
sT	Switch, toggle w/ thermal overload protection
▢	Push button
▱	Buzzer
▬	Light or power panel
▥	Telephone cabinet
Ⓙ	Junction box
▭⌐	Disconnect switch-FSS: fused safety switch; NFSS: Non-fused safety switch

——	Conduit, concealed in ceiling or wall
– – – –	Conduit, concealed in floor or wall
- - - - - -	Conduit exposed
—×—	Flexible metallic armored cable
⟶	Home run to panel - number of arrowheads indicates number of circuits. Note: any circuit without further designation indicates a two-wire circuit. For a greater number of wires, read as follows: ⊬⊬⊬3 wires, ⊬⊬⊬⊬4 wires, etc.
—T—	Telephone conduit
—TV—	Television-antenna conduit
—S—⊬	Sound-system conduit-number of crossmarks indicates number of pairs of conductors.
Ⓕ	Fan coil unit connection
○	Motor connection
M.H.	Mounting height
▣F	Fire-alarm striking station
Ⓖ	Fire-alarm gong
Ⓓ	Fire detector
ⓈⒹ	Smoke Detector
Ⓑ	Program bell
Ⓨ	Yard gong
Ⓒ	Clock
Ⓜ	Microphone, wall-mounted
▣M	Microphone, floor-mounted
▷S	Speaker, wall-mounted
Ⓢ	Speaker, recessed
Ⓥ	Volume control
▲	Telephone outlet, wall
⏢	Telephone outlet, floor

Figure E-8: Electrical drawing symbols.

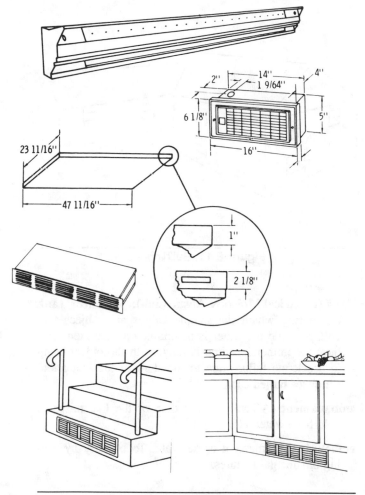

Figure E-9: Electric heating units.

electric water valve: Solenoid type (electrically operated) valve used to turn water flow on and off.

electrician: A person familiar with the theory and practical application of electricity.

electricity: Relating to the flow or presence of charged particles; a fundamental physical force or energy.

electrocution: Death caused by electrical current through the heart, usually in excess of 50 ma.

electrode: A conductor through which current transfers to another material.

electrolysis: The production of chemical changes by the passage of current from an electrode to an electrolyte or vice versa.

electrolyte: A liquid or solid that conducts electricity by the flow of ions.

electrolytic condenser-capacitor: Plate or surface capable of storing small electrical charges. Common electrolytic condensers are formed by rolling thin sheets of foil between insulating materials. Condenser capacity is expressed in microfarads.

electromagnet: A device consisting of a ferromagnetic core and a coil that produces appreciable magnetic effects only when an electric current exists in the coil.

electromotive force (emf) voltage: Electrical force that causes current (free electrons) to flow or move in an electrical circuit. Unit of measurement is the volt.

electron: The subatomic particle that carries the unit negative charge of electricity.

electronegative gas: A type of insulating gas used in pressure cables; such as SF6.

electron emission: The release of electrons from the surface of a material into surrounding space due to heat, light, high voltage, or other causes.

electronics: The science dealing with the development and application of devices and systems involving the flow of electrons in vacuum, gaseous media, and semiconductors.

electro-osmosis: The movement of fluids through diaphragms because of electric current.

electroplating: Depositing a metal in an adherent form upon an object using electrolysis.

electropneumatic: An electrically controlled pneumatic device.

electrostatics: Electrical charges at rest in the frame of reference.

electrotherapy: The use of electricity in treatment of disease.

electrothermics: Direct transformations of electric and heat energy.

Figure E-10: Elizabethan style of architecture.

Figure E-11: Building ell.

electrotinning: Depositing tin on an object.

elevation: Drawing showing the projection of a building on a vertical plane.

Elizabethan: Relating to the time of Elizabeth, Queen of England. See Fig. E-10.

ell: A wing or addition to a building forming an angle with the main structure. See Fig. E-11.

ellipse: A plane curve, such that the sum of the distance from any point of the curve of two fixed points is a constant.

ellipsoid: A solid, every plane section of which is an ellipse or a circle.

elliptical arch: An arch, elliptical in form, being described from three centers. See Fig. E-12.

elm: A coarse, open-grained wood much given to warping; is very sparingly used in engineering work.

elongation: 1) The fractional increase in length of a material stressed in tension. 2) The amount of stretch of a material in a given length before breaking.

EMA (Electrical Moisture Absorption): A water tank test during which the sample cables are subjected to voltage while the water is maintained at rated temperature; the immersion time is long, with the object being to accelerate failure due to moisture in the insulation; simulates buried cable.

embankment: A bank, mound dike or the like raised to hold back water, etc.

embellish: To ornament or decorate. To beautify by adding ornamental features.

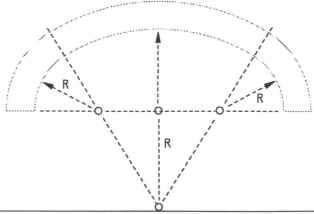

Figure E-12: Elliptical arch.

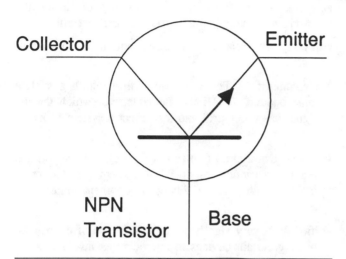

Figure E-13: Parts of a transistor.

emery: A species of corundum composed of oxide of alumina, iron, silica, and a small portion of lime. It is used as an abrasive.

emery cloth: Powdered emery glued on thin cloth, used for removing file marks and for polishing metallic surfaces.

emery wheel: A wheel made of emery. It is revolved at a high speed and is used for grinding.

EMI: Electromagnetic interference.

eminent domain: The right of the people or government to take private property for public use upon payment of compensation.

emitter: The part of a transistor that emits electrons. See Fig. E-13.

emulsifying agent: A material that increases the stability of an emulsion.

emulsion: The colloidal suspension of one liquid in another liquid, such as oil in water for lubrication.

enamel: A material applied as a paint which dries with a hard, glossy finish.

enameled brick: Brick with a glazed or enamel-like surface.

enameled wire: Wire insulated with a thin baked-on varnish enamel, used in coils to allow the maximum number of turns in a given space.

encased knot: A knot whose growth rings are not intergrown and homogeneous with the growth rings of the piece it is in.

enclosed: Surrounded by a case that will prevent anyone from accidentally touching live parts.

encroachment: A building, part of building, or obstruction which intrudes upon or invades a highway or sidewalk or trespasses upon property of another.

encumbrance: A claim, lien, charge, or liability attached to and binding upon real property, such as a judgement, unpaid taxes, or a right of way; defined in law as any right to, or interest in, land which may subsist in another to the diminution of its value, but consistent with the passing of the fee.

end bearing pile: Pile acting like a column; the point has a solid bearing in rock or other dense material.

end-grain: That face of a piece of timber exposed by the cutting of its fibers transversely.

end-lap joint: A corner joint formed by halving both pieces for a distance equal to their widths. See Fig. E-14.

endless saw: A band saw. See Fig. E-15 on page 96.

endnailing: Nailing into the end of a board, which results in very poor withdrawal resistance.

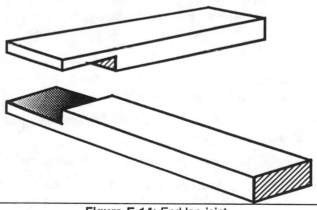

Figure E-14: End-lap joint.

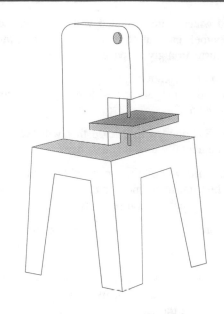

Figure E-15: Endless or band saw.

energy: A body is said to possess energy when it is capable of doing work or of overcoming resistance. Energy may be either kinetic or potential.

engine: A machine or mechanism by which generated power is applied to doing work, as the steam, gas, or gasoline engine.

engineer: An expert in design, construction, and development in the fields of electricity, mining mechanics, building, etc.

engineering: The art and science relating to expert planning and constructing in various fields of industry.

English bond: See Dutch bond.

English cross bond: Same as Dutch bond. A bond in which the courses are alternately headers and stretchers.

engraving: 1) The act or art of cutting designs on a plate. 2) An engraved design.

ensemble: The work considered as a whole, not a part.

entablature: The uppermost member of a classical order sometimes considered as all that portion above the column. See Fig. E-16.

entresol: A low floor between two higher floors, the lower one usually being a ground floor; mezzanine.

entryway: A passage for affording entrance into a building.

environment: 1) The universe within which a system must operate. 2) All the elements over which the designer has no control and that affect a system or its inputs.

EPA (Environmental Protection Agency): The federal regulatory agency responsible for keeping and improving the quality of our living environment—mainly air and water.

epitaxial: A very significant thin film type of deposit for making certain devices in microcircuits involving a realignment of molecules.

EPRI (Electric Power Research Institute): An organization to develop and manage a technology for improving electric power production, distribution and utilization; sponsored by electric utilities.

equilibrium: Properties are time constant.

equipment: A general term including material, fittings, devices appliances, fixtures, apparatus, and the like used as part of, or in connection with, an electrical installation.

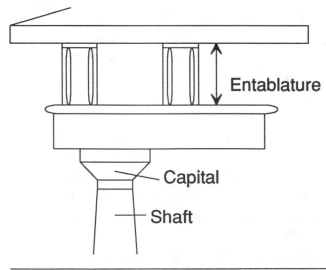

Figure E-16: Entablature.

equipotential: Having the same voltage at all points.

equity: The interest or value which an owner has in real estate over and above the mortgage against it; system of legal rules administered by courts of chancery.

equity of redemption: Right of original owner to reclaim property sold through foreclosure proceedings on a mortgage, by payment of debt, interest and costs.

equivalent: Numbers or quantities numerically equal to each other but expressed in different terms.

equivalent circuit: An arrangement of circuit elements that has characteristics over a range of interest electrically equivalent to those of a different circuit or device.

erecting: The final building up, or putting together in position.

erosion: Destruction by abrasive action of fluids.

error: Fault, mistake, omission, inaccuracy. Deviation from that which is correct or desired.

escalator: A moving stairway such as is used in stores, railroad stations, etc.

escheat: Reversion of property to the sovereign state owing to lack of any heirs capable of inheriting.

escrow: A deed delivered to a third person for the grantee to be held by him until the fulfillment or performance of some act or condition.

escutcheon: The plate about a keyhole or the one to which a door knocker is attached. See Fig. E-17.

escutcheon pins: Small roundhead nails, usually brass, used for attaching escutcheon plates.

estate: The degree, quantity, nature and extent of interest which a person has in real property.

estate in reversion: The residue of an estate left in the grantor, to commence in possession after the termination of some particular estate granted by him. In a lease, the lessor has the estate in reversion after the lease is terminated.

essex board measure: Means for the rapid calculation of board feet. See Fig. E-18.

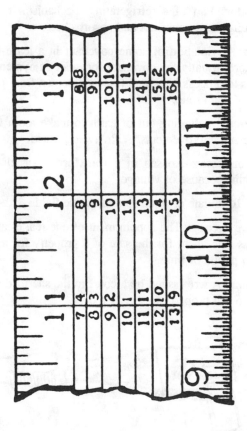

Figure E-18: Essex method of board measure, found on most framing squares. Courtesy The Stanley Works.

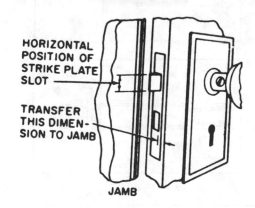

HORIZONTAL POSITION OF STRIKE PLATE SLOT

TRANSFER THIS DIMENSION TO JAMB

JAMB

Figure E-17: Escutcheon.

estimating: Calculating the amount of material required for a piece of work; also the labor required and determining the value of the finished product.

etching: Revealing structural details of a metal surface using chemical or electrolytic action.

ethics: That branch of moral science, which treats of the duties which a member of a professional or craft owes to the public, to his client, and to the other members of the profession.

ETL: Electrical Testing Laboratory.

et ux.: Abbreviation for et uxor, meaning "and wife."

evaporation: A term applied to the changing of a liquid to a gas, heat is absorbed in this process.

evaporator: Part of a refrigerating mechanism in which the refrigerant vaporizes and absorbs heat.

eviction: A violation of some covenant in a lease by the landlord, usually the covenant for quiet enjoyment; also refers to process instituted to oust a person from possession of real estate.

excavation: A digging out of earth to make room for engineering improvements. A cavity so formed.

excitation losses: Losses in a transformer or electrical machine because of voltage.

excite: To initiate or develop a magnetic field.

exclusive agency: The appointment of one real estate broker as sole agent for the sale of a property for a designated period of time.

execution: A writ issued by a court to the sheriff directing him to sell property to satisfy a debt.

executor: A person named in a will to carry out its provisions.

expanded: That which is increased in size without increasing the substance.

expansion bit: A boring bit having a cutter or cutters arranged to permit radial adjustment, to enable one tool to bore holes of different diameters. See Fig. E-19.

expansion bolt: Bolt with a casing arranged to wedge the bolt into a masonry wall to provide an anchor. See Fig. E-20

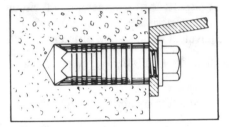

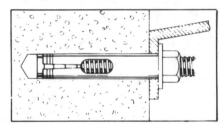

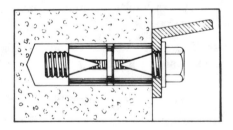

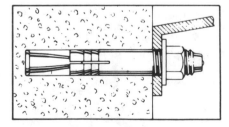

Figure E-19: Expansion bit.

Figure E-20: Types of expansion bolts.

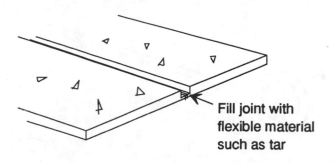

Figure E-21: Expansion joint.

expansion joint: Joint between two adjoining concrete members arranged to permit expansion and contraction with temperature changes. See Fig. E-21.

expansion, thermal: The fractional change in unit length per unit temperature change.

expansion valve: A device in a refrigerating system that maintains a pressure difference between the high side and low side and is operated by pressure.

experiment: A test or trial for the purpose of determining the results obtainable under certain known conditions.

explosion proof: Designed and constructed to withstand an internal explosion without creating an external explosion or fire. See Fig. E-22.

explosive: That which is liable to explode or to violently burst forth from within by force.

exponent: A small number or symbol placed above and to the right of a mathematical quantity, to indicate the number of times the quantity is to be taken as a factor.

exponential: Pertaining to the mathematical expression; $y = aebx$.

exposed (as applied to live parts): Live parts that a person could inadvertently touch or approach nearer than a safe distance. This term is applied to parts not suitably guarded, isolated, insulated.

exposed (as applied to wiring method): Not concealed; externally operable; capable of being operated without exposing the operator to contact with live parts.

extender: A substance added to a plastic to reduce the amount of the primary resin required per unit area.

extending: Carrying out farther than the original point or limit.

exterior: The outside of, as a whole or in part, as the exterior of a building; an exterior wall.

exterior finish: A type of finish suited to outside work.

external thread: The thread on the outside of a screw, bolt, or pipe.

extrados: The outside curve of an arch. See Fig. E-23.

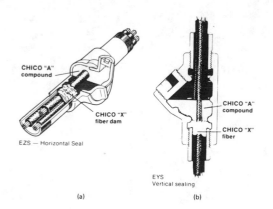

Figure E-22: Typical explosionproof fittings.

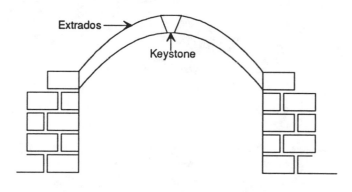

Figure E-23: Extrados.

extraction: The transfer of a material from a substance to a liquid in contact with the substance.

extraneous: Not essential; foreign.

extrude: To form materials to a given cross section by forcing through a die.

extruder types: a) Strip—uses strips of compound. b) Powder/pellet—uses compound in powder or pellet form.

extrusion of metal: A process by which hot or cold metal is forced under high pressure through an opening to produce a desired shape.

eye: Opening on the head of a pounding or cutting tools—hammer or axe—into which a handle is inserted.

eyebolt: A bolt provided with a hole or eye at one end, instead of the usual head. The eye receives a pin, stud, or hook, that takes the pull of the bolt.

eye dormer: A dormer that has a gable roof. See *dormer*.

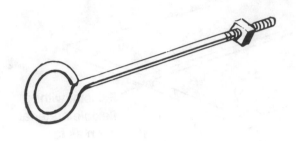

Figure E-24: Eyebolt.

eyelet: Something used on printed circuit boards to make reliable connections from one side of the board to the other.

eye protection: Protective devices such as glasses, goggles, face shields, welding helmets, etc. to protect the eyes from adverse conditions.

F

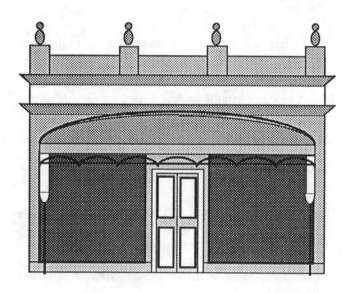

Figure F-1: The front elevation of a building is called *facade*.

fabrication: The act of building or putting together. Forming into a whole by uniting or assembling of parts.

facade: The front elevation or exterior face of a building. See Fig. F-1.

face: An operation that machines the sides or ends of the piece.

facebrick: Brick of the better quality, such as is used on walls prominently exposed.

face mark: A mark placed on one surface of a piece of wood which is being worked, in order that it may be identified as the face from which other surfaces are trued.

face mold: The full size or scale drawing or diagram of the curved portions of a sloping handrail. The drawing gives the true dimension and shape of the top of the handrail.

facenailing: Nailing perpendicular to the initial surface being penetrated. Also termed direct nailing. See Fig. F-2.

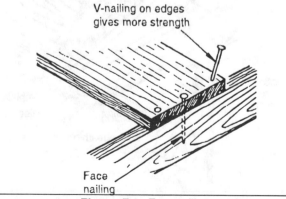

Figure F-2: Facenailing.

facies: A geologic term designating a portion of a rock formation that differs in mineral composition from other portions of the rock, as a shale facies interbedded with or grading into sandstone; a facies of diorite enclosed in a body of granite.

facilitate: To promote easier operation or completion.

facsimile: An exact copy or reproduction.

factor of safety: Radio of ultimate strength of material to maximum permissible stress in use.

Fahrenheit: Gabriel Fahrenheit, a German physicist (1686-1736). His name is given to the commonly used thermometer scale in which the freezing point is 32 degrees, and the boiling point 212 degrees.

fail-safe control: Device that opens a circuit when the sensing element fails to operate.

failure: The inability of materials and structures to endure or accomplish the work for which they were selected and designed.

faldstool: A desk at which the litany is read, as in the Church of England; also a type of folding stool or chair.

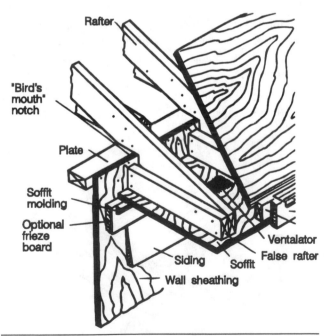

Figure F-3: Application of false rafters.

Figure F-4: Typical fan blower. Courtesy Emerson Electric.

false rafter: A short extension added to a main rafter over a cornice; especially where there is a change of roof line. See Fig. F-3.

fan: A radial or axial flow device used for moving or producing artificial currents of air.

fan blower: A rotating fan for producing a current of air. It may be used for carrying off fumes as of chemicals, for ventilating, and for forced draft in furnaces. See Fig. F-4:

"Fannie Mae": The secondary mortgage market. It provides a market for mortgages held by primary lenders, such as banks and savings and loan associations and provides the primary market with a ready market for mortgages, so as to permit greater turnover of money for loans.

FAO: This symbol on a mechanical drawing means that the piece is machined or finished all over.

farad: The basic unit of capacitance.

fascia: A flat board, band, or face, used by itself or, more often, in combination with moldings, generally located at the outer face of the cornice. See Fig. F-5.

fascia backer: The main structural support member to which the fascia is nailed.

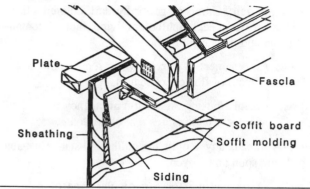

Figure F-5: Fascia used on cornice molding.

faucet: A fixture attached to the end of a pipe having a spout and valve to permit the drawing off and control of flow of liquids. See Fig. F-6.

faun: A legendary demigod, represented by a half goat and half man, used largely as a decorative detail in work of the Adam period.

favus: Diaper detail, resembling the cells in a honeycomb.

feather-edge: A keen edge, tapering off to nothing.

feedback: The process of transferring energy from the output circuit, of a device back to its input.

feeder: The conductors between the main service equipment and the branch circuit panelboards. See Fig. F-7.

feed pipe: A main line pipe; one which carries a supply directly to the point of use, or to secondary lines.

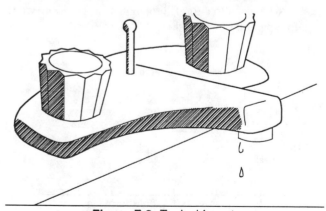

Figure F-6: Typical faucet.

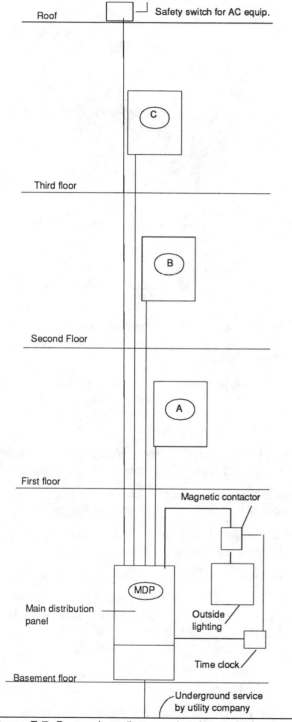

Figure F-7: Power riser diagram showing main service equipment, subpanels, and feeders.

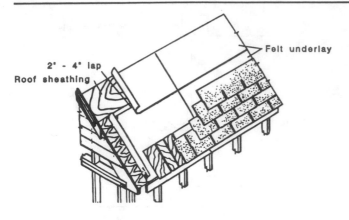

Figure F-8: Felt paper used as sheathing paper under roof shingles.

fee simple: The largest estate or ownership in real property; free from all manner of conditions or encumbrances.

feldspar: Any one of a group of rock-forming minerals which consist of silicates of aluminum with potassium, sodium, or calcium.

felt papers: Used as sheathing papers on roofs and side walls for protection against dampness, heat, and cold. Those used for roofing are often impregnated with tar, asphalt, or chemical compounds. See Fig. F-8.

female: The recessed portion of any piece of work into which another part fits is called the female portion.

female thread: A thread which is cut in a hole or on a hollow surface.

fender: A metal guard, often quite decorative, placed before an open fireplace to protect the floor.

fender pile: Outside row of piles that protects a pier or wharf from damage by ships.

fenestral: In early times, a frame on which oiled paper or thin cloth was fastened to keep out wind and rain, before or when windows were not glazed.

fenestration: The design or arrangement of the windows of a building.

feretory: A shrine, either portable or fixed, in which the relics of saints are kept.

ferromagnesian: Group of generally dark-colored silicate minerals (pyroxene, amphibole, olivine, biotite) containing iron and magnesium.

festoon: An ornament of carved work, representing a garland or wreath of flowers or leaves, or both.

F.H.A.: Federal Housing Authority; an agency of the Federal Government that insures real estate loans.

fiber: A thread; any tough, threadlike substance capable of being spun and woven.

Fiberlic: A trade name for a particular type of building board.

fibrous: As applied to the structure of metals; the opposite of granular.

fiddleback: Name given to a chair in which the back panel is somewhat similar to a fiddle in shape.

field magnet: The electromagnet by which the magnetic field of force is produced in a generator.

figure: In wood, the mottled, streaked, or wavy grain.

filament: A cathode in the form of a metal wire in an electron tube.

file: A hard steel instrument, made in various shapes and sizes, for smoothing wood or metal. See Fig. F-9.

file card: A kind of brush fitted with short, fine wires; used for cleaning files. See Fig. F-10.

filigree: Delicate ornamental work, used chiefly in decorating gold and silver.

filler: A composition for filling holes or pores in wood before painting or varnishing.

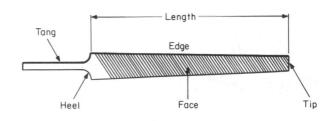

Figure F-9: Typical metal file.

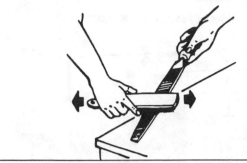

Figure F-10: File card. *Courtesy Sears.*

fillet: The rounded corner or portion that joins two surfaces which are at an angle to each other. See Fig. F-11.

filling in: The process of building in the center of the wall between the face and back.

fillister: A plane for making grooves.

filter: A porous article through which a gas or liquid is passed to separate out matter in suspension; a circuit or devices that pass one frequency or frequency band while blocking others, or vice versa.

finder's fee: A fee or commission paid to a broker for obtaining a mortgage loan for a client or for referring a mortgage loan to a broker. It may also refer to a commission paid to a broker for locating a property.

finger: A narrow, projecting piece used as a guide or index in various kinds of work.

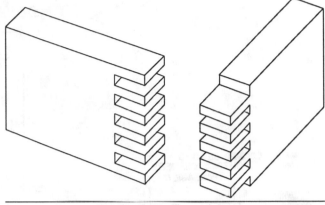

Figure F-12: Finger joint.

finger joint: Composed of five tongues or fingers interlocking; used on table brackets. See Fig. F-12.

finial: An ornament at the top of a spire or steeple. A terminating or crowning detail. See Fig. F-13.

finish plaster: Final or white coat of plaster. See Fig. F-14 on page 106.

finished string: The end string of a stair, secured to the rough carriage; cut, mitered, dressed, and often finished with a bead or molding.

finishing: Completing or bringing to an end. Perfecting finally.

Fillet

Figure F-11: Fillet.

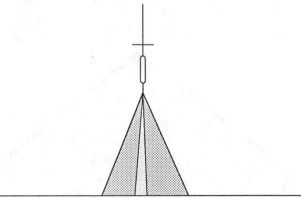

Figure F-13: Spire with finial at top.

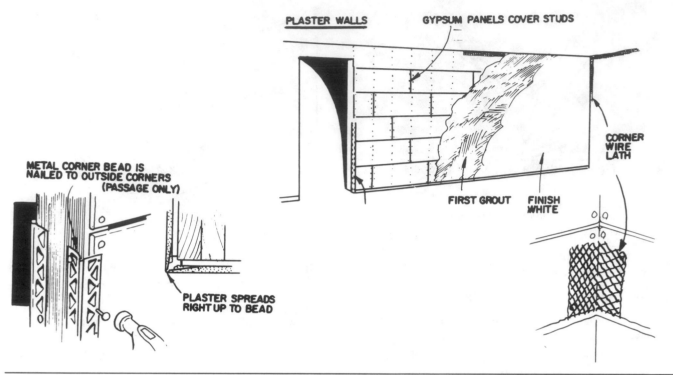

Figure F-14: Principles of plastering.

Fink truss: A type of roof truss commonly used for short spans. It is very economical due to the shortness of its struts. See Fig. F-15.

firebrick: Brick made to withstand high temperatures that is used for lining chimneys, incinerators, and similar structures.

fire clay: Clay capable of withstanding high temperature; its quality is due to the large amount of silica and small amount of fluxing agents. It is usually light in color and is used for firebrick, retorts, furnace linings, etc.

fire irons: Metal utensils for a fireplace, usually matching the fender. See Fig. F-16.

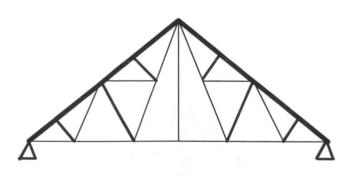

Figure F-15: Fink truss.

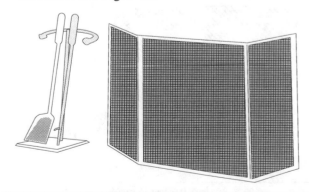

Figure F-16: Fire irons and fire screen.

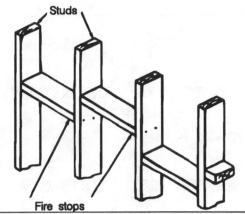

Figure F-17: Building structure showing firestops.

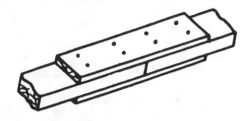

Figure F-18: Details of a fished joint.

fireproof: To build with a minimum amount of combustible material.

fireproof wood: Chemically treated wood; fire-resistive, used where incombustible materials are required.

fire-rated doors: Doors designed to resist standard fire tests and labeled for identification.

fire-resistance rating: The time in hours the material or construction will withstand fire exposure as determined by certain standards.

fire screen: A framelike piece of furniture, used for protection against the heat of the fireplace and flying sparks. See Fig. F-16.

firestops: Filling in air passages, or passages through which flames might travel, preferby with some incombustible material, although standard lumber is frequently used. See Fig. F-17.

firm commitment: A commitment by the F.H.A. to insure a mortgage on specified property with a specified mortgagor.

fish: To fish wire or cable means to pull it through conduit, raceway or other confined spaces, like walls or ceilings.

fish tape: A flexible metal tape for fishing through conduits or other raceway to pull in wires or cables; also made in non-metallic form of "rigid rope" for hand fishing of raceways.

fished joint: When a stud or other piece is to be lengthened, an extra piece may be butted against it longitudinally and the joint covered by two pieces which are nailed or bolted to opposite sides. See Fig. F-18.

fitment: Any article made and fixed to a wall or room, including paneling, chimney pieces, and fitted furniture.

fitting: An accessory such as a locknut, bushing, or other part of a wiring system that is intended primarily to perform a mechanical rather than an electrical function. See Fig. F-19.

Figure F-19: Typical electrical fittings.

Figure F-20: Fixture splice.

fixture: An article that was once personalty, but has become real estate by reason of its permanent attachment in or to the improvement.

fixture supply: Water supply pipe that connects a fixture to a branch water supply pipe or directly to a main water supply pipe.

fixture wire: Usually 16 or 18 gauge, solid or stranded and insulated. It is used for wiring electric fixtures. See Fig. F-20.

fixture unit, drainage (dfu): A measure of the probable discharge into the drainage system by various plumbing fixtures. In general, on small systems, one dfu approximates one cubic foot of water a minute.

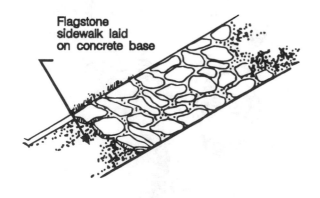

Figure F-21: Application of flagstone.

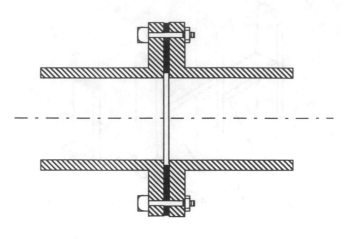

Figure F-22: Cross-section of a flange union.

fixture unit, water supply (sfu): A measure of the probable water demand by various plumbing fixtures.

flagstone: Flat stones, from 1 to 4 inches thick, used for rustic walks, steps, and floors. See Fig. F-21.

flamboyant: A name applied to the ornamentation of a certain period of French architecture.

flame retardant: 1) Does not support or convey flame. 2) An additive for rubber or plastic that enhances its flame resistance.

flange union: A pair of flanges to be threaded onto the ends of pipes to be joined. The flanges are bolted together when the pipes are joined. See Fig. F-22.

flank: The side of an arch.

flapper valve: The type of valve used in refrigeration compressors that allows gaseous refrigerants to flow in only one direction.

flare: Copper tubing is often connected to parts of a refrigerating system by flared fittings. They require that the end of the tube be expanded at about a 45-degree angle.

flashing: Metal strips placed around roof openings to provide water tightness. See Figure F-23.

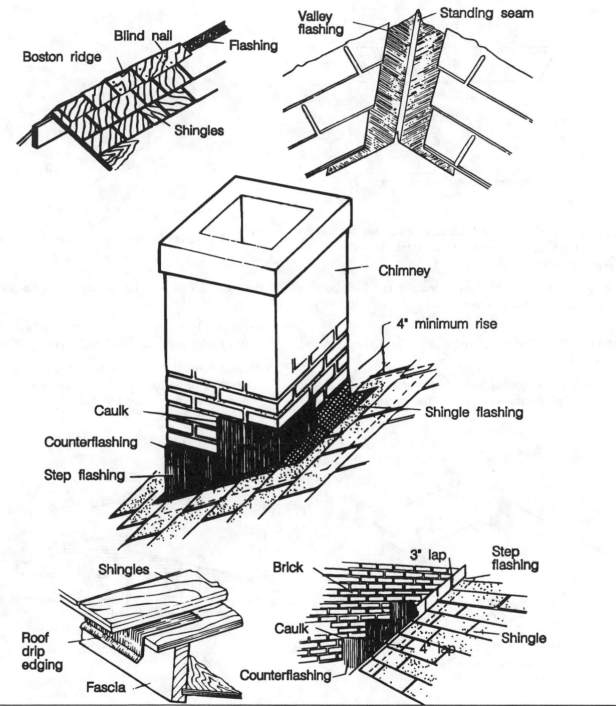

Figure F-23: A variety of flashing applications.

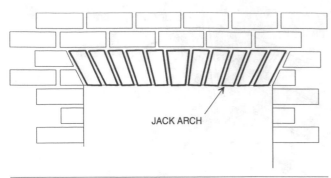

Figure 24: Flat or jack arch.

flash point: Lowest temperature at which vapors above a volatile combustible substance ignite in air when exposed to flame.

flat arch, or jack arch: A construction in which both the soffits and extrados are flat. Frequently used in fireplace construction. See Fig. F-24.

flat molding: Thin, flat strips used for finishing woodwork.

flat paint: An interior paint that contains a high proportion of pigment, and dries to a flat or lusterless finish.

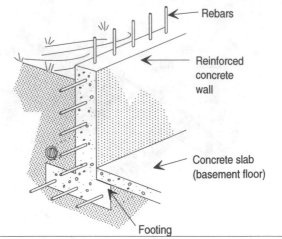

Figure 26: Reinforced concrete construction.

flat roof: A roof with just enough pitch for drainage. See Fig. F-25.

flat skylight: A skylight, the entire surface of which is flat, having only sufficient pitch to carry off water.

flat slab construction: Reinforced concrete floor construction of uniform thickness; eliminates the drops of beams and girders. See Fig. F-26.

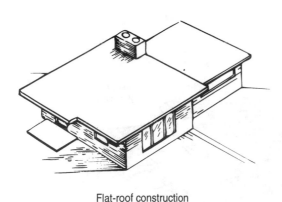

Flat-roof construction

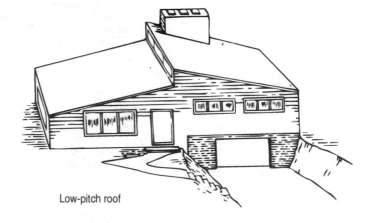

Low-pitch roof

Figure 25: Flat roof (left) and low-pitch roof for comparison.

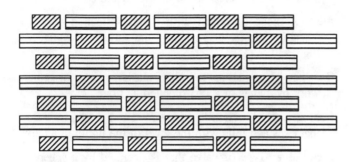

Figure F-27: Flemish and double Flemish bond.

flatting: A veneering process used on buckled veneers.

flaw: A crack or fracture in a casting or forging. In general, any defect which may eventually cause failure.

fleam: The angle of bevel of the edge of a saw tooth with respect to the plane of the blade.

Flemish bond: Pattern of bonding in brickwork consisting of alternate headers and stretchers in the same course. See Fig. F-27.

Flemish bond, double: When both the inner and outer surfaces of an exposed wall are laid in Flemish bond, all headers being true or full headers, the bond is termed "double Flemish bond." See Fig. F-27.

Flemish garden bond: Consists of three stretchers followed by a header in each course. The headers in each course center between the stretchers in the course above and below. See *garden bond*.

fleur'-de-lis': The royal insignia of France. It is widely used as a decorative unit. See Fig. F-28.

flier: A stair tread that is of uniform width throughout its length.

flight of stairs: The series of steps between floors or landings. Two flights of stairs may be broken by a landing. See Fig. F-29.

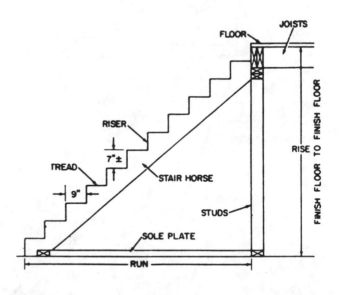

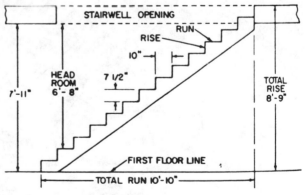

Figure F-28: Fleur'-de-lis

Figure F-29: Flight of stairs.

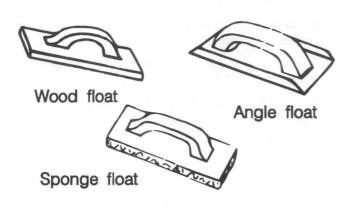

Wood float

Angle float

Sponge float

Figure F-30: Several types of floats used in masonary work.

flitch beam: Built-up beam consisting of a steel plate sandwiched between wood members and bolted.

float: A piece of board with handle attached used for spreading plaster or stucco on the surface of walls. See Fig. F-30.

floating: The equal spreading of plaster, stucco, or cement work by means of a board called a "float."

float switch: A switch that is opened and closed by a float that rises and falls with the level of the liquid in a tank.

float valve: Type of valve that is operated by a sphere or pan which floats on a liquid surface and controls the level of liquid.

flooding: Act of filling a space with a liquid.

flood level rim: The edge of a fixture receptacle from which water overflows.

floor: That portion of a structure or building on which one walks. Also a story of a building is often referred to as a floor, as ground floor, first floor, etc. See Fig. F-31 for details of floor framing.

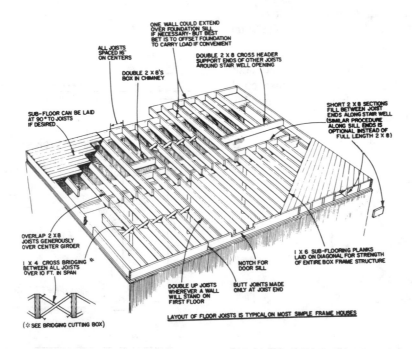

Figure F-31: Details of floor framing.

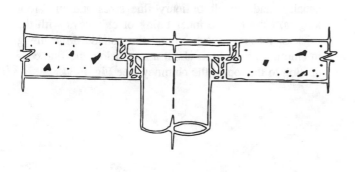

Figure F-32: Typical floor drain.

floor drain: A fixture used to drain water from floors into the plumbing system. In homes, floor drains are usually located in the laundry and near the water tank, and are fitted with a deep seal trap. See Fig. F-32.

floor plan: A drawing which shows the length and breadth of a building and of the rooms which it contains. A separate plan is made for each floor. See Fig. F-33.

flowering dogwood: A small, ornamental tree bearing white flowers. Wood is very hard and tough; valuable for golfstick heads, tool handles, and scales.

flow meter: Instrument used to measure velocity or volume of fluid movement.

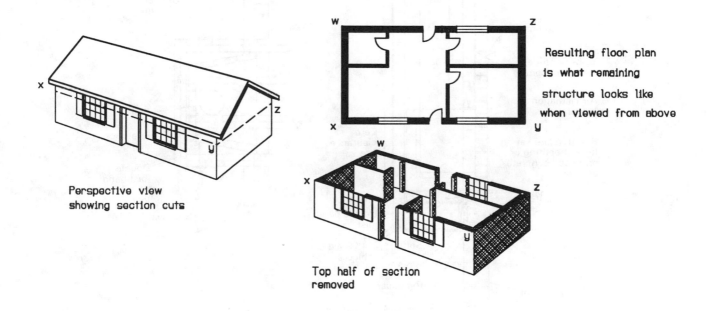

Perspective view showing section cuts

Resulting floor plan is what remaining structure looks like when viewed from above

Top half of section removed

Figure F-33: Principles of floor-plan layout.

fluctuation: A variation, an irregular change of movement.

flue: The space or passage in a chimney through which smoke, gas, or fumes ascend. Each such passage is called a flue, which together with any others and the surrounding masonry make up the chimney. See Fig. F-34.

flue lining: Fire clay or terra cotta pipe, round or square, usually made in all ordinary flue sizes and in 2-foot lengths, used for the inner lining of chimneys with the brick or masonry work around the outside. Flue lining in chimneys runs from about a foot below the flue connection to the top of the chimney. See Fig. F-34.

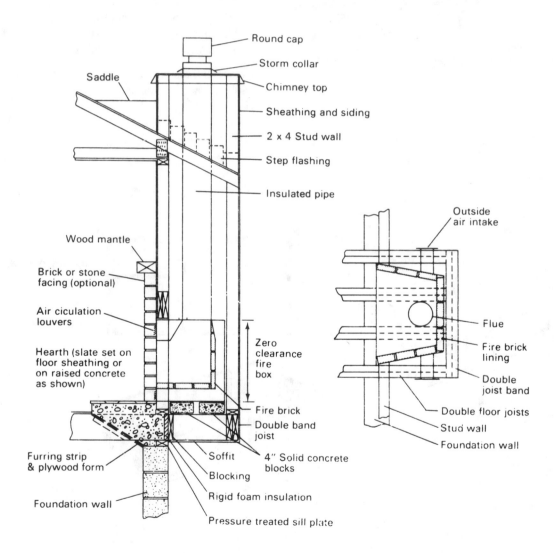

Figure F-34: Cross-section of chimney showing all details of construction. *Courtesy U.S. Dept. of Agriculture.*

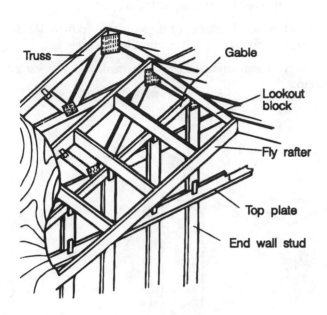

Figure F-35: Details of framing showing fly rafter.

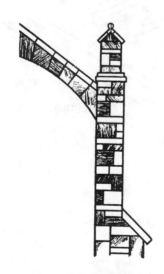

Figure F-36: Flying buttress.

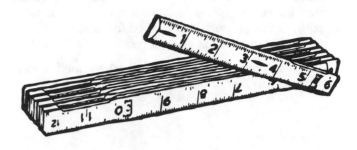

Figure F-37: Folding rule.

fluid: A substance that yields to any force tending to change its form without changing its volume; consisting of particles which move and change in shape but do not separate.

flush: Parts are said to be flush when their surfaces are on the same level.

flush valve: A valve used for flushing a plumbing fixture by using water directly from the water-supply pipes or in connection with a special flush tank.

flute: A concave channel as in a column, baluster, leg, frieze, etc.

flux: Any substance or mixture used to promote the fusion of metals or minerals.

fly rafters: End rafters of the roof overhang supported by sheathing and lookouts. See Fig. F-35.

flying buttress: Consists of a detached pier or buttress connected with a wall some distance from it by a portion of an arch, thus distributing the roof thrust. Has frequent application in church architecture. See Fig. F-36.

foil: A leaflike division in architectural ornamentation, used in groups of three, a trefoil, of four, a quatrefoil, etc.

folding rule: A collapsible instrument used for measuring. See Fig. F-37.

foliation: The banding or lamination of metamorphic rocks arising from parallelism of mineral grains (distinct from stratification in sedimentary rocks).

footing: A concrete section in a rectangular form, wider than the bottom of the foundation wall or pier it supports. With a pressure-treated wood foundation, a gravel footing may be used in place of concrete. See Fig. F-38.

forced convection: Movement of fluid by mechanical force such as fans or pumps.

foreclosure: A court process instituted by a mortgagee or lien creditor to defeat any interest or redemption which the debtor-owner may have in the property.

foreman: A man in charge of a group of workmen. He is usually responsible to a superintendent or manager.

foreshore: Land between high-water mark and low-water mark.

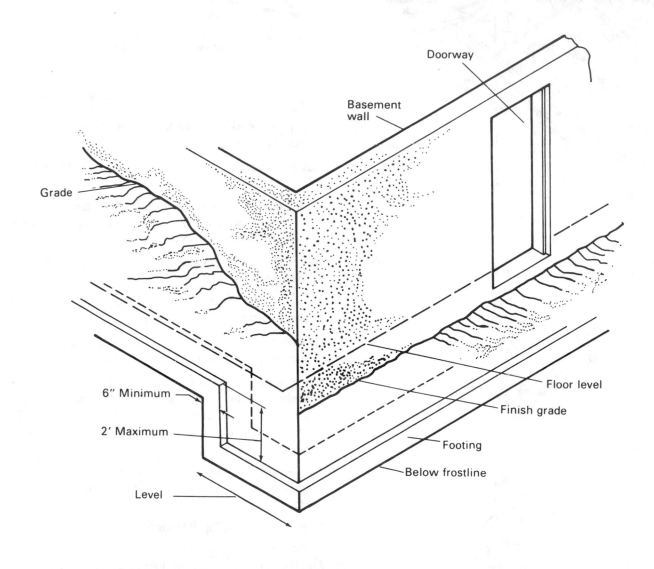

Figure F-38: Details of Foundations and related components.

furniture: Useful and decorative movable articles, such as tables, chairs, etc., placed in a building.

furred: Provided with wood strips so as to form an air space between the walls and the plastering.

furring: The leveling up or building out of a part of a wall or ceiling by wood strips, etc.

furring strips: Pieces of wood attached to a surface, as for lathing.

fuse: Electrical safety device consisting of a strip of fusible metal in a circuit that melts when the current is overloaded.

fuse block: A porcelain or slate base to which are fastened fuse clips or other contacts for holding fuses.

fuse clips: The spring part of a cutout or switch that holds the ferrules of a cartridge fuse.

fuse link: The fusible part of a cartridge case.

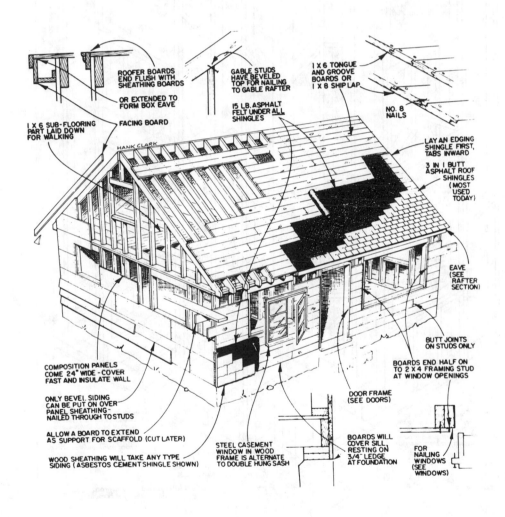

Figure F-39: Details of house framing.

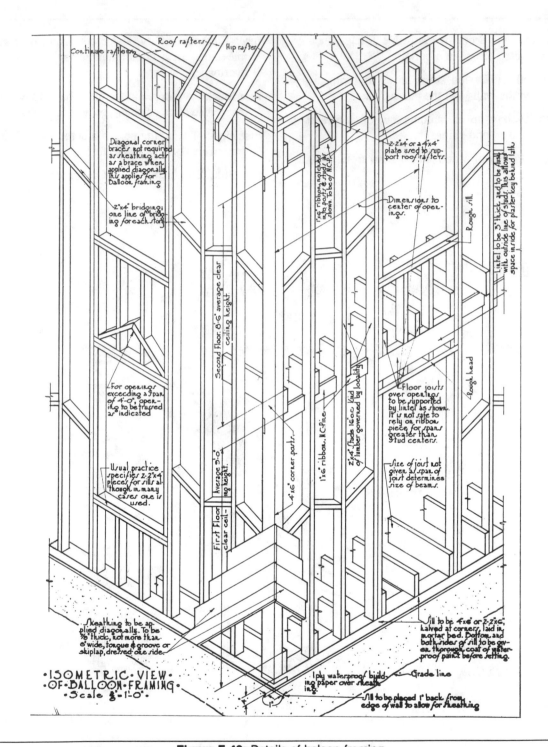

Continue rafters

Roof rafters

Hip rafters

Diagonal corner braces not required as sheathing acts as a brace when applied diagonally. This applies for balloon framing

2"x4" bridging one line of bridging for each story

For openings exceeding a span of 4'-0", opening to be trussed as indicated

Usual practice specifies 2-2"x4" pieces for sills although in many cases one is used.

Sheathing to be applied diagonally. To be 7/8" thick, not more than 6" wide, tongue & groove or shiplap, dressed one side.

Second Floor 8'-6" average clear ceiling height

First Floor Average 9'-0" clear ceiling height.

4"x6" corner posts.

1"x6" ribbon, N.C. Pine.

2"x4" studs 16"o.c. Kind of lumber governed by locality.

1"x6 ribbon, notched into posts & studs as shown. To be of N.C.P.

2-2"x4 or a 4"x4" plate used to support roof rafters.

Dimensions to center of openings.

Floor joists over openings to be supported by lintel as shown. It is not safe to rely on ribbon piece for spans greater than stud centers.

Size of joist not given as span of joist determines size of beams.

Rough sill

Lintel to be 3" thick and to be flush with outside line of stud. This allows space inside for plaster key behind lath.

Rough head

Sill to be 4"x6" or 2-2"x6, halved at corners, laid in mortar bed. Bottom and both sides of sill to be given a thorough coat of waterproof paint before setting.

1 ply waterproof building paper over sheathing

Grade line

Sill to be placed 1" back from edge of wall to allow for sheathing

·ISOMETRIC·VIEW·
·OF·BALLOON·FRAMING·
·Scale 3/4"=1'-0".

Figure F-40: Details of boloon framing.

framing: The skeleton work of a structure. The act of erecting the same.

framing, balloon: A system of framing in which all exterior studs extend in one piece from the sill plate to the roof plate. See Fig. F-40 on page 118.

framing, ladder: Framing for the roof overhang at a gable. Cross pieces are used similar to a ladder to support the overhang.

framing, platform: A system of framing in which floor joists of each story rest on the top plates of the story below or on the foundation sill for the first story, and the bearing walls and partitions rest on the subfloor of each story. See Fig. F-39.

framing square: Right-angle tool used for laying out carpentry cuts. The long arm is the "blade," the short arm is the "tongue," and the corner is the "heel." Four tables are inscribed on the square for calculating rafter

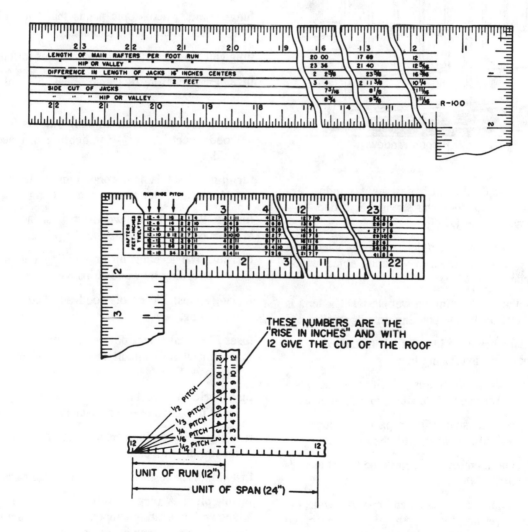

Figure F-41: characteristics, scales, and application of framing square.

Figure F-42: French window.

French window: A double-sash casement window extending down to the floor and serving as a door to a porch or terrace. See F-42.

frequency: The number of complete cycles an alternating electric current, sound wave, or vibrating object undergoes per second.

fresco: The method of painting on wet plaster. The term is often incorrectly applied to painting on a dry wall.

fret: Ornamental work done in relief, characterized by angular interlocked or interlacing lines.

friction: The resistance to motion which is set up when two moving surfaces come in contact with each other.

friction pile: Pile with supporting capacity produced by friction with the soil in contact with the pile.

frieze: A horizontal member connecting the top of the siding with the soffit of the cornice.

front foot: A standard of measurement, one foot wide, extending from street line for a depth, generally conceded to be 100 feet.

frost line: The depth of frost penetration in soil. This depth varies in different parts of the country.

full frame: The old fashioned mortised-and-tenoned frame, in which every joint was mortised and tenoned. Rarely used at the present time.

full size: Drawings are full size when made to the actual size of the work which they represent. The term "full size" is usually written on the drawing.

fuming: Aging wood by the use of chemicals.

fungi, wood: Microscopic plants that live in damp wood and cause m old, stain, and decay.

fungicide: A chemical that is poisonous to fungi.

furlong: A measure of length equal to $\frac{1}{8}$ of a mile, 660 ft.

furnace: A heating unit, utilizing various fuels—gas, oil, wood, electric, etc.—for heating a space, room, or building.

furniture: Useful and decorative movable articles, such as tables, chairs, etc., placed in a building.

furred: Provided with wood strips so as to form an air space between the walls and the plastering.

furring: The leveling up or building out of a part of a wall or ceiling by wood strips, etc.

furring strips: Pieces of wood attached to a surface, as for lathing.

fuse: Electrical safety device consisting of a strip of fusible metal in a circuit that melts when the current is overloaded.

fuse block: A porcelain or slate base to which are fastened fuse clips or other contacts for holding fuses.

fuse clips: The spring part of a cutout or switch that holds the ferrules of a cartridge fuse.

fuse link: The fusible part of a cartridge case.

fusible plug: A plug or fitting made with a metal of a known low melting temperature; used as a safety device to release pressures in case of fire.

G

gable: The triangular end of an exterior wall above the eaves. See Fig. G-1.

gable end: An end wall having a gable. See Fig. G-1.

gable molding: The molding used to finish a gable.

gable roof: A ridge roof terminating in a gable. See Fig. G-1.

gain: The mortise or notch cut out of a timber to receive the end of a beam.

gallery: 1) An elevated floor in a large audience room, usually projecting from the walls, supported by columns from below, or hung from above, or both; generally equipped with seats. 2) A room in which works of art are displayed.

Figure G-2: Gambrel roof.

gallon: A unit of liquid measure containing 4 quarts, 8 pints, 231 cubic inches.

galvanize: To coat iron with zinc. It is not usually an electrical process, but consists simply of dipping the iron in molten zinc.

gambrel roof: A roof having its slope broken by an obtuse angle. See Fig. G-2.

gang switch: A unit of two or more switches to give control of two or more circuits from one point. The entire mechanism is mounted in one box under one cover.

Figure G-1: Gable roof.

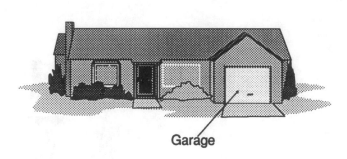

Figure G-3: Residence with one-car garage attached.

garage: A building or portion of a building in which one or more self-propelled vehicles carrying volatile, flammable liquid for fuel or power are kept.

garden bond: Consists of three stretchers in each course followed by a header, although this bond may have from two to five stretchers between headers. See Fig. G-4.

garderobe: A wardrobe. A private room, as a bed chamber.

gargoyle: A stone spout, grotesquely carved, projecting at the upper part of a building.

gas: Vapor phase or state of a substance.

gas pliers: Pliers used for gripping small pipe or round objects. See Fig. G-5.

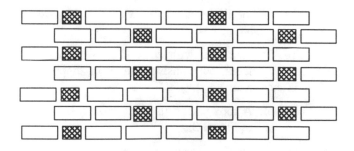

Figure G-4: Garden bond brick pattern.

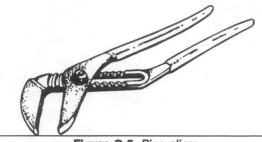

Figure G-5: Pipe pliers.

gate: A device that makes an electronic circuit operable for a short time.

gate valve: A valve whose action depends on the motion of a wedge-shaped gate between the inlet and outlet openings. See Fig. G-6.

geometrical stair: A stair which returns on itself with winders or with winders and a landing built around a comparatively narrow well. The balustrade follows the curve without newel posts at the turns. Often called a spiral stair.

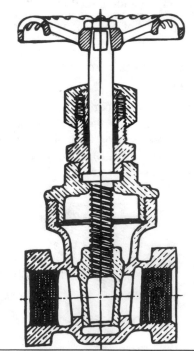

Figure G-6: Gate valve.

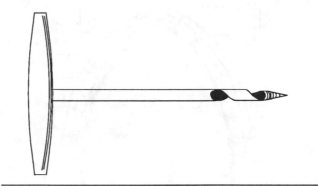

Figure G-7: Gimlet.

Georgia pine: Dark in color, with well-marked grain. It is often used as a finishing material. The wood is hard, heavy, and strong, but decays rapidly in damp places. It is of resinous nature and does not take paint well.

German siding: A type of weatherboard on which the upper portion of the exposed face is finished with a concave curve, and the lower part of the back face is rebated.

GFCI (Ground-Fault Circuit-Interrupter): A protective device that detects abnormal current flowing to ground and then interrupts the circuit.

GFPE: (Ground-Fault Protection of Equipment): A system intended to provide protection of equipment from damaging line-to-ground fault currents by operating to cause a disconnecting means to open all ungrounded conductors of the faulted circuit.

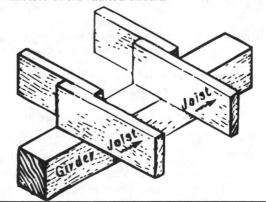

Figure G-8: Application of girders.

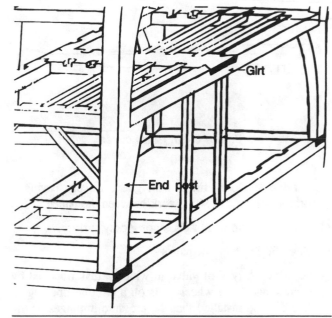

Figure G-9: Girt positioned for building support.

gimlet: A small wood-boring tool with the handle attached at right angles to the bit. See Fig. G-7.

gingerbread work: Elaborate, gaudy or overornamented work as in the trim of a house.

girder: A beam, either timber or steel, used for supporting a super structure. See Fig. G-8.

girt: 1) Heavy timber framed into corner posts as support for the building. See Fig. G-9. 2) The circumference of round timber.

girt strip: See ledger board.

glass: A hard, brittle substance made by melting together sand or silica with lime, potash, soda, or lead oxide.

glass cutter: Any device used for cutting glass to size, usually a diamond or a small rotary wheel set in a handle. See Fig. G-10 on page 124.

glazed: Equipped with window panes.

glazed brick: Brick having a glassy surface made by fusing on a glazing material.

Figure G-10: Glass cutter.

glazed doors: Doors fitted with glass, usually having a pattern or lattice of woodwork between the panes.

glazed tile: A tile with a glassy or glossy surface.

glazing: Fitting window panes.

globe valve: A type of valve in which a disk operated by a screw and hand wheel seats on a circular opening. It should be so installed that the pressure impinges on the underside of the seat. See Fig. G-11.

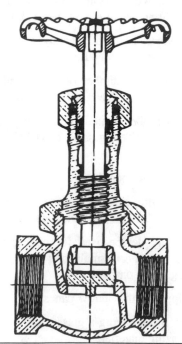

Figure G-11: Globe valve.

Figure G-12: Gothic arch.

gloss paint or enamel: A paint or enamel that contains a relatively low proportion of pigment and dries to a sheen or luster.

glueline, exterior: Waterproof glue at the interface of two veneers of plywood.

glyph: A short, vertical groove.

Goncalo Alves: A tall slender tree found in Brazil. It is a hardwood with general color and grain similar to mahogany but marked with light and dark longitudinal streaks. Used for small table tops, borders, and panel sections.

gooseneck: The curved or bent section of the handrail on a stair.

Gothic: A particular style of classic architecture or ornament.

Gothic arch: A pointed arch. One usually high and narrow, coming to a point at the center. See Fig. G-12.

gouge: Cutting chisel which has either a concave or convex cutting surface. See Fig. G-13.

government anchor: A V-shaped anchor usually made of ½ round bars to secure the steel beam to masonry.

grade: 1) The level of the ground around a building. 2) The fall or slope of a line in reference to the horizontal. In laying plumbing pipes, grade is usually expressed as fall in fractions of an inch per foot of pipe length; that is, ¼″ fall means a fall of ¼″ per foot of length.

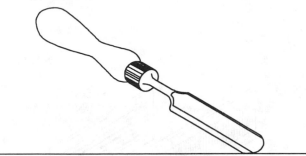

Figure G-13: Typical gouge.

grade beam: Horizontal, reinforced concrete beam between two supporting piers at or below ground supporting a wall or structure.

grading: The leveling of a plot of ground, or sloping it to a desired angle. Also, the sloping of a sidewalk to secure proper drainage.

gradual load: A load gradually applied to a structure and which furnishes the most favorable conditions of stress.

graduate: To divide into regular steps or grades, as a scale.

graduation: The method or system of dividing a graduated scale; also one of the equal divisions or one of the dividing lines of such a scale.

grain: Relating to the direction or arrangement of wood fibers; working a piece of wood longitudinally may be with or against the grain; transversely is spoken of as cross grain. See Fig. G-14.

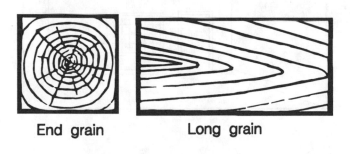

End grain   Long grain

Figure G-14: End and long grain.

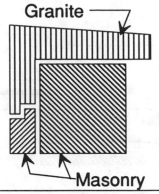

Figure G-14: Granite coping.

grain, edge or vertical: Edge-grain lumber has been sawed parallel to the pith of the log and approximately at right angles to the growth rings.

grain, flat: Flat-grain lumber has been sawed parallel to the pith of the log and approximately tangential to the growth rings.

graining: A painting process applied to cheap woods to imitate oak, walnut, and other better woods.

graining comb: A comb-like tool made of steel or leather. Used by wood finishers for obtaining grain effects.

granite: A rock composed of quartz, feldspar, and mica. It is very hard and takes a high polish. Used extensively in building work and for monuments. See Fig. G-14.

graphic: Illustrating ideas by pictures or diagrams.

graphic arts: A broad term which embraces every branch of pictorial representation.

grating: A gratelike arrangement of bars used to cover an opening. Also used for forming platforms in engine rooms, stair landings, on fire escapes, etc.

gravel: A natural mixture of sand and pebbles.

gravity: That force which draws all bodies toward the center of the earth or to its surface.

gravity water system: Any water system in which pressure is obtained by gravity.

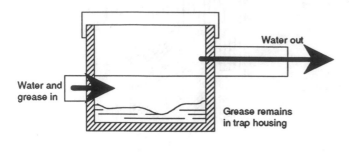
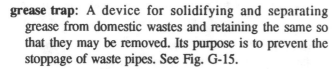

Figure G-15: Grease trap.

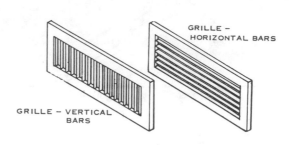

Figure G-16: Various types of grilles used in heating, ventilating, and air-conditioning systems.

grease trap: A device for solidifying and separating grease from domestic wastes and retaining the same so that they may be removed. Its purpose is to prevent the stoppage of waste pipes. See Fig. G-15.

green wood: Timber from which the sap has not been removed by seasoning and drying.

grid: An electrode having one or more openings for the passage of electrons or ions.

grid leak: A resistor of high ohmic value connected between the control grid and the cathode in a grid-leak capacitor detector circuit and used for automatic biasing.

grillage: Steel framework in a foundation designed to spread a concentrated load over a wider area; generally enclosed in concrete.

grille: An openwork of metal or wood, plain or ornamental, used to cover an opening at the end of an air passageway. See Fig. G-16.

grind: To sharpen, to reduce to size, or to remove material by contact with a rotating abrasive wheel.

grinder: Any appliance or device on which work is done by grinding.

grindstone: A revolving stone against which tools and materials are abraded by grinding. Grindstones are natural sandstone.

groin: The line of intersection of two vaults where they cross each other.

groined ceiling: Arched ceiling consisting of two intersecting curved, planes. See Fig. G-17.

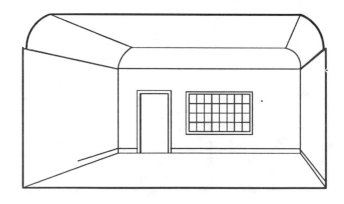

Figure G-17: Groined ceiling.

groined vaulting: The system of covering a building or passageways with stone vaults which cross and intersect.

grommet: A metallic eyelet, used principally in awnings and flags. Also a ring of candle wicking used as a water tight gasket or washer around bolts and studs.

groove: A depressed or sunken channel, usually small; so designated, whether some other part does or does not fit into it.

ground: 1) Nailing strip set in or attached to a brick or stone wall. 2) A large conducting body (as the earth) used as a common return for an electric circuit and as an arbitrary zero of potential.

ground clamp: A clamp used for attaching a wire or other conductor to a pipe to; make a good electrical connection. See Fig. G-18.

ground clip: A spring clip used to secure a bonding conductor to an outlet box. See Fig. G-18.

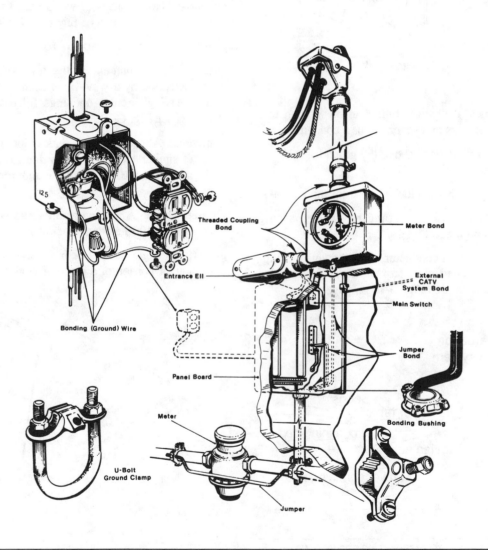

Figure G-18: National Electrical Code regulations for grounding.

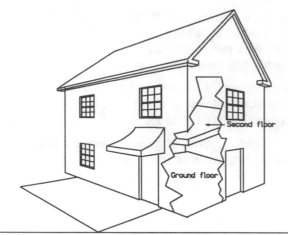

Figure G-19: Ground floor.

ground coil: A heat exchanger buried in the ground that may be used either as an evaporator or a condenser.

ground floor: The first floor above the ground level; usually the main floor. See Fig. G-19.

ground joist: A joist that is blocked up from the ground.

ground water: Water that is standing in or passing through the ground. See Fig. G-20.

grounded conductor: A conductor used to connect equipment or the grounded circuit of a wiring system to a grounding electrode. See Fig. G-18 on page 127.

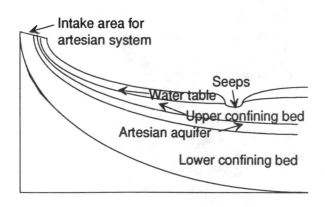

Figure G-20: Geological cross-section showing water table.

Figure G-21: Guilloche.

grounds: Narrow strips of wood nailed to walls as guides to plastering and as a nailing base for interior trim.

grout: A fluid cement mixture for filling crevices.

guilloche: An ornament in the form of bands or strings characterized by a wavy interlacement of the main motifs, leaving circular openings filled with round ornaments. See Fig. G-21.

gumwood: A medium hard, dark-colored wood with grain effect similar to Circassian walnut; used extensively in the manufacture of furniture and for interior trim in buildings.

Gunters chain: A surveyor's measure 66 ft. in length. It consists of 100 links each being 7.92 in. long.

gusset: An angle bracket or brace used to stiffen a corner or angular portion of a piece of work. See Fig. G-22.

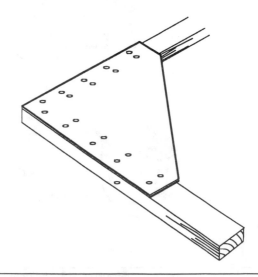

Figure G-22: Practical application of a gusset.

guttae: Small cylindrical or cone-shaped pendants used to ornament the Doric entablature.

gutter: 1) A shallow channel or conduit of metal or vinyl set below and along the eaves of a house to catch and carry off rainwater from the roof. See Fig. G-23. 2) The space provided along the sides and at the top and bottom of enclosures for switches, panels, and other apparatus, to provide for arranging conductors that terminate at the lugs or terminals of the enclosed equipment. Gutter is also used to refer to a rectangular sheet metal enclosure with removable side, used for splicing and tapping wires at distribution centers and motor control layouts. Usually called "auxiliary gutter." See Fig. G-24.

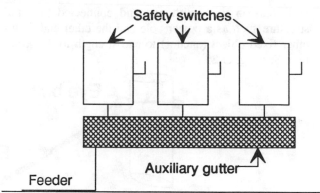

Figure G-24: Auxiliary gutter used in an electrical distribution system.

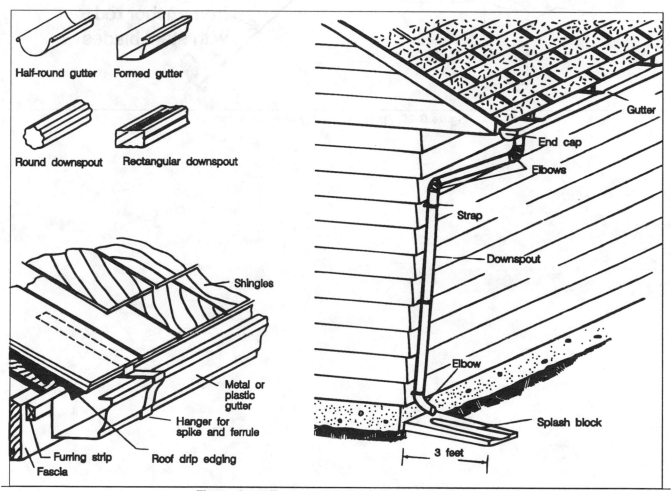

Figure G-23: Types and applications of roof gutters.

guy: A tension wire with one end connected to a tall structure such as a power pole and the other end to another fixed object (anchor) to add strength to the structure. See Fig. G-25 .

gypsum: Hydrous sulphate of calcium, colorless when pure. When it is heated slowly, part of the water is driven off, the resulting product being known as "plaster of Paris."

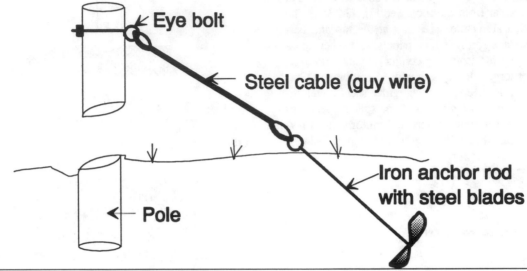

Figure G-25: Application of guy wire and anchor.

H

habendum clause: The "To Have and To Hold" clause which defines or limits the quantity of the estate granted in the premises of the deed.

hack: To cut roughly in an unworkmanlike manner.

hack saw: A light-framed saw used for cutting metal, operated by power or by hand. See Fig. H-1.

hack-saw blade: Replaceable blades used in hack saws; made in various sizes (teeth-per-inch) for different applications. See Fig. H-2.

haft: A handle as of a knife or an awl.

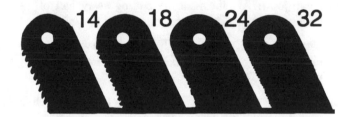

Teeth Per Inch	Uses
14	Iron, soft steel, rails
18	Tool steel, iron pipe, light angle iron
24	Copper, brass, medium tubing, sheet metal
32	Thing tubing, thin sheet metal
Note	Never allow the teeth to straddle the work.

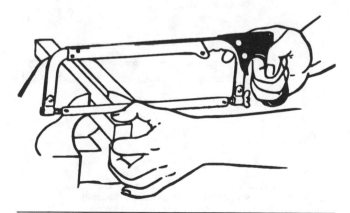

Figure H-1: Starting a cut with a hack saw.

Figure H-2: Hack-saw blades and their application.

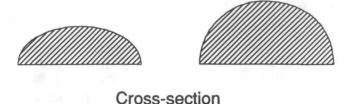

Cross-section

Figure H-3: Half-round file.

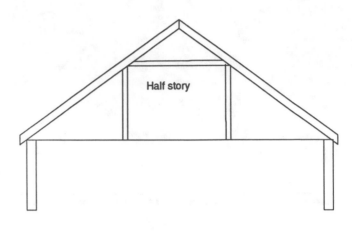

Half story

Figure H-4: Half story.

half-back bench saw: Gives both the advantage of a stiff cutting edge and the ability to cut entirely through the work. The stiffening bar extends over only a portion of the blade length thus combining the action of both the handsaw and the backsaw.

half-lap joint: A joint made by cutting away half of the thickness of the pieces to be joined. See *lap joint*.

half-moon stake: A stake with the top curved and beveled to an edge. Its principal use is in circular flanging operations.

half pattern: One of the halves of a pattern which is parted through the center for convenience of molding.

half-round file: A file which is flat on one side and curved on the other. The amount of convexity never equals a semicircle. See Fig. H-3.

half section: In mechanical drawing, a sectional view which terminates at the center line, showing an external view on one side of the center line, and on the other side an interior view.

half story: That part of a pitched roof structure directly under the roof, having a finished ceiling and floor and some side wall. See Fig. H-4.

halving: The making of a joint by cutting away half of the thickness from the face of one piece and the other half from the back of the piece to be fitted to it, so that when the two are put together the outer surfaces will be flush. See *lap joint*.

hammer: An instrument or tool used for striking blows in metal working, driving nails, etc. Hammers are of various kinds, each bearing a name according to the purpose it is to serve. See Fig. H-5.

handbook: A book of reference to be carried in the hand. A manual or guide, usually containing a compilation of data and formulas.

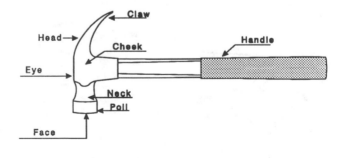

Figure H-5: Typical hammer used on building construction projects and their respective parts.

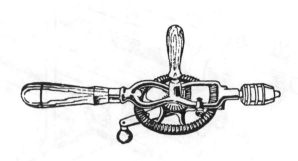

Figure H-6: Hand drill.

hand drill: A drilling machine operated by hand. See Fig. H-6.

hand feed: The feeding by hand of the cutting tools of machines of various kinds.

hand file: This file has parallel sides but is tapered in thickness. It is double cut with various degrees of fineness. Its principal use is in finishing flat surfaces. See *file*; also Fig. H-7.

hand money: Same as an earnest money deposit.

handrail: A rail, as on a stair or at the edge of a gallery, placed for convenient grasping by the hand. See Fig. H-8.

handrail wreath: Curved section of a stair rail.

handsaw: An ordinary one-handled saw, either rip or cross-cut, used by woodworkers. See Fig. H-9.

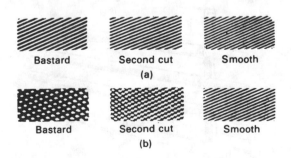

Bastard   Second cut   Smooth
(a)

Bastard   Second cut   Smooth
(b)

Figure H-7: Hand file cuts.

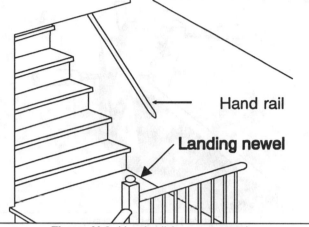

Hand rail

Landing newel

Figure H-8: Hand rail for use on stairs.

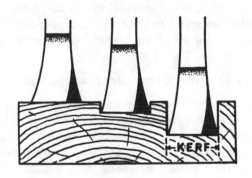

Rip teeth

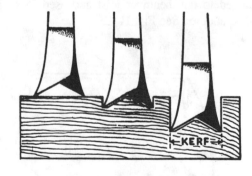

Cross-cut teeth

Figure H-9: Types of common hand saw cuts.

Figure H-10: Hand screw.

hand screw: A woodworkers' clamp consisting of two parallel jaws and two screws. Clamping action is obtained by means of the screws, one operating through each jaw. See Fig. H-10.

hand tools: Tools that are guided and operated by hand.

handy box: The commonly used, single-gang outlet box used for surface mounting to enclose wall switches or receptacles, on concrete or cinder block construction of industrial and commercial buildings, non-gangable, also made for recessed mounting, also known as "utility boxes". Fig. H-11.

hanger: A strap of iron or steel used as a drop support, attached to one beam or joist and used to support the end of another. See Fig. H-12.

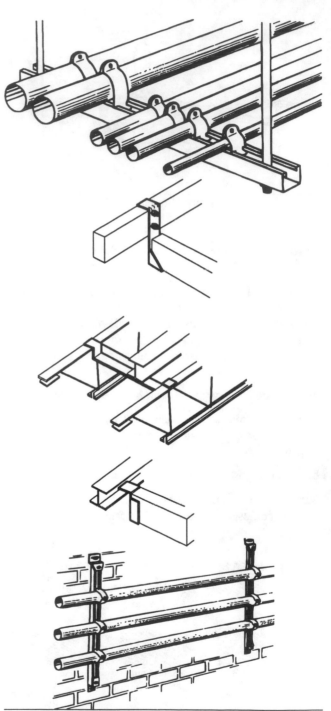

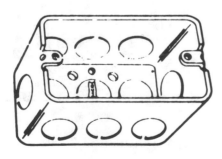

Figue H-11: Handy box.

Figure H-12: Variety of hangers used on building construction projects.

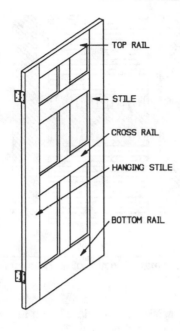

TOP RAIL

STILE

CROSS RAIL

HANGING STILE

BOTTOM RAIL

Figure H-13: Principle parts of a door, showing hanging stile.

hanger bolt: Consists of a lag screw at one end and a machine-bolt thread and nut at the other. It is used for attaching hangers to woodwork.

hanging stile: That vertical part of a door or of a casement window to which the hinges are fastened. See Fig. H-13.

hard brass: Brass which has not been annealed after drawing or rolling; used for springs, etc.

hard drawn: A relative measure of temper; drawn to obtain maximum strength.

hardness: The resistance a body offers to being scratched or worn by another substance.

hardpan: A layer of rock under soft soil.

hard water: Water containing a large quantity of compounds of calcium and magnesium in solution.

hardwood: Wood which is a close-grained, dense, and heavy, as oak, hickory, ash, and beech. See Fig. H-14.

Type of Wood	Uses
American beech	Lumber, veneer, flooring, crates.
American elm	Containers, furniture, barrels, and kegs.
Black cherry	Lumber and furniture.
Black Walnut	Furniture and interior trim.
Cottonwood	Lumber, veneer, containers, and furniture.
Hickory	Tool handles.
Mahogany	Quality furniture and veneer plywood.
Mahogany, Philippine	Lumber and veneer plywood for furniture, built-ins, and paneling.
Pecan	Lumber and veneer for furniture.
Red oak	Lumber for flooring.
Red wood	Boards for outdoor decks and trim.
Rock elm	Containers and furniture.
Sugar maple	Lumber and veneer for furniture and interior paneling and trim.
Sweet gum	Lumber, veneer, plywood.
White ash	Tool handles with some use in furniture making.
White oak	Lumber for flooring and trim.
Yellow birch	Lumber, veneer, general millwork.
Yellow popular	Lumber, veneer, for furniture and interior finishes.

Figure H-14: Principle hardwoods used in building construction, along with their applications.

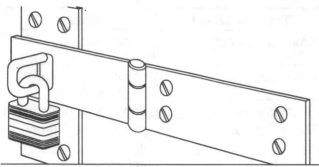

Figure H-15: Typical hasp.

H-clip: A metal clip into which edges of adjacent plywood sheets are inserted to hold edges in alignment. See *plywood clip*.

hasp: A fastening as for a door, usually passing over a staple, and secured by a peg or padlock. See Fig. H-15.

hatchway: An opening covered by a trap door, as in a roof, to permit easy access for repairs; or in a ceiling to give entrance to an attic. See Fig. H-16.

haunch: The shoulder of an arch.

hawk: A small, square board with handle underneath, used to hold plaster or mortar.

H beam: A steel beam whose section is like the letter H. See Fig. H-17.

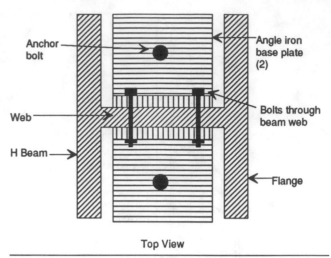

Top View

Figure H-17: Characteristics of an H beam used for column support.

header: 1) A brick or stone placed with its end toward the face of a wall. See Fig. H-18. 2) A beam placed perpendicular to joists, to which joists are nailed in framing for chimneys, stairways, or other openings. See Fig. H-19. 3) A wood lintel.

heading bond: Pattern of brick bonding formed with headers. See Fig. H-20.

head pressure: Pressure that exists in the condensing side of a refrigerating system.

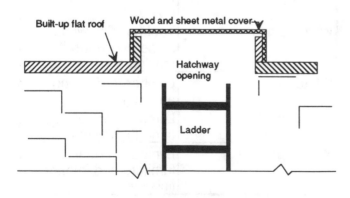

Figure H-16: Hatchway leading to building roof.

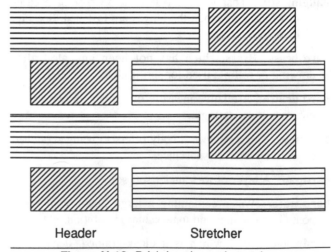

Header   Stretcher

Figure H-18: Brick header and stretcher.

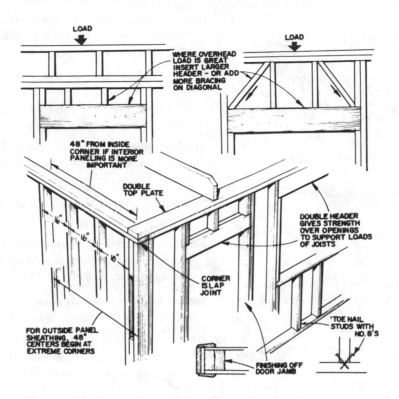

Figure H-19: House framing details showing headers.

head room: The vertical space between a stair and the ceiling or stair above.

hearth: The floor of a fireplace including that portion in front of the fireplace. See Fig. H-21.

Figure H-20: Heading bond brick pattern.

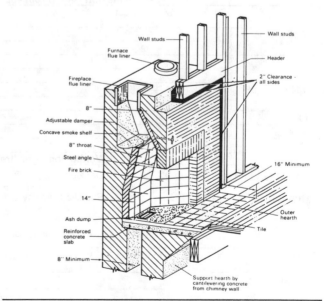

Figure H-21: Fireplace with hearth.

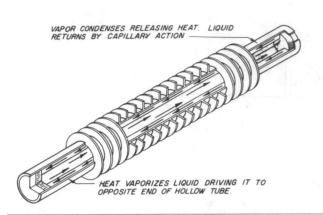

VAPOR CONDENSES RELEASING HEAT. LIQUID
RETURNS BY CAPILLARY ACTION

HEAT VAPORIZES LIQUID DRIVING IT TO
OPPOSITE END OF HOLLOW TUBE.

Figure H-22: One type of heat exchanger.

heartwood: The wood extending from the pith to the sap-wood, the cells of which no longer participate in the life process of the tree.

heat: Added energy that causes substances to rise in temperature; energy associated with random motion of molecules.

heat exchanger: Device used to transfer heat from a warm or hot surface to a cold or cooler surface. Evaporators and condensers re heat exchangers. See Fig. H-22.

heating values: Amount of heat that may be obtained by burning a fuel, usually expressed in Btu per pound or gallon.

heat load: Amount of heat, measured in Btu, that is removed during a period of 24 hours.

heat loss calculations: Calculations used to determine heat loss in a particular building or area; used to size heating system. See Fig. H-23.

heat of compression: Mechanical energy of pressure transformed into heat energy.

HEATING LOAD		
1. Design Conditions	**Dry Bulb (F)**	**Specific Humidity gr./lb.**
Outside		
Inside		
Difference		
ITEMS		**HEAD LOAD (BTUH)**
2. Transmission Gain (From appropriate tables)		
Sq. Ft. X Factor X		Dry Bulb Temp. Diff.
Windows		=
		=
		=
		=
Walls		=
		=
Roof		=
Floor		=
Other		=
3. Ventilation or Infiltration (from appropriate tables)		
Sensible Load CFM X Dry Bulf Temp. Diff. X 1.08 (Factor)		=
Humidification Load CFM X Specific Humidity Diff. X 0.67 (Factor)		=
4. Duct Heat Loss (from appropriate tables)		
Heat Loss ____ X Factor for Insulation Thickness ____ X Duct = Length (Ft.) ____ ÷ 100		
5. Total Heating Load		=

Figure H-23: Principles of calculating heat loss in a building or area.

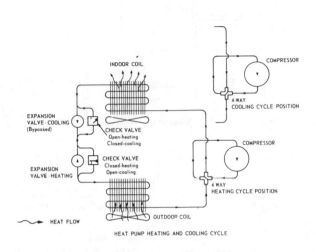

Figure H-24: Operating characteristics of a typical heat pump.

heat pressure control: Pressure-operated control that opens an electrical circuit if high side pressure becomes excessive.

heat pump: A compression cycle system used to supply heat to a temperature—controlled spaced, which can also remove heat from the same area. See Fig. H-24.

heat transfer: Movement of heat from one body or substance to another. Heat may be transferred by radiation, conduction, convection, or a combination of these.

heavy joist: Timber over four inches and less than six inches in thickness and eight inches or over in width.

heel: The end of a rafter or beam which rests on the wall plate. See Fig. H-25.

heel wedges: Triangular shaped pieces of wood that can be driven into gaps between rough framing and finished items, such as window frames, to provide a solid backing for these items.

helium: One of the rarest elements. It exists in certain uranium minerals and to a small extent in the air. Its chief source is in the natural gas of southwestern United States.

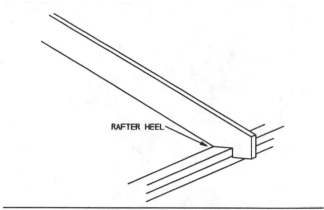

Figure H-25: Rafter heel.

helix: A curve, as would be obtained by winding a thread around a cylinder in such a manner that there would be a uniform amount of advance with each revolution.

hematite: An important iron ore.

hemlock: An inexpensive wood somewhat like spruce in appearance; used extensively for framing.

henry: The basic unit of inductance.

hereditaments: The largest classification of property; includes lands, tenements, and incorporeal property, such as rights of way.

hermaphrodite caliper: A caliper in which one leg is pointed as in a pair of dividers, the other being slightly hooked as in the ordinary outside caliper. See Fig. H-26.

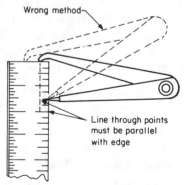

Figure H-26: Hermaphrodite caliper.

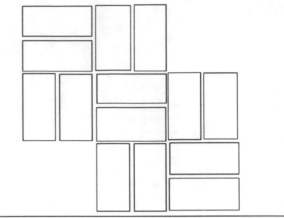

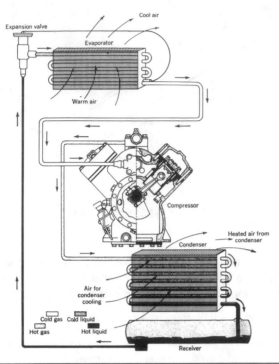

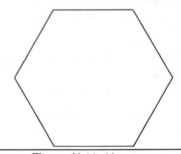

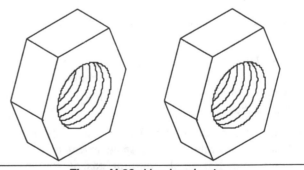

Figure H-27: Operating priniples of a refrigeration system showing hermetic motor, high-pressure cutout and other principle parts.

hermetic motor: Compressor drive motor sealed within the same casing that contains the compressor.

hermetic system: Refrigeration system that has a compressor driven by a motor contained in a compressor dome or housing.

high pressure cutout: Electrical control switch operated by the high side pressure that automatically opens an electrical circuit if too high head pressure or condensing pressure is reached. See Fig. H-27.

herringbone: The name given to masonry or brick when laid up in a zigzag pattern. See Fig. H-28.

herringbone bond: A zigzag arrangement of bricks or tile, in which the end of one brick is laid at right angles against the side of a second brick.

hewing: The dressing of timber by chopping or by blows from an edged tool.

Figure H-28: Herringbone bond brick pattern.

hexagon: A plane figure having six sides and six angles. All sides of a regular hexagon are equal. All angles are equal; their sum totals 720 degrees. See Fig. H-29.

hex head: A common shop expression referring to screws and bolts with hexagonal heads. See Fig. H-30.

Figure H-29: Hexagon.

Figure H-30: Hex head nuts.

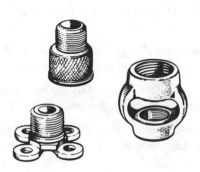

Figure H-31: Hickey fitting.

hickey: 1) A small threaded fitting of brass or iron placed in a fixture assembly between the stem and the support to provide an outlet for the wires coming out of the fixture stem. See Fig. H-31. 2) A pipe bending device used for bending conduit or other pipe to the desired shape. See Fig. H-32.

hickory: A very hard and tough wood, hard to work; used extensively in work which requires bending.

hidden surface line: A line consisting of short dashes. It is used to indicate the surface of a hidden part.

high flashing point: When oil will ignite only at a very high temperature, it is said to have a high flashing point.

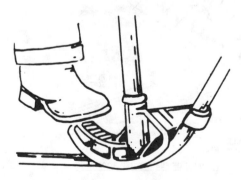

Figue H-32: Hickey used to bend rigid steel conduit.

high gloss: Paint which dries with a lustrous, enamel-like finish.

high side: Parts of a refrigerating system that are under condensing or high side pressure. See Fig. H-27.

high speed steel: Steel containing tungsten or molybdenum, which has been added to increase its efficiency for cutting tools. Such tools can be operated at much higher speeds without injury than can the ordinary carbon-steel tools.

high tension bolts: High strength steel bolts tightened with calibrated wrenches to high tension; used as a substitute for conventional rivets in steel frame structures.

hinge: A mechanical device, consisting primarily of a pin and two plates, which may be attached to the door and door frame to permit opening and closing of the door, or for use under any similar condition. See Fig. H-33.

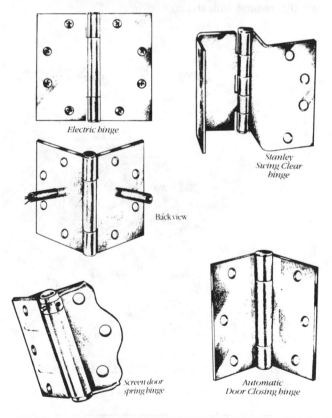

Figure H-33: Various types of hinges.

hip: The external angle formed by the meeting of two sloping sides of a roof. See Fig. H-34.

hip jack rafter: Short rafter extending from the plate to the hip ridge. See Fig. H-34.

hip rafters: The rafters that form the hip of a roof; distinguished from ridge. See Fig. H-34

hip roof: A roof which rises with equal angles from all four sides of a building. See Fig. H-34.

hod: A long-handled receptacle for carrying bricks and mortar.

holdover tenant: A tenant who remains in possession of leased property after the expiration of the lease term.

hollow tile: A building material used extensively for both exterior walls and partitions. It is made in a variety of forms and sizes. When used for an outside wall, it is usually covered with stucco.

holly: Usually a small, slow-growing tree. Wood is white, hard, and close grained. Used for piano keys, inlay work, and interior finish.

hologen: "Salt former." Applied to the family of elements consisting of fluorine, chlorine, bromine, and iodine.

homestead: Real estate occupied by the owner as a home; the owner enjoys special rights and privileges.

hone or oilstone: A stone used for whetting edged tools, to give the clean, fine edge necessary for clean cutting. See Fig. H-35.

honey locust: Tree of medium size. It is a flower and pod-bearing tree. Wood is hard and strong, very durable in contact with soil. Used for fence posts, rails, and wheel hubs.

hook bolt: A bolt which, instead of having a head, has the unthreaded end bent U shaped, or straight at right angles to the body of the bolt. See Fig. H-36.

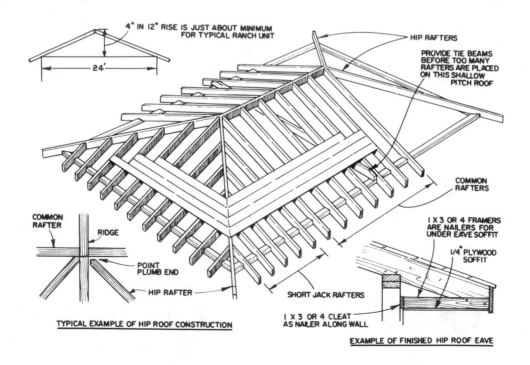

Figure H-34: Details of hip roof framing.

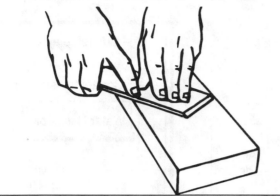

Figure H-35: Honing operation on an oilstone.

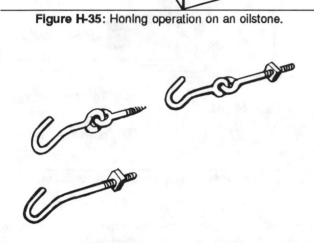

Figure H-36: Typical hook bolts.

hook joint: A dustproof joint for doors of showcases.

horse: A trestle. Also one of the slanting supports of a set of steps to which are attached the treads and risers.

hopper: A box or receiver used for the purpose of feeding supplies of materials to machines or furnaces of various kinds.

hopper window: A window that is hinged at the bottom to swing inward. See Fig. H-37.

hot gas bypass: Piping system in a refrigerating unit that moves hot refrigerant gas from a condenser into the low pressure side.

hot junction: That part of the thermoelectric circuit that releases heat.

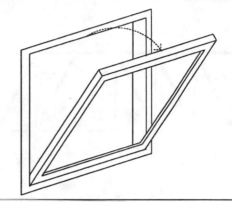

Figure H-37: Hopper window.

hot wire: A resistance wire in an electrical relay that expands when heated and contracts when cooled.

house drain: That part of the lowest horizontal piping of a plumbing system which receives the discharge from soil, waste, and other drainage pipes inside of any building and conveys the same to the house sewer. See Fig. H-38.

housed string: A stair string with vertical and horizontal grooves cut on the inside to receive the ends of the risers and treads. Wedges covered with glue are generally used to hold the risers and treads in place in the grooves.

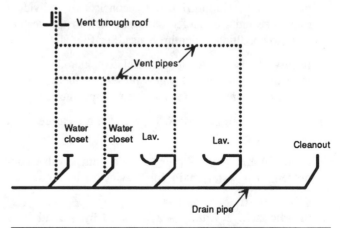

Figure H-38: Riser diagram of a house drain.

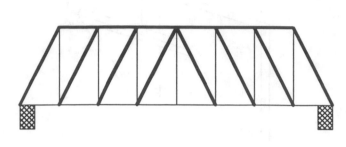

Figure H-39: Howe truss.

housing for the elderly: A project designed specially for older persons (62 years or over) which provides living unit accommodations, and common social and activities space, and facilities for health and nursing services for residents.

howe truss: A form of truss used both in roof and bridge construction; especially adapted to wood and steel construction. See Fig. H-39.

hue: The particular shade of a color.

H.U.D.: Department of Housing and Urban Development.

humidifier: A device designed to increase he humidity within a room or a house by means of the discharge of water vapor. Humidifiers may consists of individual room-size units or larger units attached to the heating plant to condition the entire house. See Fig. H-40.

humidity: Moisture, dampness. Relative humidity is the ratio of the quantity of vapor present in the air to the greatest amount possible at a given temperature.

hutch: 1) A small or dark room; a chest; a measure. 2) A place for storing.

hydrant: A fire plug; a plug or pipe with a valve connected to a water main for service in extinguishing fires.

hydraulic jack: A lifting jack, actuated by a small force pump enclosed within it, and operated by a lever from the outside.

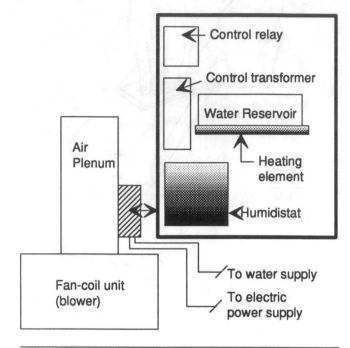

Figure H-40: Principle parts of a humidifier.

hydraulic lime: Limestones which have sufficient impurities to give the calcined product hydraulic properties but not enough to take up all the lime present make hydraulic limes when burned.

hydraulics: The science of liquids, especially of water in motion.

hydrometer: An instrument for determining the specific gravity of liquids. See Fig. H-41.

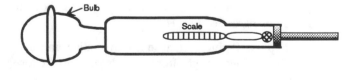

Figure H-41: Hydrometer.

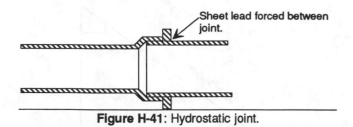

Figure H-41: Hydrostatic joint.

hydrostatic joint: Used in large water mains, in which sheet lead is forced tightly into the bell of a pipe by means of the hydrostatic pressure of a liquid. See Fig. H-41.

hydronic: Type of heating system that circulates a heated fluid, usually water, through baseboard coils. Circulating pump is usually controlled by a thermostat. See Fig. H-42.

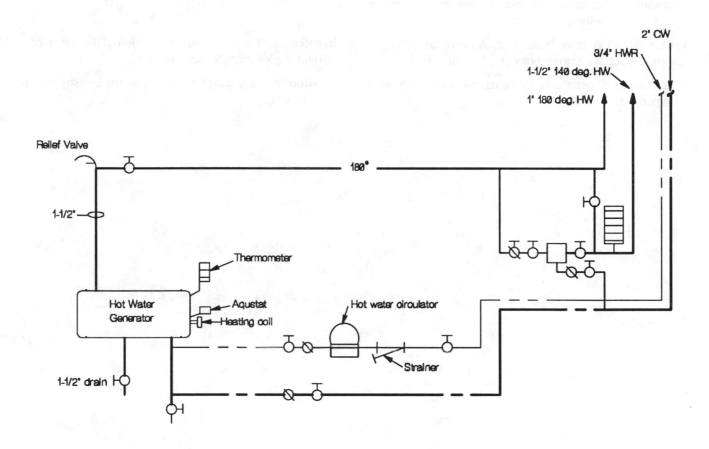

Figure H-42: Hydronic heating system diagram.

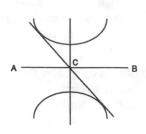

Figure H-43: Hyperbola.

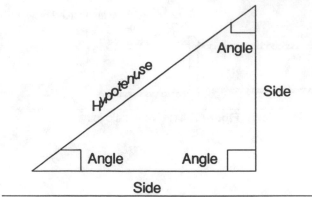

Figure H-44: Triangle showing hypotenuse.

hygrometer: An instrument used to measure the degree of moisture in the atmosphere.

hyperbola: 1) A conic section. 2) A curve consisting of two distinct and similar branches. See Fig. H-43.

hyperbolic: Of, pertaining to, or having a the shape of a hyperbola.

hypotenuse: The diagonal which joins the sides of a right-angled triangle. See Fig. H-44.

hypothesis: An assumption as a basis for investigation or reasoning.

I

I^2r: The power losses in a device or apparatus due to resistance heating.

I^2t: Relating to the heating effect of a current (amps-squared) for a specified time (seconds), under specified conditions.

IACS (International Annealed Copper Standard): Refined copper for electrical conductors; 100% conductivity at 20C for 1m x 1mm2 has 1/58 ohm resistivity, 8.89 grams per mm2 density, 0.000 017 per degree C coefficient of linear expansion, 1/254.45 per degree C coefficient of variation of resistance; however, NBS suggests the use of 8.93 for density (1977).

IAEI: International Association of Electrical Inspectors.

I beam: A rolled steel beam or built-up beam shaped like the letter I. I-beams are used for long spans as basement beams or over wide wall openings, such as a double garage door, when wall and roof loads are imposed on the opening. See Fig. I-1.

IBEW: International Brotherhood of Electrical Workers.

IC (Pronounced "eye see"): 1) Refers to interrupting capacity of any device required to break current (switch, circuit breaker, fuse, etc.), taken from the initial letters I and C, is the amount of current which the device can interrupt without damage to itself. 2) Electronics: integrated circuit.

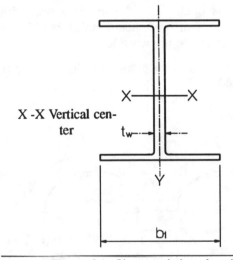

Figure I-1: Characteristics of an I beam.

ID: Inside diameter.

identified: Marked to be recognized as grounded.

IEC: International Electrochemical Commission.

IEEE: Institute of Electrical and Electronics Engineers.

igneous rock: Rocks formed under the action of heat, as basalt, quartz, granite, etc.

ignition transformer: A transformer designed to provide a high voltage current.

IIR (Isobutylene Isoprene Rubber): Butyl synthetic rubber.

illustrate: To explain by means of photos, figures, examples, comparisons, and the like; to make clear.

imaginary number: Quanity or value that involves the square root of a negative quantity and is unreal.

impair: To lessen in quantity or quality; to deteriorate.

impedance: The opposition to current flow in an ac circuit; impedance includes resistance (R), capacitive reactance (X_C) and inductive reactance (X_L); it is measured in ohms.

impedance matching: Matching source and load impedance for optimum energy transfer with minimum distortion.

imperviousness: Is obtained by laying up a wall with paving; using cement or cement-lime mortar. These bricks should always be dry when laid.

impetus: Momentum; the force with which any body is driven or impelled.

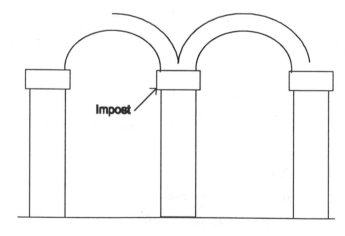

Figure I-2: Impost

impost: The top member of a pillar on which the arch rests. See Fig. I-2.

impregnated: Lumber for outside use is impregnated with various chemicals to enable it to resist the decomposing influences of the atmosphere.

impulse: A surge of unidirectional polarity.

in phase: The condition existing when waves pass through their maximum and minimum values of like polarity at the same instant.

incandescence: Glowing due to heat.

incandescent: That which gives light or glows at a white heat.

incandescent lamp: An electric lamp or bulb containing a thin wire or filament of infusible conducting material.

incarnadine: Shades of color from red to flesh.

inching: Momentary activation of machinery used for inspection or maintenance.

incinerator: A container for burning trash.

incise: To carve; cut in; engrave.

inclined plane: A plane that is inclined to the plane of the horizon, the angle that it makes being its inclination.

incombustible material: Material that will not ignite or actively support combustion in a surrounding temperature of 1200 degrees Fahrenheit during an exposure of five minutes also, material that will not melt when the temperature of the material is maintained at 900 degrees Fahrenheit for at least five minutes.

increment: The amount whick a varying quantity increases between two of its stages.

incrustation: The forming of scale on the interior portions of steam boilers.

indefinite: Not precise; unsettled; uncertain.

indenture: A formal written instrument made between two or more persons in different interests; name comes from practice of indenting or cutting the deed on the top or side in a waving line.

indicate: Point out; show; to point out very briefly.

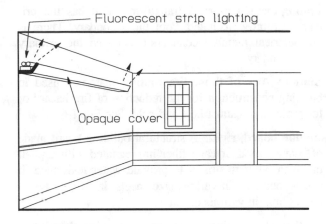

Fluorescent strip lighting

Opaque cover

Figure I-3: One type of indirect lighting.

indirect lighting: Lighting effect obtained by focusing the light against the ceiling or some other surface from which it is diffused in the area to be lighted. See Fig. I-3.

indoor: Not suitable for exposure to the weather.

induced voltage: A voltage set up by a varying magnetic field linked with a wire, coil, or circuit.

inductance: The creation of a voltage due to a time-varying current; the opposition to current change, causing current changes to lag behind voltage changes; units of measure; henry.

induction: The production of magnetization or electrification in a body by the mere proximity of magnetized or electrified bodies, or of an electric current in a conductor by the variation of the magnetic field in its vicinity.

induction coil: Essentially a transformer with an open magnetic circuit, in which an alternating current of high voltage is induced in the secondary by means of a pulsating direct current in the primary.

induction heater: The heating of a conducting material in a varying electromagnetic field due to the material's internal losses.

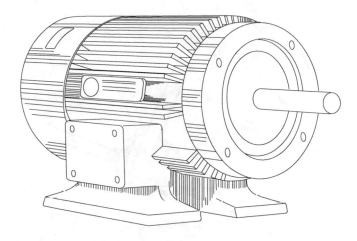

Figure I-4: Typical induction motor.

induction machine: An asynchronous ac machine to change phase or frequency by converting energy—from electrical to mechanical, then from mechanical to electrical.

induction motor: An ac motor that does not run exactly in step with the alternations. Currents supplied are led through the stator coils only; the rotor is rotated by currents induced by the varying field set up by the stator coils. See Fig. I-4.

inductive reactance: The opposition to the flow of an electrical current in a circuit that consists of turns of wire. The opposition is greater if the turns are wound on an iron core. The measure of resistance to the flow of alternating current through a coil.

inductivity: The capacity or power for induction.

inductor: A device having winding(s) with or without a magnetic core for creating inductance in a circuit.

industrial waste: The liquid waste resulting from the processes employed in industry.

inert: Will not readily combine with anything.

infrared lamp: An electrical device that emits infrared rays—invisible rays just beyond red in the visible spectrum.

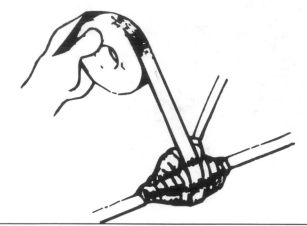

Figure I-5: Insulating tape.

infrared radiation: Radiant energy given off by heated bodies which transmits heat and will pass through glass.

inflation: 1) The act of expanding by filling with air or gas. 2) Establishing of false standard of value.

infrigement of patent: The unlicensed manufacture, sale, or use of a thing pateted.

inglenook: A fireside corner.

ink: The material used for legends and color coding.

inspector: One whose duty is to check a product with regard to requirements or to pass judgement on quality and quantity of work perform.

installment contract: Purchase of real estate upon an installment basis; upon default, payments are forfeited.

instantaneous value: The value of a variable quantity at a given instant.

instrument: A device for measuring the value of the quantity under observation.

insulated: Separated from other conducting surfaces by a substance permanently offering a high resistance to the passage of energy through the substance.

insulating tape: Adhesive tape made nonconducting by being saturated with an insulating compound or manufactured from a nonconducting material; used for covering wire joints and exposed parts. See Fig. I-5.

insulating transformer: A transformer that has the primary carefully insulated from the secondary. There is no electrical metallic connection between the primary and secondary.

insulation: Any of those fireproofing materials used in building construction for the reduction of fire hazard or for protection against heat and cold. See Fig. I-6.

insulation board, rigid: A structural building board made of coarse wood or cane fiber impregnated with asphalt or given other treatment to provide water resistance. It can be obtained in various size sheets, in various thicknesses, and in various densities.

insulation, class rating: A temperature rating descriptive of classes of insulations for which various tests are made to distinguish the materials; not related necessarily to operating temperatures.

insulation, electrical: A medium in which it is possible to maintain an electrical field with little supply of energy from additional sources; the energy required to produce the electric field is fully recoverable only in a complete vacuum (the ideal dielectric) when the field or applied voltage is removed: used to a) save space b) enhance safety c) improve appearance.

Inustation	Material	Use
Batts	Fiberglass and rock wool	Unfinished attics and underside of floors
Blanket	Fiberglass and rock wool	Same as for batts
Rigid board	Polystyrene, urethane, fiberglass	Exterior wall sheathing, floor slab perimeter, mansory walls
Loose fill	Fiberglass, celulose, and rock wool	Unfinished attic floors or blown into finished walls and other contained areas

Figure I-6: Types of insulation.

insulation fall-in: The filling of strand interstices, especially the inner interstices, which may contribute to connection failures.

insulation level (cable): The thickness of insulation for circuits having ground fault detectors which interrupt fault currents within 1) 1 minute = 100% level 2) 1 hour = 133% level 3) over 1 hour = 173% level.

insulation, thermal: Substance used to retard or slow the flow of heat through a wall or partition.

intake belt course: A belt course in which the molded face is so cut that it serves as an intake between the varying thicknesses of two walls.

integral: Built into or self-contained; that is, an electric heater with an integral thermostat.

integrated circuit: A circuit in which different types of devices such as resistors, capacitors, and transistors are made from a single piece of material and then connected to form a circuit. See Fig. I-7:

integrated electrical system: An industrial wiring system in which:

- An orderly shutdown is required to minimize personnel hazard and equipment damage.

- The conditions of maintenance and supervision assure that qualified persons will service the system.

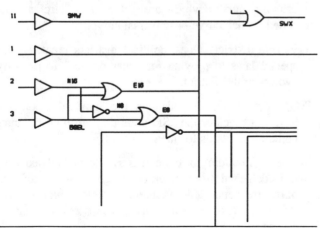

Figure I-7: Schematic diagram of an integrated circuit.

- Effective safeguards, acceptable to the authority having jurisdiction, are established and maintained.

integrator: Any device producing an output proportionate to the integral of one variable with respect to a second variable; the second is usually time.

intercalated tapes: Two or more tapes of different materials helically wound and overlapping on a cable to separate the materials.

interconnected system: Operating with two or more power systems connected thru tie lines.

interface: 1) A shared boundary. 2) (nuclear power); a junction between Class 1E and other equipment of systems.

interference: Extraneous signals or power which are undesired.

interior finish: The general effect of the inside finishing of a building.

interlock: A safety device to insure that a piece of apparatus will not operate until certain conditions have been satisfied.

interpolate: To estimate an intermediate between two values in a sequence.

interpole: A small field pole placed between the main field poles and electrically connected in series with the armature of an electric rotating machine.

interrupter: A device that opens and closes ae electrical circuit at very frequent intervals.

interrupting time: The sum of the opening time and arcing time of a circuit opening device.

interstice: The space or void between assembled conductors and within the overall circumference of the assembly.

intrinsically safe: Incapable of releasing sufficient electrical or thermal energy under normal or abnormal conditions to cause ignition of a specific hazardous atmospheric mixture in its most ignitable concentration.

inverted arch: Arch in which the keystone is at the lowest point of the arch.

inverter: An item which changes dc to ac.

ion: An electrically charged atom or radical.

ionic: Pertaining to that order of Greek architecture characterized by scroll-like ornaments of the capital.

ionic order: Style of architecture developed by the Ionians. Its columns are fluted and surmounted by a capital in which scrolls form an important feature of decoration.

ionization: 1) The process or the result of any process by which a neutral atom or molecule acquires charge. 2) A breakdown that occurs in gaseous parts of an insulation when the dielectric stress exceeds a critical value without initiating a complete breakdown of the insulation system; ionization is harmful to living tissue, and is detectable and measurable; may be evidenced by corona. An ionization smoke detector is shown in Fig. I-8.

ionization factor: This is the difference between percent dissipation factors at two specified values of electrical stress; the lower of the two stresses is usually so selected that the effect of the ionization on dissipation factor at this stress is negligible.

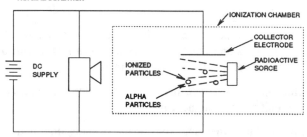

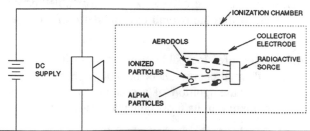

Figure I-8: Ionization smoke detector.

IPCEA (Insulated Power Cable Engineers Association): The association of cable manufacturing engineers who make nationally recognized specifications and tests for cables.

ipso facto: By the fact itself.

IR (Insulation resistance): The measurement of the dc resistance of insulating material; can be either volume of surface resistivity; extremely temperature sensitive.

IR drop: The voltage drop across a resistance due to the flow of current through the resistor.

iridescence: A multicolored appearance.

iridium: A silver-white metallic element of the platinum group.

IRK (Insulation dc resistance constant): A system to classify materials according to their resistance on a 1000 foot basis at 15.5°C (60°F).

iron: Metallic element that plays the most important role in the industrial world. It is obtained fro ores in combination with other substances. Iron is marketed as cast, wrought, malleable, and steel.

ironwork: The term relates to the use of iron for ornamental purposes.

irradiation, atomic: Bombardment with a variety of subatomic particles; usually caused changes in physical properties.

irregular curve: A drafting instrument used as an aid in drawing curves which are not arcs of circles. Also called French curve or universal curve.

irrigation district: Quasi-political districts created under special laws to provide for water services to property; owners in the district.

ISO: International Organization for Standardization who have put together the "SI" units that are now the international standards for measuring.

isolated: A building or column is said to be isolated when it is detached from other buildings or when different parts are separated by incombustible materials. Not readily accessible to persons unless special means for access are used.

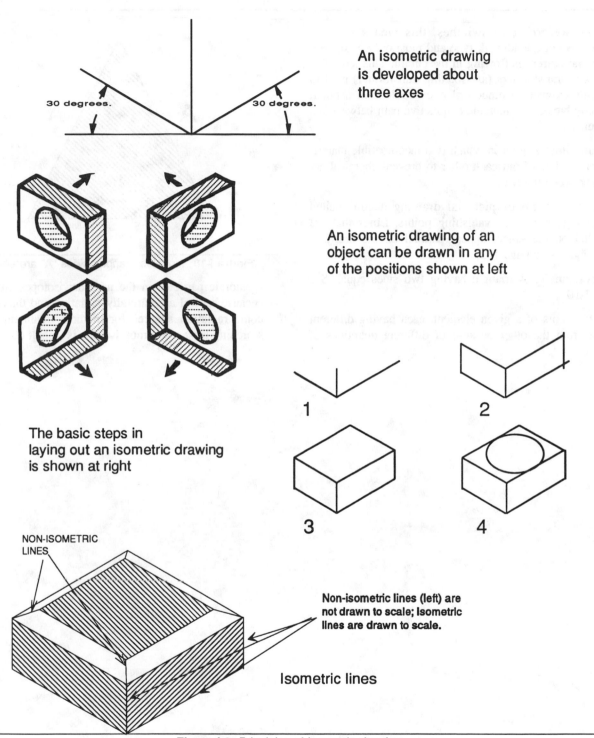

An isometric drawing is developed about three axes

30 degrees.   30 degrees.

An isometric drawing of an object can be drawn in any of the positions shown at left

The basic steps in laying out an isometric drawing is shown at right

1

2

3

4

NON-ISOMETRIC LINES

Non-isometric lines (left) are not drawn to scale; Isometric lines are drawn to scale.

Isometric lines

Figure I-9: Principles of isometric drawing.

isolating: Referring to switches, this means that the switch is not a loadbreak type and must only be opened when no current is flowing in the circuit. This term also refers to transformers (an isolating transformer) used to provide magnetic isolation of one circuit from another, thereby breaking a metallic conductive path between the circuits.

isolation joint: A joint in which two incompatible materials are isolated from each other to prevent chemical action between the two.

isometric: A form of pictorial drawing accomplished without reference to vanishing points. Lines that are parallel on the object appear parallel on the drawing. See Fig. I-9 on page 153.

isoceles triangle: A triangle having two sides equal. See Fig. I-10.

isotope: Atoms of a given element, each having different mass from the other because of different quantities of

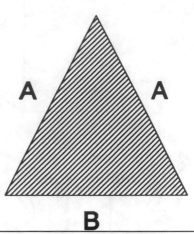

Figure I-10: Isoceles triangle; sides "A" are equal.

subatomic particles in the nucleus; isotopes are useful because several are naturally radiating and thus can become radiation sources for medical treatments or researching labs; a common isotope is Cobalt 60.

J

jack: 1) A plug-in type terminal. 2) A mechanical device used for lifting heavy loads through short distances with a minimum expenditure of manual power. See Fig. J-1.

jack arch: The ordinary flat arch; also called "French arch." See Fig. J-2.

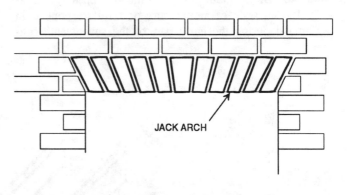

Figure J-2: Jack arch.

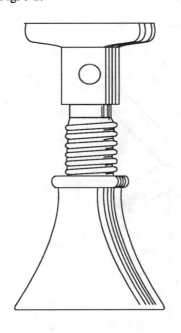

Figure J-1: Building jack.

jacket: 1) A non-metallic polymeric close fitting protective covering over cable insulation; the cable may have one or more conductors. 2) An outer casing, as around a hot-water heater, to preserve heat or cold.

jacket, conducting: An electrically conducting polymeric covering over an insulation.

jack plane: A plane used for roughing off, bringing the wood down to approximate size. See Fig. J-3 on page 160.

jack rafter: Short rafter with one end terminating against a hip or valley rafter. When one end terminates against the hip and the other against the valley rafter, it is often termed a "cripple." See Fig. J-4 on page 160.

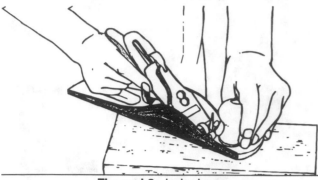

Figure J-3: Jack plane.

jackstud: A short stud that does not extend from floor to ceiling; for example, a stud that extends from the floor to a window.

jacobean: Of or relating to the time of James I of England. That style of architecture or furniture popular in England during the early part of the seventeenth century.

jalousie: A kind of window, blind or shutter made with slats or panes fixed at an angle. See Fig. J-5.

jamb: One of the upright sides of a doorway or a window frame. See Fig. J-6.

jamming: The wedging of a cable such that it can no longer be moved during installation.

japanning: A method of finishing wood or metal with baking varnish or baking japan, usually applied by dipping, after which it is baked at varying temperatures, on wood at low heat, but on metal at from 200 to 400 degrees F.

JB: Pronounced "jay bee", refers to any junction box, taken from the initial letters J and B.

jetty: A structure of stone or wood, projecting into a body of water to divert current, to protect a harbor or shore.

jewelers' saw: A saw similar to a coping saw but having a saw blade hard enough for cutting metal and other substances. See Fig. J-7:

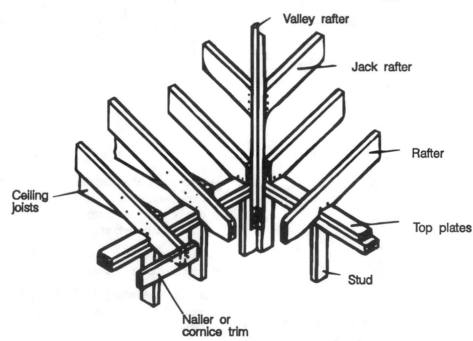

Figure J-4: Roof framing details showing the various types of rafters.

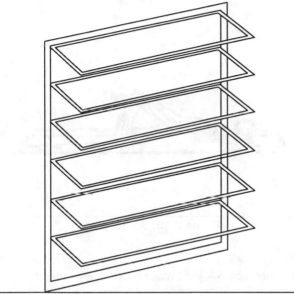

Figure J-5: Jalousie window.

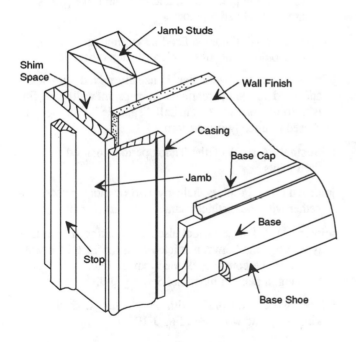

Figure J-6: Details of door framing, showing the door jamb.

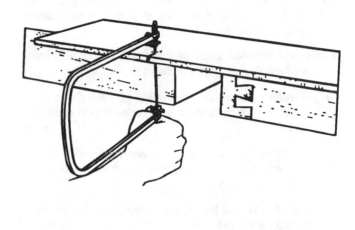

Figure J-7: Jewelers' saw.

jig saw: A narrow, thin-bladed saw, to which an up-and-down motion is imparted. See Fig. J-8.

joggle: A dowel for joining two adjacent blocks of masonry.

joiner: A woodworker who specializes in making wood joints.

Figure J-8: Jig saw.

joinery: The term relates to the various types of joints used by woodworkers.

joint: To join, fasten, or secure two or more pieces together by any of the methods well known to artisans. See Fig. J-11 on opposite page.

joint-butt: Squared ends or ends and edges adjoining each other. Popular types include:

- dovetail—Joint made by cutting pins the shape of dovetails which fit between dovetails upon another piece.

- drawboard—A mortise-and-tenon joint with holes so bored that when a pin is driven through, the joint becomes tighter.

- fished—An end butt splice strengthened by pieces nailed on the sides.

- glue—A joint held together with glue.

- halved—A joint made by cutting half the wood away from each piece so as to bring the sides flush.

- housed—A joint in which a piece is grooved to receive the piece which is to form the other part of the joint.

- lap—A joint of two pieces lapping over each other.

Figure J-10: Jointer plane.

- mortised—A joint made by cutting a hole or mortise, in one piece, and a tenon, or piece to fit each other.

- rub—A glue joint made by carefully fitting the edges together, spreading glue between them, and rubbing the pieces back and forth until the pieces are well rubbed together.

- scarfed—A timber spliced by cutting various shapes of shoulders, or jugs, which fit each other.

joint cement: A powder that is usually mixed with water and used for joint treatment in gypsum wallboard finish. Joint cement, often call "spackle," can be purchased in a ready-mixed form.

joint stool: A stool of the Tudor period, marked by mortised joints.

joint tenancy: Property held by two or more persons together with the distinct character of survivorship.

jointer: 1) A flat steel tool used for making the various types of joints between bricks upon the face of a wall, as the V, the concave, beaded, square, etc. 2) In woodworking, a planing machine. See Fig. J-9.

jointer plane: Iron plane with wood fittings, used for all kinds of plane work. See Fig. J-10.

jointing: The finishing of the exterior surface of mortar joints.

Figure J-9: Jointer.

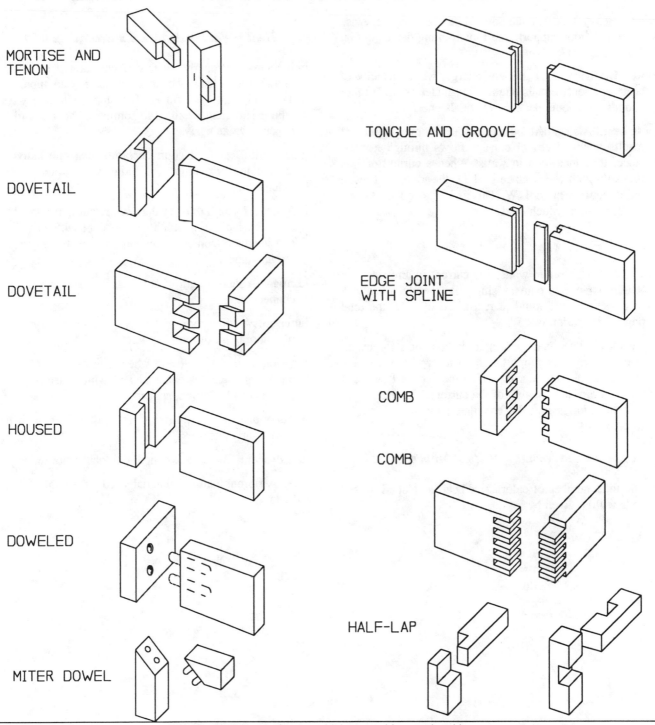

MORTISE AND
TENON

DOVETAIL

DOVETAIL

HOUSED

DOWELED

MITER DOWEL

TONGUE AND GROOVE

EDGE JOINT
WITH SPLINE

COMB

COMB

HALF-LAP

Figure J-11: Several types of joints used in woodworking.

joist: Heavy piece of planking or timber laid edgewise to form a floor support. See roof-framing details in Fig. J-4 on page 160.

joule: The derived SI unit for energy, work, quantity of heat; one joule equals one newton-meter or .73732 foot pounds. One joule per second equals one watt.

joule heat: Also called joule effect, the thermal effect that results when electrical current flows through a resistance. It is measured in watts. When a current of 1 A flows through a resistance of 1 Ω, the joule heat given off is equivalent to 1 W. This is explained by Ohm's law for power, which reads

$$P - I^2R$$

where P is power in watts, I is current in amperes, and R is resistance in ohms. Using this formula, when 2 A of current flow through a resistance of 1 Ω, the total power dissipation is 4 W.

Joule's law: The heat generated in a conductor by an electric current is proportional to the resistance of the conductor, the time during which the current flows, and the square of the strength of the current. The quantity of heat in calories may be calculated by the use of the equation

Calories per second = Volts x Ampere x 0.24

The total number of calories or heat developed in seconds will be given by

Heat = Volts x Amperes x Seconds x 0.24

journeyman: Properly, one who has gained a thorough knowledge of his or her trade by serving an apprenticeship, although the term is often applied to any worker who is sufficiently skilled to command the standard rate of journeyman's pay.

judgment: Decree of court declaring that one individual is indebted to another and fixing the amount of such indebtedness.

judgment d.s.b.: D.s.b. is the abbreviation for the Latin debitum sine brevi, which means "debt without writ." It is a judgment confessed by authority of the language in the instrument.

jumper: A short length of conductor, usually a temporary connection.

junction: A connection of two or more conductors. Place of union; point of meeting; joint.

junction box: An enclosure where one or more electrical raceways or cables enter, and in which electrical conductors can be, or are, spliced.

junior mortgage: A mortgage second in lien to a previous mortgage.

jurisdiction: The area assigned to a local labor union.

jute: A fibrous natural material used as a cable filler or bedding.

ka: KiloAmpere.

kc: Kilocycle, use kiloHertz.

kcmil: One thousand circular mils. 250 kcmil is a conductor or wire size of 250,000 circular mils; 500 kcmil is 500,000 circular mils, etc.

kel: Kilogram.

kellastone: A stucco with crushed finish.

kelvin (K): The basic SI unit of temperature :1/273.16 of thermodynamic temperature of the triple freezing point of water.

kelvin double bridge: A special bridge that is used for measuring very low electrical resistance (0.1 ohm or less). The arrangement of the bridge reduces the effects of contact resistance, which causes significant error when such low resistances are connected to conventional resistance bridges.

Kelvin's temperature: Any temperature in the absolute scale. This is a temperature scale with its zero point at -273.1°C, or absolute zero. The unit of thermodynamic temperature is the kelvin, and its symbol is K.

kerf: A cut or incision made by a saw in a piece of wood. See *Kerfing*.

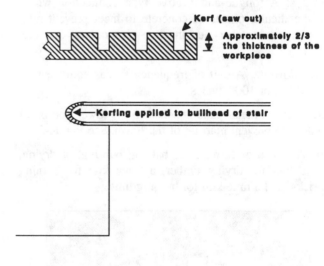

Figure K-1: Principles of kerfing.

kerfing: The process of cutting grooves or kerfs across a board to make it flexible for bending. The kerfs are cut about two thirds of the thickness of the piece. The bullnose of a stair is frequently bent by kerfing. See Fig. K-1.

Figure K-2: Keyhole saw.

keyhole saw: A small tapered-blade saw used for cutting keyholes, small openings for fishing wires, and the like. See Fig. K-2.

keystone: The uppermost stone of an arch which locks its members together. See Fig. K-3.

keyways: A tongue-and groove type connection where perpendicular planes of concrete to meet prevent relative movement between the two components. See Fig. K-4.

kHz: kiloHertz. A unit of frequency that is equivalent to 1000 Hz, or 1000 cycles.

kick plate: A metal plate attached to the lower portion of a door to prevent marring of the finish. See Fig. K-5.

kiln: An oven or furnace for baking, burning, or drying, as a kiln for drying limber, a brick kiln for burning bricks, and a lime kiln for burning lime.

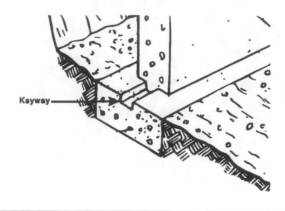

Figure K-4: Keyway in footing to tie-in poured foundation walls.

kiln dried: Lumber that has been dried by means of controlled heat and humidity, in ovens or kilns, to specified ranges of moisture content.

kilo: 10^3 or 1000.

kilocycle: One thousand cycles.

kilogram (kg): The basic SI unit for mass; an arbitrary unit represented by an artifact kept in Paris, France.

kilometer: A metric unit of linear measurement equal to 1000 meters.

kilovolt ampere: One thousand volt-amperes.

kilowatt: Unit of electrical power equal to 1000 watts.

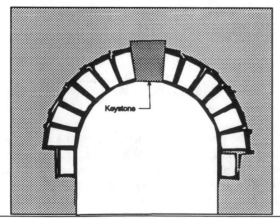

Figure K-3: Keystone.

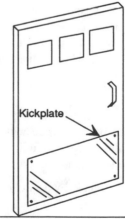

Figue K-5: Kickplate on bottom portion of door.

kilowatt-ft: The product of load in kilowatts and the circuits distance over which a load is carried in feet; used to compute voltage drop.

kilowatt hour: The work performed by one kilowatt of electric power in one hour. The unit on which the price of electrical energy is based.

kinetic energy: Energy by virtue of motion. Whenever work is accomplished on an object, energy is consumed (changed from one kind to another). If no energy is available, no work can be performed. Thus, energy is the ability to do work. One form of energy is that which is contained by an object in motion. In driving a nail into a block of wood, a hammer is set in motion in the direction of the nail. As the hammer strikes the nail, motion of the hammer is converted into work as the nail is driven into the wood. This energy is called kinetic energy.

king post: The central upright pieces in a roof truss against which the rafters abut, and which supports the tie beam. See Fig. K-6.

Kirchoff's Laws: 1) The algebraic sum of the currents at any point in a circuit is zero. That is,

$$I_1 + I_2 + I_3 + ... = 0$$

where I_1, I_2, I_3, etc., are the currents entering and leaving the junction. Currents entering the junction are assumed to be positive, while currents leaving the junction are negative. When solving a problem using this equation, the currents must be placed into the equation with the proper polarity signs attached.
2) The algebraic sum of the product of the current and the impedance in each conductor in a circuit is equal to the electromotive force in the circuit.

knee: The bend in a response curve that is most often an indication of the onset of saturation or cutoff. The point on the curve that represents the knee marks an abrupt change.

kneeler: A stone cut to provide a change of direction. See Fig. K-7.

knee wall: A short wall extending from the floor to the roof in the second story of a 1½-story building. See Fig. K-8.

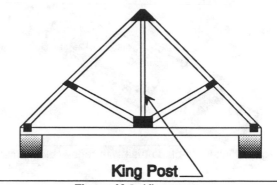
Figure K-6: King truss.

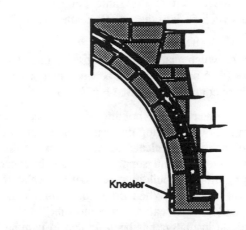
Figure K-7: Kneeler cut to change direction in a stone wall.

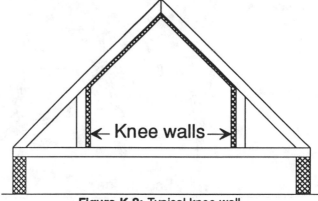
Figure K-8: Typical knee wall.

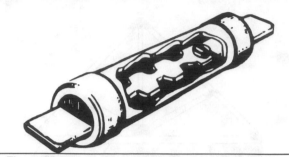

Figure K-9: Knife-blade fuse with renewable element.

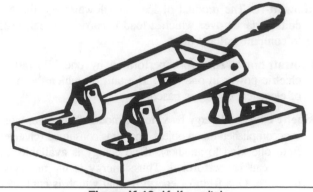

Figure K-10: Knife switch.

knife-blade fuse: A fuse having end connections that resemble the blades of a knife switch and that fit into the contacts of the cutout in the same manner that the blades of a switch fit into the switch contacts. See Fig. K-9.

knife switch: A switch that opens or closes a circuit by the contact of one or more blades between two or more flat surfaces or contact blades. See Fig. K-10.

knob: A porcelain device for holding electrical conductors in place. For open wiring on insulators, the knob is used in conjunction with cleats, tubes and flexible tubing for the protection and support of single insulated conductors run in or on buildings, and not concealed by the building structure. When concealed, this wiring method is called "concealed knob-and-tube" wiring. Open wiring on insulators has been limited to farm and industrial applications; concealed knob-and-tube wiring is allowed only as an extension of an existing system or else with special permission from the inspecting authorities.

knockout: A portion of an enclosure designed to be readily removed for installation of a raceway or cable connector. Sometimes called "concentric" or "eccentric."

knot: 1) A fault in timber, supposed to be caused by a branch or offshoot when the tree is growing. A live knot is one that cannot be knocked out. A dead knot is one which is loose and can be separated from the timber. 2) Intertwining of two or more sections of rope to join them together. See Fig. K-11.

KO: Pronounced "kay oh"; a knockout, the partially cut opening in boxes, panel cabinets and other enclosures, which can be easily knocked out with a screw driver and hammer to provide a clean hole for connecting conduit, cable or some fittings. See knockout.

kVA: Kilovolts times Ampere. Also referenced as 1000 volt-amperes.

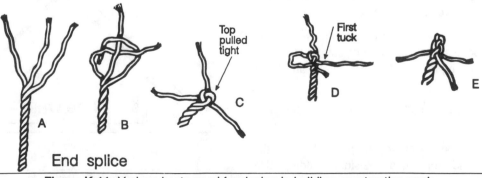

End splice

Figure K-11: Various knots used for rigging in building construction work.

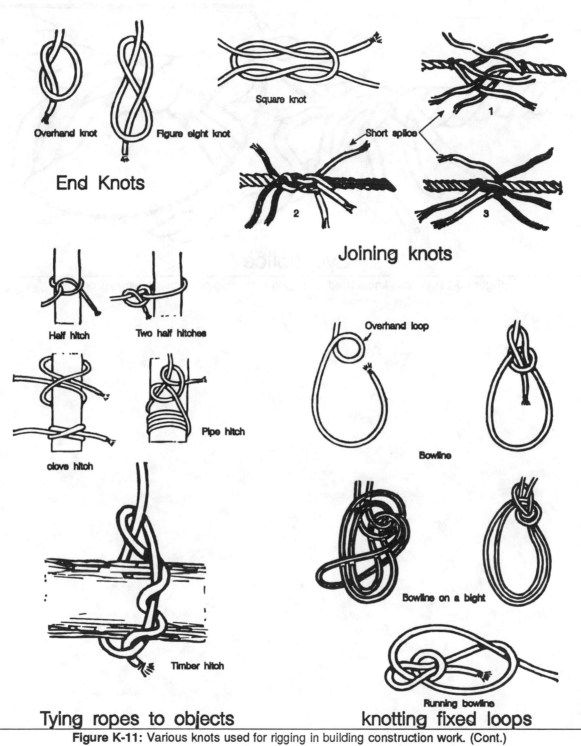

End Knots
Overhand knot   Figure eight knot

Square knot

Short splice

Joining knots

Half hitch   Two half hitches

clove hitch   Pipe hitch

Timber hitch

Overhand loop

Bowline

Bowline on a bight

Running bowline

Tying ropes to objects   **knotting fixed loops**

Figure K-11: Various knots used for rigging in building construction work. (Cont.)

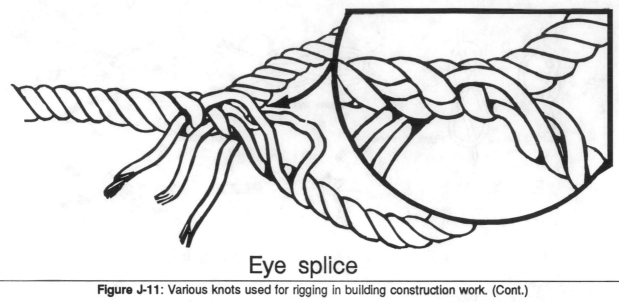

Eye splice

Figure J-11: Various knots used for rigging in building construction work. (Cont.)

L

LA: Lightning arrestor.

label: A projecting molding or dripstone over an opening in a wall.

labeled: Items to which a label, trademark or other identifying mark of nationally recognized testing labs has been attached to identify the items as having been tested and meet appropriate standards.

laboratory: A place where scientific tests, experiments, analysis, etc. are carried on.

lac: Lac is the secretion of the lac insect; shellac is a product manufactured from it. The name is often applied to quick-drying wood finishes.

lacewood: Sometimes called silky oak. Native of Australia. An inexpensive but decorative wood marked with small evenly distributed silky spots. Used generally in small surfaces and inlays.

laches: Delay or negligence in asserting one's rights.

lacquer: A protective coating or finish that dries to form a film by evaporation of a volatile constituent.

lacunar: A paneled or coffered ceiling.

ladder: An aid to climbing. Usually made of two parallel uprights connected by regularly spaced rungs. See Fig. L-1.

Figure L-1: Several types of ladders used on building construction projects.

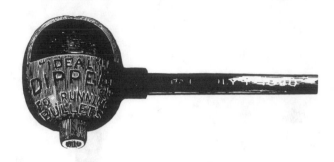

Figure L-2: Ladle.

ladle: Receptacle used for taking the molten metal from a cupola, in transporting it, and in pouring it into molds. Ladles are of various shapes and sizes with capacities from 25 pounds to 100 tons. See Fig. L-2.

La Farge cement: An imported nonstaining by-product cement produced during the calcination of hydraulic lime. It develops slightly less strength than Portland cements.

lagging: The wood covering for a reel.

lag screw: A square-headed, heavy wood screw. It must be tightened down with a wrench as its head is not slotted. See Fig. L-3.

lally column: Concrete-filled cylindrical steel structural column. See Fig. L-4.

lambert: A unit of luminance = 1/candle per cm^2 and therefore equal to the luminance of a perfectly diffusing surface emitting or reflecting light at the rate of one lumen per square centimeter.

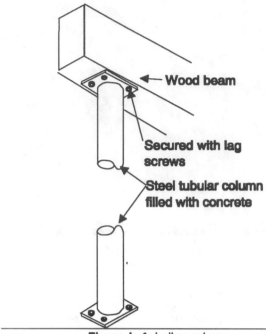

Figure L-4: Lalley column.

lambert cosine law: The intensity from a source of perfectly diffusing distribution in proportional to the cosine of the angle between the direction of emission and the normal to that surface. A source that obeys the law will appear equally bright from an direction.

lambert emitter: A source having the same luminance or brightness from any angle of view and therefore obeying the lambert cosine law.

lambrequin: An ornamental drapery or short decorative hanging, pendant from a shelf or from the casing above a window.

laminate: To build up wood in layers, each layer being a lamination or ply. The construction of plywood.

laminated core: An assembly of steel sheets for use as an element of magnetic circuit; the assembly has the property of reducing eddy-current losses.

laminated wood: Wood built up of piles or laminations that have been joined either with glue or with mechanical fasteners. The piles usually are too thick to be classified as veneer, and the grain of all piles is parallel.

Figure L-3: Lag screw.

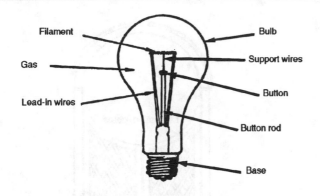

Figure L-5: Parts of an incandescent lamp.

lamp: A device to convert electrical energy to radiant energy, normally visible light: usually only 10-20% is converted to light. See Fig. L-5.

land contract: A contract for the purchase of real estate upon an installment basis; upon payment of last installment, deed is delivered to purchaser.

land economics: Branch of the science of economics which deals with the classification, ownership, and utilization of land and buildings erected thereon.

landing: A platform introduced at some point in a stair run; used to change direction of stairs or to break the run. See Fig. L-6.

landing newel: The post placed at the landing point of a stair and supporting the baluster. See Fig. L-7.

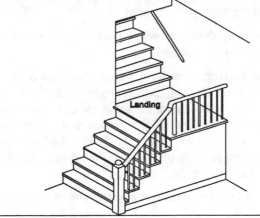

Figure L-6: One type of stair landing.

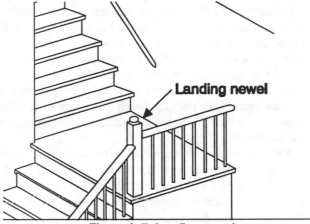

Figure L-7: Landing newel.

landing tread: The front end of a stair landing. Usually it is so built that the front edge has the thickness and finish of a stair trend and the back has the thickness of the flooring of the landing.

landlord: One who rents property to another.

landscape panel: A wood panel with a horizontal grain.

lap: The relative position of applied tape edges; "closed butt lap"—tapes just touching; "open butt" or "negative lap"—tapes not touching; "positive lap" or "lap"—tapes overlapping.

lap joint: A joint produced by the overlapping of contiguous faces of wood or metal. See Fig. L-8.

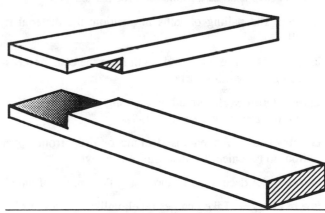

Figure L-8: Lap joint.

lap winding: A parallel winding that has an even number of segments and slots. The number of bars, however, is equal to or a multiple of the number of slots. It is known as a parallel winding because there are as many circuits in parallel as there are poles. Thus, a four-pole motor has four circuits in parallel, and a six-pole motor has six circuits in parallel through the armature. This requires that there be as many brushes as there are poles for the motor.

Since there are so many paths in parallel, a larger current can pass through the winding because it has several paths to travel. This, in turn, reduces the amount of voltage needed in the armature to push the current through so many paths.

larch: A medium-size, cone-bearing, deciduous tree. Wood is heavy, hard, and strong, white to red. Used for telephone poles, fence posts, and in some types of boat building.

large knot: A sound knot more than $1\frac{1}{2}$ inches in diameter.

lastic: A substance which at a certain temperature exhibits the physical properties of rubber.

latent heat: Heat given off or absorbed in a process (as vaporization or fusion) other than a change in temperature.

lateral: Relating to the side, or crosswise of the length.

lateral thrust: The pressure of a load that extends to the sides of a masonry wall.

lath: A strip used as a foundation for plaster.

lathing: The nailing of laths in position; the material itself.

latitude: Distance north and south from the equator measured on the earth's surface.

lattice: Open work formed by crossing or interlacing laths or other thin strips. See Fig. L-9:

laundry chute: An enclosed chute or drop from upper floor to basement for disposal of laundry.

laureling: A decorative feature using the laurel-leaf motif.

law of charges: Like charges repel, unlike charges attract.

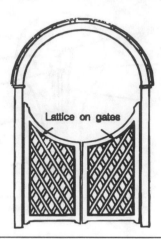

Figure L-9: Lattice work.

law of magnetism: Like poles repel, unlike poles attract.

lay: The axial length of one turn of the helix of any element in a cable.

lay direction: Direction of helical lay when viewed from the end of the cable.

lay length: Distance along the axis for one turn of a helical element.

lay up: To place materials together in the relative positions they will have in the finished building.

LDD (luminaire dirt depeciation): A factor utilized in illumination computations to predict the reduction in initial illumination based upon the type of lamp and the relamping program to be used.

lead (leed): A short connecting wire brought out from a device or apparatus.

lead-acid battery: The most widely used type of storage battery, which has an EMF (voltage) of 2.2 V per cell. In its charged condition, the active materials in the lead-acid battery are lead dioxide (sometimes referred to as lead peroxide) and spongy lead. The lead dioxide is used as the positive plate, while the spongy lead forms the negative plate.

lead squeeze: The amount of compression of a cable by a lead sheath.

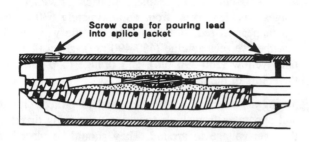

Figure L-10: Lead sheath used on high-voltage cable.

leading: Applying a lead sheath to electrical cable. See Fig. L-10.

leakage: Undesirable conduction of current.

leakage distance: The shortest distance along an insulation surface between conductors.

leakage resistance: The ohmic value of the path between two electrodes that are insulated from each other.

lean-to roof: A wing or extension to a building, with a roof sloping only in one direction.

lease: A contract, written or oral, for the possession of lands and tenements on the one hand and a recompense of rent or other income, on the other hand.

leasehold: An estate in realty held under a lease.

ledge: A shelflike projection as from a wall.

ledger board: Same as ribbon strip; attached to studding to carry joists. See Fig. L-11.

leg: A portion of a circuit, such as a switch leg or switch loop.

legal description: A description recognized by law, which is sufficient to locate and identify the property without oral testimony.

legend, embossed: Molded letters and numbers in the jacket surface, letters may be raised or embedded.

let-in brace: A board nominally 1 inch thick applied diagonally into notched studs.

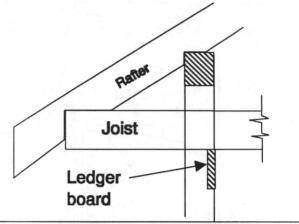

Figure L-11: Application of ledger board.

Lenz's Law: "In all cases the induced current is in such a direction as to oppose the motion which generates it." That which states that whenever the value of an electric current is changed in a circuit, it creates an electromotive force by virtue of the energy stored up in its magnetic field, which opposes the change.

lessee: A person to whom property is rented under a lease.

lessor: Landlord.

level: Horizontal; in a horizontal plane. Also, a tool for testing with regard to the horizontal. See Fig. L-12.

lewis: A dovetailed tenon inserted in a heavy stone for the purpose of attaching hoisting apparatus.

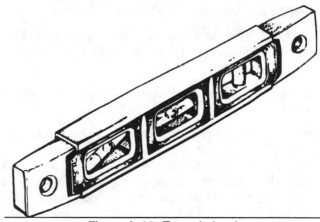

Figure L-12: Torpedo level.

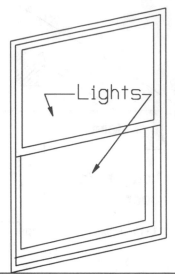

Figure L-13: Parts of a windows showing the windows lights.

lewis bolt: An anchor bolt; a bolt having a jagged and tapered tail. It is used for insertion into masonry, where it is held with lead.

liability: The condition of being responsible for a possible or actual loss, penalty, expense, or burden.

license: A privilege or right granted by the state, county, or city to operate a business. An authority to go upon or use another person's land or property, without possessing any estate therein.

license year: Period specified in license law for license; usually different from calendar year.

lien: A hold or claim which one person has upon property of another as security for a debt or charge; judgements, mortgages, taxes.

life estate: An estate or interest held during the term of some certain person's life.

light: 1) A pane of glass; a division in a sash for a single pane of glass. See Fig. L-13. 2) Radiant energy lying within a wavelength that spreads from 100 to 10,000 nm. The human eye can perceive light radiations in a frequency range of between 450 and 700 nm. The color of light is determined by its wavelength. Energy at the short-wave end of the visible spectrum, from 380 to about 450 nm, produces the sensation of violet. The longest visible waves, from approximately 630 to 760 nm, appear as red. Between these lie the wavelengths that the eye sees as blue (450-490 nm), green (490-560 nm), yellow (560-590 nm), and orange (590-630 nm), the colors of the rainbow.

lightning arrestor: A device designed to protect circuits and apparatus from high transient voltage by diverting the over-voltage to ground. They should be placed on upward projections such as chimneys, towers, and the like. On flat roofs, rods should be placed 50 ft on center, and on edges of flat roofs and ridges of pitched roofs, about 25 ft. on center. The rods should project from 10 to 60 in. above flat roofs and ridges, and from 10 to 14 in. above upward projections. See surge arrester.

light quantity: A measure of the total light emitted by a source of light falling on a surface. It may be expressed in lumens or watts. If all the light were of green-yellow color, at the peak of the spectral luminosity curve, 1 W would be equal to 681 lm. This gives some idea of the relative luminous efficiency of common light sources. For instance, a 100-W incandescent lamp emits about 1600 lm, only 2.4 W as light and the balance as heat. A 40-W fluorescent lamp emits about 3100 lm, 4.5 W as light and the balance as heat. From this, it can be seen that the fluorescent lamp gives four to five times as much light as the incandescent on a watt-for-watt basis.

light relay: A photoelectric device that triggers a relay in accordance with fluctuations in the intensity of a light beam. These may be purely electronic in nature or electromechanical, depending on the individual circuit. Often, light relays use an active or passive photoelectric device in the base circuit of a transistor, which is connected in series with an electromechanical relay. When light strikes the photocell, the transistor is driven to saturation and conducts current through the relay coil, causing it to change states.

light wave: A stream of electromagnetic energy that falls into the light spectrum. This lies between 100 and 10,000 nm. Visible light (that which can be seen by the human eye) lies within a very narrow band width of this spectrum at a frequency range of about 500 to 600 nm.

LIM (Laboratory Inspection Manual): A summary of specified values for quality testing.

limestone: A very commonly used stone for buildings of the better types; also used for making lime.

limit: A boundary of a controlled variable.

limit control: Control used to open or close electrical circuits as temperature or pressure limits are reached.

limiter: A device in which some characteristic of the output is automatically prevented from exceeding in predetermined value.

linden: A wood, same as "basswood."

line: A circuit between two points; ropes used during overhead construction.

line, bull: A rope for large loads.

line, finger: A rope attached to a device on a pole when a device is hung, so further conductor installation can be done from the ground.

line, pilot: A small rope strung first.

lineal foot: A foot in length as distinguished from square foot or cubic foot.

linear: Arranged in a line.

linearity: When the effect is directly proportional to the cause.

line, tag: A rope to guide devices being hoisted.

Lintel: The horizontal top piece over a doorway or window opening. See Fig. L-14.

liquid absorbent: A chemical in liquid form that has the property to absorb moisture.

liquid line: The tube that carries liquid refrigerant from the condenser or liquid receiver to the refrigerant control mechanism.

liquid receiver: Cylinder connected to a condenser outlet for storage of liquid refrigerant in a system.

lis pendens: Suit pending; usually recorded so as to give constructive notice of pending litigation.

Lissajous Figure: A special case of an x-y plot in which the signals applied to both axes are sinusoidal functions; useful for determining phase and harmonic relationships.

listed: Items in a list published by a nationally recognized independent lab that makes periodic tests. Fig. L-15. See labeled.

listing: Oral or written employment of broker by owner to sell or lease real estate.

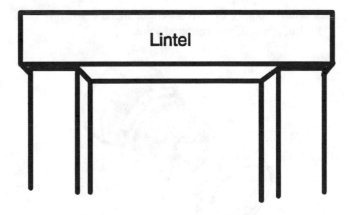

Figure L-14: Lintel application.

Figure L-15: Various labels used to indicate that materials or an apparatus is "listed."

liter: Metric unit of volume that equals 61.03 cubic inches.

littoral: Belonging to shore as of sea or Great Lakes; corresponds to riparian rights.

live: Energized.

live-front: Any panel or other switching and protection assembly, such as switchboard or motor control center, which has exposed electrically energized parts on its front, presenting the possibility of contact by personnel.

live load: The load, expressed in pounds per square foot, of people, furniture, snow, etc., that are in addition to the weight of the structure itself.

LMP: Low Molecular Weight Polyethylene.

load: 1) A device that receives power. 2) The power delivered to such a device.

load-break: Referring to switches or other control devices, this phrase means that the device is capable of safely interrupting load current—to distinguish such devices from other disconnect devices which are not rated for breaking load current and must be opened only after the load current has been broken by some other switching device.

load center: An assembly of circuit breakers or switches. See Fig. L-16.

load factor: The ratio of the average to the peak load over a certain period. The time may be either the normal number of operating hours per day or 24 h, as generally used by the power companies. The average load is equal to the kilowatt-hours used in the specified time, as measured by a watthour meter, divided by the number of hours. The maximum load is the highest load at one time, as measured by some form of maximum-demand or curve-drawing watthour meter.

load losses: Those losses incidental to providing power.

LOCA (Loss of Coolant Accident) (nuclear power): The test to simulate nuclear reactor accident exhibited by high radiation, high temperature, etc.

location, damp: A location subject to a moderate amount of moisture such as some basements, barns, cold-storage, warehouses, and the like.

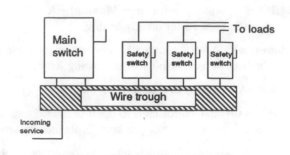

Figure L-16: Diagram of an electric load center.

location, dry: A location not normally subject to dampness or wetness; a location classified as dry may be temporarily subject to dampness or wetness, as in the case of a building under construction.

location, wet: A location subject to saturation with water or other liquids.

locked rotor: When the circuits of a motor are energized but the rotor is not turning.

locking stile: That part of a door to which the lock is attached.

locknut: A fitting for securing cable and conduit connectors to outlet boxes, panelboards, and other apparatus; also used on threaded metal conduit. See Fig. L-18.

lockout: To keep a circuit locked open. To "tag out" a circuit as per OSHA rules.

lock seam: Joining of two sheets of metal consisting of a folded, pressed, and soldered joint.

Figure L-17: Application of locknut.

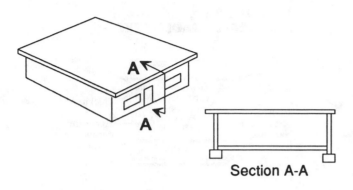

Section A-A

Figure L-18: Logitudinal section taken through an object.

lock washer: A thin washer whose action is similar to that of a compression spring. Frequently used for the same purpose as a lock nut. See Fig. L-17.

logarithm: The exponent that indicates the power to which a number is raised to produce a given number.

logarithmic: Pertaining to the function y = log x.

longitudinal section: A section taken through a part in the direction of its length. See Fig. L-18.

lookout: A short wood bracket or cantilever to support an overhang portion of a roof, usually concealed from view by a soffit.

Figure L-19: One type of louver.

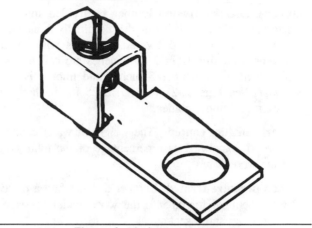

Figure L-20: Lug connector.

looping-in: Avoiding splices by looping wire through device connections instead of cutting the wire.

loss: Power expended without doing useful work.

louver: An opening with a series of horizontal slats arranged to permit ventilation but to exclude rain, sunlight, or vision. See Fig. L-19.

lozenge molding: A molding characterized by lozenge-shaped ornament; used in connection with Norman architecture.

lug: A device for terminating a conductor to facilitate the mechanical connection. Fig. L-20.

lug sill: A type of sill longer than the width of the window opening in stone or brick walls. Such a sill differs from a "slip sill" in that its ends are "let in" to the wall.

lumber: Timber cut to size in marketable form, as boards, planks, etc.

lumber, boards: Lumber less than 2 inches thick and 2 or more inches wide.

lumber, dimension: Lumber from 2 inches to, but not including, 5 inches thick and 2 or more inches wide. Includes joists, rafter, studs, plank, and small timbers.

lumber, dressed size: The dimension of lumber after shrinking from green dimension and after machining to size of pattern.

lumber grades: Softwood lumber is divided into three size categories: 1) finish grades nominally 1 inch thick and thicker are called *boards*. 2) pieces with nominal thickness ranging from 2 to 4 inches are called *dimension lumber*. 3) lumber 5 inches and thicker is called *timber*. See Figs. L-21 through L-25 for further information on lumber grades.

lumber, moisture content: The weight of water contained in wood, expressed as a percentage of the total weight of the wood.

lumber, pressure-treated: Lumber that has had a preservative chemical forced into the wood under pressure to resist decay and insect attack.

lumber, shiplap: Lumber that has been milled along the edge to make a close rabbeted or lapped joint.

lumber, timbers: Lumber 5 or more inches in least dimension. Includes beams, stringers, posts, caps, sills, girders, and purlins.

lumen: The derived SI unit for luminous flux.

A, *Northern Hardwood and Pine Manufacturers Association, Inc.;*

B, *Pacific Lumber Inspection Bureau, Inc.;*

C, *Southern Pine Inspection Bureau;*

D, *Canadian Lumberman's Association.*

Figure L-21: Examples of symbols of lumber logos of quality control agencies.

TYPICAL APA REGISTERED TRADEMARKS

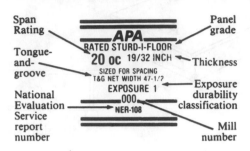

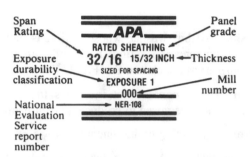

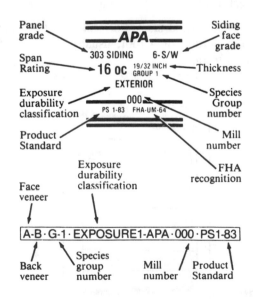

Figure L-22: Typical American Plywood Association (APA) markings designating performance standards.

	For material 2 to 4 inches thick by 2 to 4 inches wide				For material 2 to 4 inches thick by 5 inches or more wide			
	Extreme fiber in bending (f)		Modulus of elasticity (E)	Horizontal shear (H)	Extreme fiber in bending (f)		Modulus of elasticity (E)	Horizontal shear (H)
	Single member uses	Repetitive member uses			Single member uses	Repetitive member uses		
Southern pine (surfaced dry, used at 19% maximum moisture content)								
Select structural	2,000	2,300	1,400,000	90	1,700	1,950	1,400,000	90
No. 1	1,700	1,950	1,400,000	90	1,450	1,650	1,400,000	90
No. 2	1,400	1,600	1,300,000	90	1,200	1,350	1,300,000	90
No. 3	775	875	1,100,000	90	700	800	1,100,000	90
Appearance	1,700	1,950	1,400,000	90	1,450	1,650	1,400,000	90
Stud	775	875	1,100,000	90	700	800	1,100,000	90
Construction	1,000	1,150	1,100,000	90	—	—	—	—
Standard	550	650	1,100,000	90	—	—	—	—
Utility	275	300	1,100,000	90	—	—	—	—
Spruce–pine–fir (surfaced dry or green, used at 19% maximum moisture content)								
Select structural	1,450	1,650	1,500,000	70	1,250	1,450	1,500,000	70
No. 1	1,200	1,400	1,500,000	70	1,050	1,200	1,500,000	70
No. 2	1,000	1,150	1,300,000	70	875	1,000	1,300,000	70
No. 3	550	650	1,200,000	70	500	575	1,200,000	70
Appearance	1,200	1,400	1,500,000	70	1,050	1,200	1,500,000	70
Stud	550	650	1,200,000	70	500	575	1,200,000	70
Construction	725	850	1,200,000	70	—	—	—	—
Standard	400	475	1,200,000	70	—	—	—	—
Utility	175	225	1,200,000	70	—	—	—	—
Hem–fir (surfaced dry or green, used at 19% maximum moisture content)								
Select structural	1,650	1,900	1,500,000	75	1,400	1,650	1,500,000	75
No. 1	1,400	1,600	1,500,000	75	1,200	1,400	1,500,000	75
No. 2	1,150	1,350	1,400,000	75	1,000	1,150	1,400,000	75
No. 3	650	725	1,200,000	75	575	675	1,200,000	75
Appearance	1,400	1,600	1,500,000	75	1,200	1,400	1,500,000	75
Stud	650	725	1,200,000	75	575	675	1,200,000	75
Construction	825	975	1,200,000	75	—	—	—	—
Standard	475	550	1,200,000	75	—	—	—	—
Utility	225	250	1,200,000	75	—	—	—	—
Douglas–fir south (surfaced dry or green, used at 19% maximum moisture content)								
Select structural	2,000	2,300	1,400,000	90	1,700	1,950	1,400,000	90
No. 1	1,700	1,950	1,400,000	90	1,450	1,650	1,400,000	90
No. 2	1,400	1,600	1,300,000	90	1,200	1,350	1,300,000	90
No. 3	775	875	1,100,000	90	700	800	1,100,000	90
Appearance	1,700	1,950	1,400,000	90	1,450	1,650	1,400,000	90
Stud	775	875	1,100,000	90	700	800	1,100,000	90
Construction	1,000	1,150	1,100,000	90	—	—	—	—
Standard	550	650	1,100,000	90	—	—	—	—
Utility	275	300	1,100,000	90	—	—	—	—

Source: National Forest Products Association. National Design Specification: Wood Construction, Supplement to the 1986 Edition. Table 4A.
The species and grades selected from Table 4a and reproduced here reflect those most commonly used for framing.

Figure F-23: Design values for visualy graded structural lumber.

Materials Being Joined	Nailing Method	No.	Nail Size	Nailing Procedures
Header to joists	Endnail	3	12d	Joist hangers advised
Joist to sill or girder	Toenail	2	8d	
Header and stringer joist to sill	Toenail		8d	24 inches on center
Subfloor, boards				
1 by 6 inch and smaller		2	8d	To each joist
1 by 8 inch		3	8d	To each joist
Subfloor, plywood				
At edges			8d	6 inches on center
At intermediate joists			8d	8 inches on center
Subfloor (2 by 6 inch, tongue and groove to joist or girder)	Blind-nail (casing) and facenail	2	12d	
Soleplate to stud, horizontal assembly	Endnail			
Top plate to stud	Endnail	2	12d	
Soleplate to joist or blocking	Facenail		12d	24 inches on center, maximum
Doubled studs	Facenail		12d	24 inches on center, staggered
End stud of intersecting wall to exterior wall stud	Facenail		12d	24 inches on center
Upper top plate to lower top plate	Facenail		12d	24 inches on center
Upper top plate, laps and intersections	Facenail		12d	
Header, two pieces			12d	12 inches on center, staggered
Ceiling joist to top wall plates	Toenail	2	8d	
Ceiling joist laps at partition	Facenail	3	12d	
Rafter to top plate	Toenail	2	8d	
Rafter to ceiling joist	Facenail	3	12d	
Rafter to valley or hip rafter	Toenail	3	8d	

Figure F-24: Lumber nailing schedules.

Materials Being Joined	Nailing Method	No.	Nail Size	Nailing Procedures
Ridge board to rafter	Endnail	3	12d	
Rafter to rafter through ridge board	Toenail	3	12d	
Collar beam to rafter				
2-inch member	Facenail	2	12d	
1-inch member	Facenail	3	8d	
1-inch diagonal let-in brace to each stud and plate (4 nails at top)		2	8d	
Built-up corner studs				
Studs to blocking	Facenail	2	10d/12d	Each side
Intersectng stud to corner studs	Facenail	2	12d	24 inches on center, staggered, each layer
Wall sheathing:				
1 by 8 or less, horizontal	Facenail	2	8d	At each stud
Wall sheathing, vertically applied plywood				
3/8 inch and less thick	Facenail		6d	6 inches on center at edges
1/2 inch and over thick	Facenail		8d	12 inches on center intermediate
Wall sheathing, vertically applied fiberboard 1/2 inch thick	Facenail			1½-inch roofing nails, 3 inches on center at edges and 6 inches on center intermediate
Roof sheathing, boards, 4-, 6-, 8-inch width	Facenail	2	8d	At each rafter
Roofsheathing, plywood 3/8 inch and less thick	Facenail		6d	6 inches on center at edges and 12 inches on center intermediate
Plywood, 1/2 inch and over thick	Facenail		8d	

Figure F-24: Lumber nailing schedules. *(Cont.)*

Grades	Use Category	Description
Construction, Standard, & Utility	Light framing 2 by 2 through 4 by 4	For use where high strength values are not required (studs, plates, sills, cripples, blocking, etc.)
Stud	Studs, 2 by 4 through 4 by 6, 10 feet and shorter	An optional all-purpose grade limited to 10 feet and shorter. Characteristics affecting strength and stiffness values re limited so that the grade is suitable for all stud uses, including load-bearing walls.
Select Structural, No. 1, No. 2, & No. 3	Structural light framing	Designed to fit engineering applications where higher bending/strength ratios are needed in light framing sizes (for trusses, concrete pier wall forms, etc.)
Select Structural	Structural joists and planks	For joists, rafters, and general framing.

Figure L-25: Dimension lumber grades and sizes.

lumen maintenance: A term applied to data usually to the flux through a unit solid angle (sterandian) from a uniform point source of one candela (candle).

lumen method: A method of calculation to predict the average illumination on a work plane in a room. The maximum and minimum values of illumination will depend upon type of luminaire selected and their placement within a room. All calculations assume that the principles of good lighting layout are applied and that the conditions such as voltage, temperature, etc, will permit luminaires to provide their rated output.

luminaire: A complete lighting unit that includes the lamp, sockets, and equipment for controlling light, such as reflectors and diffusers. On electric discharge lighting, the luminaire also includes a ballast. The common term used for luminaire is lighting fixture, or in some cases, simple fixture.

luminaire classification: Luminaires are classified for general lighting by the C.I.E. in accordance with their percentage of total luminaire output emitted above and below horizontals.

luminance: Intensity in a given direction divided by projected area, as intercepted by the air. It is subjective intensity and ranges from very dim to very bright. Luminance is expressed as candelas per square inch in a certain direction. Candelas per square inch may be put into more convenient form by multiplying by 452, giving luminance in footlamberts. Another way of looking at luminance is in relation to illumination and the reflection factor. For a nonspecular surface

Luminance = Illumination x Reflection factor

or

$$L = E \, x \, R$$

where E equals footcandles, R equals reflection factor, and L equals footlamberts.

To illustrate, if E = 100 fc and R = 50%, then L = 100 x 0.50 = 50 footlamberts.

luminance coefficients: A coefficient, similar to the Coefficient of Utilization, used to redict the average luminance or room surfaces.

lus: The derived SI unit for illuminance.

lux: The unit of illumination when the meter is the unit of length; equal to one lumen per square meter. The phot is the unit of illumination per square centimeter.

LV: Low voltage.

LWBR (nuclear): Light water breeder reactor.

M

machine: An item to transmit and modify force or motion, normally to do work.

machine rating: The amount of power a machine can deliver without overheating.

machine screw: A very commonly used type of screw with clear-cut threads and a variety of head shapes. It may be used either with or without a nut. See Fig. M-1.

magnet: A body that produces a magnetic field external to itself; magnets attract iron particles.

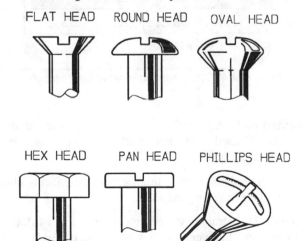

Figure M-1: Types of machine screws.

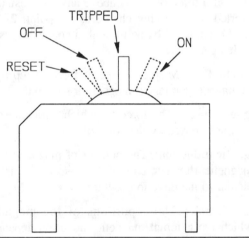

Figure M-2: Operating characteristics of a magnetic circuit breaker.

magnetic circuit breaker: An electromagnetic device for opening a circuit. See Fig. M-2.

magnetic coil: The winding of an electromagnet. A coil of wire wound in one direction, producing a dense magnetic field capable of attracting iron or steel when carrying a current of electricity.

magnetic cutout: A device for breaking an electrical circuit by means of an electromagnet, instead of by fusing a part of the circuit.

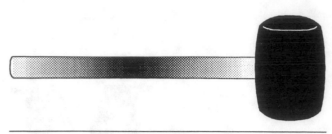

Figure M-3: Rubber mallet.

magnetic density: The number of lines of force or induction per unit area taken perpendicular to the induction. In free space, flux density and field intensity are the same numerically, but within magnetic material, the two are quite different.

magnetic field: 1) A magnetic field is said to exist at a point if a force over and above any electrostatic force is exerted on a moving charge at the point. 2) The force field established by ac through a conductor, especially a coiled conductor.

magnetic flux: Magnetic lines of force set up by an electromagnet, permanent magnet, or solenoid.

magnetic force: The force by which attraction and repulsion are exerted by the poles of a magnet.

magnetic induction: The number of magnetic lines or the magnetic flux per unit of cross-sectional area perpendicular to the direction of the flux.

magnetic pole: Those portions of the magnet toward which the external magnetic induction appears to converge (south) or diverge (north).

magnetic permeability: A measure of the ease with which magnetism passes through an substance.

magnetic switch: A switch operated or controlled by an electromagnet.

magnetism: 1) That property of iron, steel, and some other substances, by virtue of which, according to fixed laws, they exert forces of attraction and repulsion. 2) The science that treats the conditions and laws of magnetic force.

mahogany: True mahogany is considered the premier cabinet wood of the world. The heartwood turns dark red upon exposure to the sun. Grain produces highly attractive figures.

make and break: The term may be applied to several electrical devices. Primarily, there is a pair of contact points, one stationary, and the other operated by a cam that makes the break in a circuit between these points.

mallet: A hammer made of wood, rubber, leather, or similar material. See Fig. M-3.

manhole: A subsurface chamber, large enough for a man, to facilitate cable installation, splices, etc., in a duct bank. See Fig. M-4.

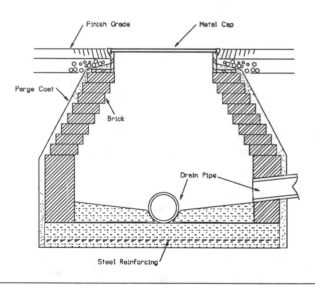

Figure M-4: Cross-section of a typical manhole.

mansard roof: A rather flat-decked, double-sloped roof, frequently used in placing an additional story on a residence. See Fig. M-5.

mantel: The shelf above a fireplace; also the facing about a fireplace including the shelf.

manual: 1) Operated by mechanical force applied directly by personal intervention. 2) A handbook of instructions.

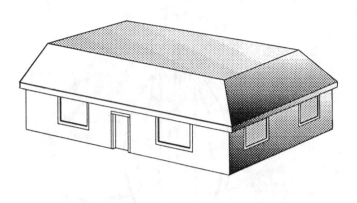

Figure M-5: Building with a mansard roof.

maple: A hard, light-colored, tough wood much used for flooring and veneers.

marker: A tape or colored thread in a cable that identifies the cable manufacturer.

market value: The highest price which a buyer, willing but not compelled to buy, would pay, and the lowest a seller, willing but not compelled to sell, would accept.

marketable title: Such a title as a court would compel a purchaser to accept; it is free from any encumbrances or clouds.

marshalling: Where a creditor has two or more funds out of which to satisfy a debt, he cannot so elect as to deprive another individual, who has but one fund, of his security.

masonry: Stone, brick, concrete, hollow-tile, concrete-block, gypsum-block, or other similar building units or materials or a combination of the same, bonded together with mortar to form a wall, pier, buttress, or similar element.

mass: The property that determines the acceleration the body will have when acted upon by a given force; unit = kilogram.

master switch: A main switch; a switch controlling the operation of other switches.

mastic: A pasty material used as a cement in such applications as setting tile or as a protective coating for thermal insulation or waterproofing.

mat: A concrete base for heavy electrical apparatus, such as transformers, motors, generators, etc.; sometimes the term includes the concrete base, a bed of crushed stone around it and an enclosing chain-link fence.

matrix: A multi-dimensional array of items.

matte surface: A surface from which reflection is predominately diffused.

maximum voltage: The highest voltage reached in each alternation of an alternating voltage.

MCM: An expression referring to conductors of sizes from 250 MCM, which stands for Thousand Circular Mils, up to 2000 MCM. The newest way of expressing thousand circular mils is kcmil; that is, 250 kcmil, 500 kcmil, etc. See Fig. M-6.

Wire Size, kcmil	Diameter, overall	Resistance, Copper	Resistance, Aluminum
250	0.575"	0.0515	0.0847
300	0.630"	0.0429	0.0707
350	0.681"	0.0367	0.0605
400	0.728"	0.0321	0.0529
500	0.813"	0.0258	0.0424
600	0.893"	0.0214	0.0353
700	0.964"	0.0184	0.0303
750	0.998"	0.0171	0.0282
800	1.03"	0.0161	0.0265
900	1.09"	0.0143	0.0235
1000	1.15"	0.0129	0.0212
1250	1.29"	0.0103	0.0169
1500	1.41"	0.00858	0.0141
1750	1.52"	0.00735	0.0121
2000	1.63"	0.00643	0.0106

Figure M-6: Specifications of various wire sizes.

mean: An intermediate value; arithmetic — sum of values divided by the quantity of the values; the average.

mechanical water absorption: A check of how much water will be absorbed by material in warm water for seven days (mg/sq. in. surface).

mechanic's lien: A species of lien created by statute which exists in favor of persons who have performed work or furnished materials in the erection or repair of a building.

meeting of minds: A mutual intention of two persons to enter into a contract affecting their legal status based on agreed-upon terms.

medium hard: A relative measure of conductor temper.

meeting rail: The horizontal wood or metal bar which divides the upper and lower sash of a window.

meg or mega: When prefixed to a unit of measurement it means one million times that unit.

megavolt: A unit of voltage equal to one million volts.

megger® A test instrument for measuring the insulation resistance of conductors and other electrical equipment; specifically, a megohm (million ohms) meter; this is a registered trade name of the James Biddle Co. See Fig. M-7.

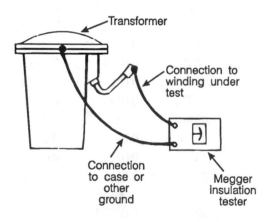

Figure M-7: Megger used to test transformer insulation.

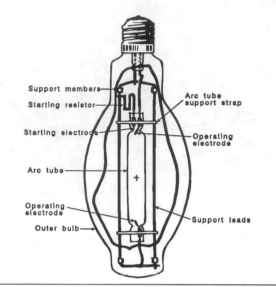

Figure M-8: Principle parts of mercury-vapor lamp.

megohm: A unit of electrical resistance equal to one million ohms.

megohmmeter: An instrument for measuring extremely high resistance.

melt index: The extrusion rate of a material through a specified orifice at specified conditions.

melting time: That time required for an overcurrent to sever a fuse.

mercury vapor lamp: A type of lamp that produces light by passing a current through mercury vapor. See Fig. M-8.

messenger: The supporting member of an aerial cable.

metal clad (MC): The cable core is enclosed in a flexible metal covering.

metal-clad switchgear: Switchgear having each power circuit device in its own metal enclosed compartment.

metal filament: The electrical conductor that glows when heated in an incandescent lamp.

meter: An instrument designed to measure; metric unit of linear measurement equal to 39.37 inches.

meter pan: A shallow metal enclosure with a round opening, through which a kilowatt hour meter is mounted, as the usual meter for measuring the amount of energy consumed by a particular building or other electrical system.

metric system: An international language of measurement. Its symbols are identical in all languages. The conversion factors in Appendix A provide conversions from metric to English and from English to metric units, along with many other useful conversions.

MFT: Thousands of feet.

mho: Reciprocal of ohm.

mica: A silicate which separates into layers and has high insulation resistance, dielectric strength, and heat resistance.

MI cable: Mineral insulated, metal sheathed cable. See Fig. M-9.

microfarad: A unit of capacity, being one millionth of a farad.

micrometer (mike): A tool for measuring linear dimensions accurately to 0.001 inch or to 0.01 mm.

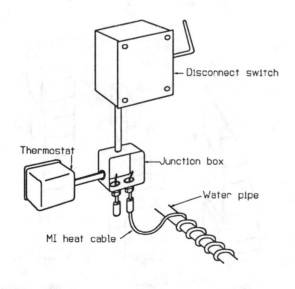

Figure M-9: Application of Type MI cable.

micro structure: The structure of polished and etched metals as revealed by a microscope at a magnification of more than ten diameters.

microwave: Radio waves of frequencies above one gigahertz.

mil: A unit used in measuring the diameter of wire, equal to 0.001 inch (25.4 micrometers).

MIL: Military specification.

mildewcide: A chemical that is poisonous specifically to mildew fungi. A Specific type of fungicide.

mill: One-tenth of one cent; the measure used to state the property tax rate. That is, a tax rate of one mill one the dollar is the same as a rate of one-tenth of one percent of the assessed value of the property.

milliampere: One thousandth of an ampere.

millivolt: One thousandth of a volt; .0001 volt.

mil scale: The heavy oxide layer formed during hot fabrication or heat treatment of metals.

millwork: Building materials made of finished wood and manufactured in millwork plants and planing mills. It includes such items as inside and outside window and door frames, blinds, porchwork, mantels, panelwork, stairways, molding, and interior trim. The term does not include flooring or siding.

minimum at a point: Specifications permit the thickness at one point to be less than the average.

minimum average: The specified average insulation or jacket thickness.

miter: The joining of two pieces at an evenly divided angle. To match angles. See Fig. M-10 on page 186.

miter box: A device used as a guide in sawing miter joints.

miter cut: A cut made at an angle so that two pieces similarly cut will form a right angle when joined.

miter joint: The joint of two pieces at an angle that is half the joining angle. For example, the miter joint at the side and head casing at a door opening is made at a 45° angle.

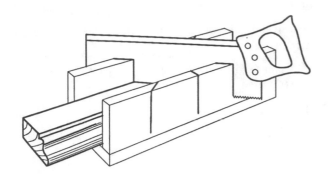

Figure M-10: Miter box and saw, cutting a miter joint.

mitering: The act of joining by the use of miter joints.

miter plane: A plane for use with a miter board.

miter-saw cut, or miter-sawing board: An appliance used to guide the saw at the desired angle.

miter square: A tool similar to the try square but having a head which permits the laying out of both 90-deg. and 45-deg. angles.

mks: Meter, kilogram, second.

mm: Millimeter; 1 meter/1000

modem: Equipment that connects data transmitting/receiving equipment to telephone lines; a word contraction of "modulator-demodulator."

modulation: 1) The varying of a "carrier" wave by a characteristic of a second "modulating" wave. To regulate by or adjust to a certain measure or proportion.

modulus of electricity: The ratio of stress (force) to strain (deformation) in a material that is elastically deformed.

mogul: A socket or receptacle (lampholder) used with large incandescent lamps of 300 watts or more.

moisture content of wood: The weight of water contained in wood, expressed as a percentage of the total weight of the wood.

moisture-repellent: So constructed or treated that moisture will not penetrate.

moisture-resistance: So constructed or treated that moisture will not readily injure.

molded case breaker: A circuit breaker enclosed in an insulating housing.

molding: An ornamental strip used in the finishing of buildings. See Fig. M-11.

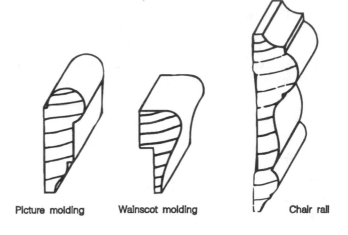

Picture molding   Wainscot molding   Chair rail

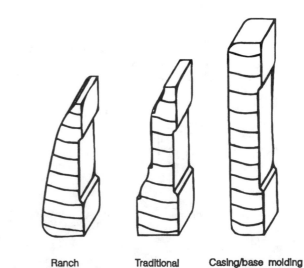

Ranch   Traditional   Casing/base molding

Figure M-11: Several types of molding used to trim buildings.

mole (mol): The basic SI unit for amount of substance; one mole is the amount of substance of a system that contains as many elementary entities as there are atoms in 0.012 kilograms of carbon 12.

molecule: The group of atoms that constitutes the smallest particle in which a compound or material can exist separately.

mortar: A mixture of cement with sand and water used as a bonding agent between bricks, stones, blocks, etc.

mortgage: A conditional transfer of real property as security for the payment of a debt or the fulfillment of some obligation.

mortgagee: A person to whom property is conveyed as security for a loan made by such person (the creditor).

mortgagee in possession: A mortgage creditor who takes over the income from the mortgaged property upon a default on the mortgage by the debtor.

mortgagor: An owner who conveys his property as security for a loan (the debtor).

mortise: A slot cut into a board, plank, or timber, usually edgewise, to receive a tenon of another board, plank, or timber to form a joint. See Fig. M-12.

motor: An apparatus to convert from electrical to mechanical energy.

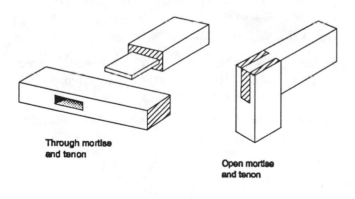

Through mortise and tenon

Open mortise and tenon

Figure M-12: Typical mortise.

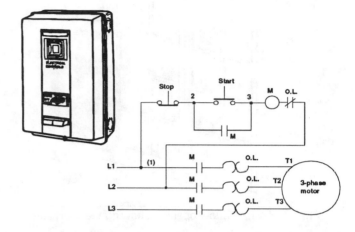

Figure M-13: Typical motor control and circuit.

motor, capacitor: A single-phase induction motor with an auxiliary starting winding connected in series with a condenser for better starting characteristics.

motor control: Device to start/stop and/or otherwise control the operation of an electric motor. See Fig. M-13.

motor control center: A grouping of motor controls such as mtoro starters.

motor effect: Movement of adjacent conductors by magnetic forces due to currents in the conductors.

mouse: Any weighted line used for dropping down between finished walls to attach to cable to pull the cable up; a type of vertical fishing between walls.

mortise: A space hollowed out, as in a piece of wood, to receive a tenon; mortise-and-tenon joint.

mortise chisel: A narrow face, heavy-bodied chisel used in cutting mortises.

mortising machine: A machine for cutting mortises in wood.

MPT: Male pipe thread.

MPX: Multiplexer.

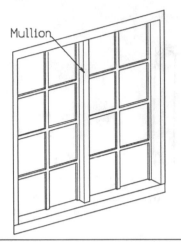

Figure M-14: Window opening showing the mullion.

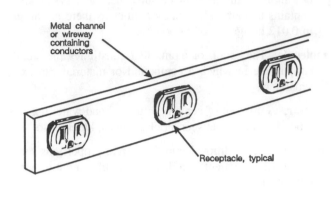

Figure M-15: Multiout assembly.

multiple listing: The arrangement among real estate board or exchange members whereby each broker brings his listings to the attention of the other members so that if a sale results, the commission is divided between the broker bringing the listing and the broker making the sale, with a small percentage going to the board or exchange.

MTW: Machine tool wire.

mudsill: A foundation timber of a structure placed directly on the ground.

mullion: The large vertical division of multiple windows.

multioutlet assembly: A type of surface or flush raceway designed to hold conductors and attachment plug receptacles. See Fig. M-15.

multiple barrier protection (nuclear power): The keeping of radioactive fission products from the public by placing multiple barriers around the reactor; for BWR these are fuel cladding, the reactor vessel, and the containment building.

multiplex: To interleave or simultaneously transmit two or more messages on a single channel.

multiplier: A known resistance used with a multimeter or a galvanometer to increase their range. See Fig. M-5.

multispeed motor: A motor capable of being driven at any one of two or more different speeds independent of the load.

muntin: The small member that divides the glass in a window frame.

mutual inductance: The condition of voltage in a second conductor because of a change in current in another adjacent conductor.

mw: Megawatt: 106 watts.

Mylar® DuPont trade name for a polyester film whose generic name is oriented polyethylene terephthalate; used for insulation, binding tapes.

myrtle: A general-purpose wood whose color makes an attractive veneer.

N/A: 1) Not available. 2) Not applicable.

Nail: A slender piece of metal, one end of which is pointed, the other end having a head, either flattened or rounded. It is a common means of fastening together several pieces of wood or other material. See chart on page 186 for examples of the most commonoly used nails for building construction.

Nail set: A small rod of steel 4 or 5 in. long with one end drawn to a taper and slightly cupped to prevent it from slipping off the head of the nail; used in sinking the head of a nail below the surface. See Fig. N-1.

NAREB: National Association of Real Estate Boards.

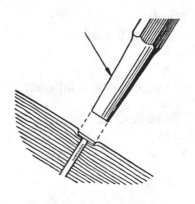

Figure N-1: Nail set.

National Electrical Code (NEC): A set of rules governing the selection of materials, quality of workmanship, and precautions for safety in the installation of electrical wiring. The NEC, originally prepared in 1897, is frequently revised to meet changing conditions, such as improved equipment and materials and new fire hazards. The code is the result of the best efforts of electrical engineers, manufacturers of electrical equipment, insurance underwriters, fire fighters, and other concerned experts throughout the country.

The NEC is published by the NFPA (National Fire Protection Association), Batterymarch Park, Quincy, MA. It contains specific rules and regulations intended to help in "the practical safeguarding of persons and property from hazards arising from the use of electricity." The NEC contains provisions considered necessary for safety. Compliance therewith and proper maintenance will result in an installation essentially free from hazard, but not necessarily efficient, convenient, or adequate for good service for future expansion of electrical use.

natural convection: Movement of a fluid or air caused by temperature change.

NBR: Nitrite-butadiene rubber; synthetic rubber.

NBS: National Bureau of Standards.

NC: Normally closed.

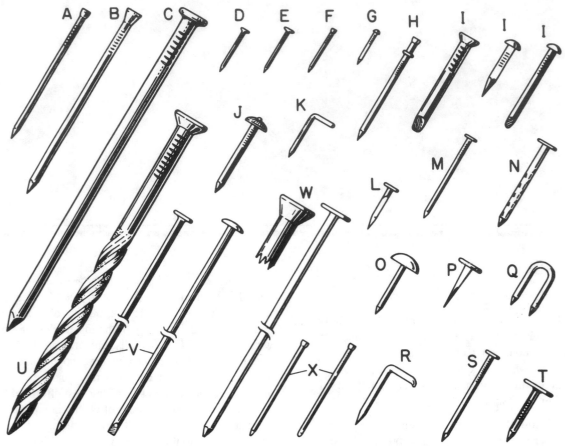

A — Finishing Nail
B — Casing Nail
C — Common Nail
D — Box Nail
E — Blued Lath Nail
F — Brad
G — Escutcheon Nail
H — Scaffold or Duplex or Double Head
I — Hinge Nails
J — Nail for Corrugated Roofing
K — Nail for Metal Lath
L — Upholsterer's Tack
M — Shingle Nail

N — Asbestos Shingle Nail (Barbed)
O — Upholsterer's Nail (Plain or Fancy)
P — Tack (Flat-Head)
Q — Staple
R — Electrician's "Staple"
S — Sheetrock Nail (Cement Coated)
T — Asphalt-Roofing Nail
U — "Screw" Nails
       (For Every Purpose — Available In Different
       "Thread" Designs)
V — Flat-Head and Round-Head Gutter Spikes
W — Round Wire Spikes
X — Flooring Brads

Figure N-2: Various types of nails used on building construction projects.

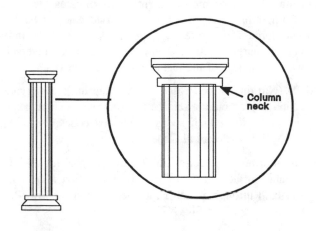

Figure N-3: Parts of a columns, showing the neck.

Neck: The upper part of the shaft of a column immediately below the capital. See Fig. N-3.

Necking: A narrow molding extending around the upper part of a column, pillar, or the like.

negative: Connected to the negative terminal of a power supply.

negative conductor: A conductor leading from the negative terminal.

negative plate: 1) In a storage cell, the spongy lead plate which, during discharge, is the negative plate or terminal. 2) In a primary cell, the carbon, copper, platinum, etc., is the negative electrode.

negative side of circuit: The conducting path of a circuit from the current-consuming device back to the source of supply.

NEMA: National Electrical Manufacturers Association.

neoprene: An oil resistant synthetic rubber used for jackets; originally a DuPont trade name, now a generic term for polychloroprene.

net listing: A price, which must be expressly agreed upon, below which the owner will not sell the property and at which price the broker will not receive a commission; the broker receives the excess over and above the net listing as his commission.

network: An aggregation of interconnected conductors consisting of feeders, mains, and services.

network limiter: A current limiting fuse for protecting a single conductor.

neutral: The element of a circuit from which other voltages are referenced with respect to magnitude and time displacement in steady state conditions.

neutral block: The neutral terminal block in a panelboard, meter enclosure, gutter or other enclosure in which circuit conductors are terminated or subdivided.

neutral wire: A grounded conductor in an electrical system that does not carry current until the system is unbalanced. Neutral conductors must have sufficient capacity for the current that they might have to carry under certain conditions. In a single-phase, three-wire system, the middle leg or neutral wire carries no current when the loads on each side between neutral are equal or balanced. If, however, the loads on the outside wires become unequal, the difference in current flows over the neutral wire.

When a 120/240-V single-phase service is used, it is highly desirable for the 120-V loads to be balanced across both sides of the service. The neutral wire can then be smaller than the two hot or ungrounded wires. The NE Code permits the reduction of the neutral to the size that will carry the maximum unbalanced load between the neutral and any one ungrounded conductor.

A general rule of thumb is to reduce the neutral by not more than two standard wire sizes. A further demand factor of 70 percent may be applied in reducing the neutral wire for that portion of the unbalanced load that is in excess of 200 A. However, if 50 percent or more of the load consists of electric-discharge lamp ballasts, the neutral will be the same size as the ungrounded conductors.

In a three-phase, four-wire system, a three-wire branch circuit consisting of two phase wires and one neutral wire has a neutral or grounded wire that carries approximately the same current as the phase conductors. Therefore, it should be the same size as the phase conductors. However, three-phase, four-wire systems generally supply a mixed load of lamps, motors, and other appliances. Motors and similar three-phase loads connected only to the phase wires cannot throw any load onto the neutral.

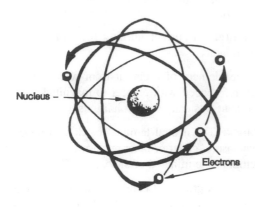

Figure N-4: Parts of an atom.

neutron: Subatomic particle contained in the nucleus of an atom. The neutron is located in the central mass, or nucleus, of the atom and possesses no electrical charge. Due to this, the neutron is not deflected by magnetic or electric fields, and its interaction with matter is mainly by collision. See Fig. N-4.

newel: A post at the top or bottom of a flight of stairs, supporting the handrail. See Fig. N-5.

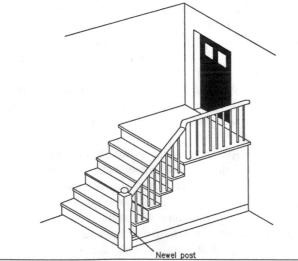

Figure N-5: Typical newel post.

newton: A unit of measurement that expresses the force of attraction or repulsion between two charged bodies. This force varies directly with the product of the individual charges, and at the same time, varies inversely with the square of the distance between charges.

NFPA (National Fire Protection Association): An organization to promote the science and improve the methods of fire protection which sponsors various codes, including the National Electrical Code.

nickel plating: The depositing of a coating of nickel on a metallic surface. Accomplished by immersion in a nickel salt bath through which an electric current of low voltage is passed. See Fig. N-6.

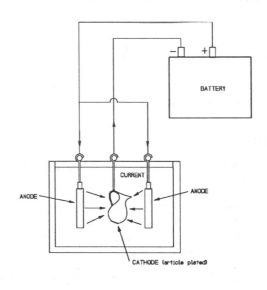

Figure N-6: Principles of nickel plating.

nineteen hundred box: A commonly used term to refer to any 2-gang 4-inch square outlet box used for two wiring devices or for one wiring device with a single-gang cover where the number of wires requires this box capacity. Fig. N-7.

nipple: A threaded pipe or conduit of less than two feet length.

Figure N-7: Nineteen-hundred box.

NO: Normally open.

node: A junction of two or more branches of a network.

no-load current: The current that flows through a device or circuit when the same device or circuit is delivering zero-output current. For example, when the primary of a transformer is connected to an ac source, a voltage appears at the secondary winding even when it is not connected to a load. While there is no current flow from the output in this condition, a small amount of current flows in the primary winding to sustain transformer operation. Here, the no-load current is the measured drain from the primary input voltage source.

nominal: Relating to a designated size that may vary from the actual.

nominal rating: The maximum constant load which may be increased for a specified amount for two hours without exceeding temperature limits specified from the previous steady state temperature conditions: usually 25 or 50 percent increase is used.

nomograph: A chart or diagram with which equations can be solved graphically by placing a straightedge on two known values and reading where the straightedge crosses the scale of the unknown value. See Fig. N-8.

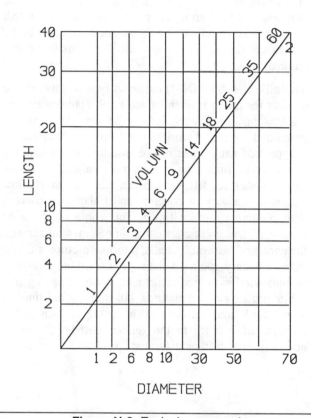

Figure N-8: Typical nomograph.

nonautomatic: Used to describe an action requiring personal intervention for its control.

Non-bearing partition: One that simply divides the space into rooms and does not carry overhead partitions or floor joists.

noncode installation: A system installed where there are no local, state, or national codes in force.

nonconductor: Any substance that does not allow electricity to pass through it.

noninductive circuit: A circuit in which the magnetic effect of the current flowing has been reduced by one of several methods to a minimum or to zero.

noninductive resistance: Resistance free from self-induction.

noninductive winding: A winding so arranged that the magnetic field set up by the current flowing in one half of the coil is neutralized by the magnetic field set up by the current flowing in the opposite directions in the second half.

nonmetallic-sheathed (NM) cable: A type of cable that is popular for use in residential and small commercial wiring systems. In general, it may be used for both exposed and concealed work in normally dry locations.

Type NM cable must not be installed where exposed to corrosive fumes or vapors, nor embedded in masonry, concrete, fill, or plaster, nor run in shallow chases in masonry or concrete and finished or covered with plaster or similar finish. This cable must not be used as a service-entrance cable, in commercial garages, theaters and assembly halls, motion picture studios, storage battery rooms, hoistways, hazardous locations, or embedded in poured cement, concrete, or aggregate.

For use in wood structures, holes are bored through wood studs and joists, and the cable is then pulled through these holes to the various outlets. The holes normally give sufficient support, providing they are not over 4 feet on center. When no stud or joist support is available, staples or some similar supports are required for the cable. The supports must not exceed 4.5 feet and must be within 12 inches of each outlet box or other termination point. See *Romex*.

normal charge: The thermal element charge that is part liquid and part gas under all operating conditions.

Nosing: That portion of a stair tread which projects beyond the riser on which it rests.

N.S.F. check: Not sufficient funds check — (not honored by bank).

NPT: National tapered pipe thread.

NR: 1) Nonreturnable reel; a reel designed for one-time use only. 2) Natural rubber.

NRC (Nuclear Regulatory Commission): The Federal agency for atomic element usage; formerly AEC.

NSD (neutral supported drop): A type of service cable.

Nylon® This is the DuPont trade name for polyhexamethylene-adipamide which is the thermoplastic used as insulation and jacketing material for various applications.

O

oak: A hard, durable, and very strong wood used for many purposes. Especially valuable in places where it may be exposed to the weather; also for furniture, flooring, and trim.

oakum: Hemp used for calking; usually made by untwisting old rope. Tarred hemp.

obelisk: A square monumental shaft with pyramidal top.

oblique: Slanting, inclined; neither vertical nor horizontal.

oblique projection: By this system one face of the object is drawn as parallel to the observer, and the faces perpendicular to this front face are drawn at an angle and to the same scale as the front face, while in a cabinet projection the lengths would be drawn at half scale. See Fig. O-1.

oblong: A rectangular figure having greater length than breadth.

obscuration: The covering power of a paint or enamel. The covering of a surface with an opaque paint or finishing material.

obsidian: A very hard, glassy, volcanic rock, usually black.

obsolete: Gone out of use.

obtuse: Not pointed or acute; greater than a right angle.

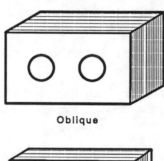

Oblique

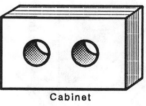

Cabinet

Figure O-1: Typical oblique projection.

obtuse angle: Greater than a right angle; exceeding 90 degrees.

octagon: A plane figure having eight sides and eight angles.

odeum: A roofed theater; that is, hall, gallery, etc.

Oersted, Hans Christian: Danish physicist noted for discovery of effects of electric current on magnetic needle.

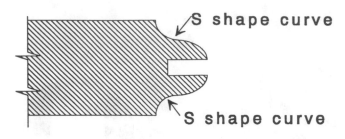

Figure O-2: Ogee molding.

offset: A recess or sunken panel in a wall.

ogee: A molding having in section a reverse curve or long S curve. See Fig. O-2.

ogive: A pointed arch.

ohm: The unit of electrical resistance or impedance; one ohm equals one volt per ampere.

ohmmeter: A type of instrument for measuring resistance in ohms.

ohm resistance: A circuit (d.c.) is said to have a resistance of one ohm when one volt (e.m.f.) will produce a current of one ampere through it.

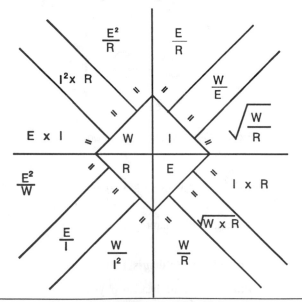

Figure O-3: Summary of Ohm's law.

Figure O-4: Oilstone used for sharpening chisel.

Ohm's law: Mathematical relationship between voltage, current, and resistance in an electric circuit. Ohm's law states that the current flowing in a circuit is proportional to the voltage and inversely proportional to the resistance or opposition. See Fig. O-3.

oil: The prefix designating the operation of a device submerged in oil to cool or quench or insulate.

oil hardening: The hardening of stel by quenching it in oil instead of water.

oil-proof: A device or apparatus designed so that the accumulation of oil or vapors will not prevent safe, successful operation.

oilstone: A smooth stone used, when moistened with oil, for sharpening tools. See Fig. O-4.

oil-tight: Construction preventing the entrance of oil or vapors not under pressure.

OL (nuclear): 1) Operating license. 2) Abbreviation for overload.

olive: A slow-growing, close-grained, heavy wood, light yellowish brown with dark brown spots and streaks.

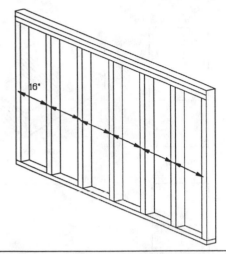

Figure O-4: Studs placed 16" on center.

on center (O.C.): The measurement of spacing for elements such as studs, rafters, and joists, from the center of one member to the center of the next. See Fig. O-5.

one-way slab: Concrete slab with reinforcing steel rods providing a bearing on two opposite sides only.

ooze: To discharge or leak out gradually.

opaque: Impervious to light; not translucent.

open circuit: A circuit which is not electrically complete and in which there is no current.

open circuit cell: Cells normally kept on open circuit for intermittent work. They exhaust quickly on closed circuit, but recover when the circuit is opened.

open shop: An establishment in which both union and nonunion workmen may be employed, as opposed to the closed shop which employs only members of trade unions.

open string stairs: A stair having a wall on one side and a balustrade or handrail on the other. The stair is so constructed that the treads and risers are visible from the side.

open washer: A washer partly cut away so that it may be slipped around a bolt without entirely removing the nut; also called "slip washer." See Fig. O-6.

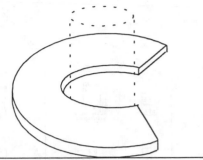

Figure O-6: Open washer.

open web joist: Steel joists built up out of light steel shapes with an open latticed web. See Fig. O-7.

open wiring: Exposed electrical conductors mounted on porcelain knobs or cleats.

operator: One who manipulates a machine or controls the working thereof.

opposite: Facing; set over against; contrary; diametrically different.

orders of architecture: There are five classic orders: Tuscan, Doric, Ionic, Corinthian, and Composite.

ordinate: The distance of any point from the axis of abscissas; also the line indicating such distance.

Oregon pine or Douglas fir: Strong, but light in weight; used extensively for masts, flagpoles, long framing, etc.
organic: Matter originating in plant or animal life, or composed of chemicals of that origin.

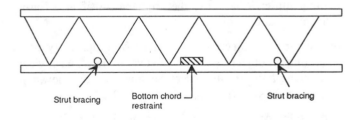

Strut bracing   Bottom chord restraint   Strut bracing

Figure O-7: Open web steel joist.

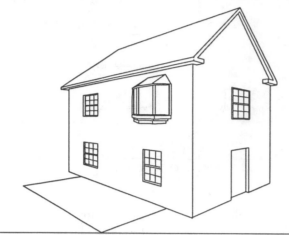

Figure O-8: Oriel.

oriel: A window built out from a wall and resting on a bracket or on corbels; distinguished from a bay window. See Fig. O-8.

oriental walnut: An Australian tree of large size. Suitable for producing great quantities of veneer with grain similar to American walnut.

oriented strand board (OSB): A type of structural flakeboard composed of layers, with each layer consisting of compressed strand-like wood particles in one direction, and with layers oriented at right angles to each other. The layers are bonded together with a phenolic resin.

orifice: Accurate size opening for controlling fluid flow.

ornamentation: The laying of stone, bricks, tiles, or other masonry, in such a manner as to form a decorative design.

orthographic: Relating to the arrangement of views in a mechanical drawing.

orthographic projection: A system of graphically presenting an object by means of several views, each view showing a face of the object, such as the front view, top view, right-side view, etc. See Fig. O-9.

orthostyle: An arrangement of columns in a straight line.

oscillation: The variation, usually with time, of the magnitude of a quantity that is alternately greater and smaller than a reference.

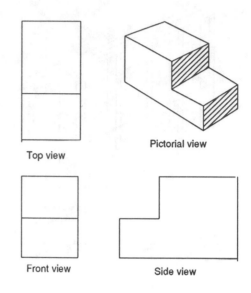

Top view   Pictorial view

Front view   Side view

Figure O-9: Simple orthograthic projection.

orthostyle: An arrangement of columns in a straight line.

oscillator: A device that produces an alternating or pulsating current or voltage electronically.

oscillograph: An instrument primarily for producing a graph of rapidly varying electrical quantities.

oscilloscope: A test instrument that shows visually on a screen the pattern representing variations in an electrical quantity. See Fig. O-10.

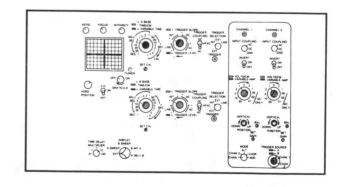

Figure Q-10: Front panel of an oscilloscope.

OSHA (Occupational Safety and Health Act): Federal Law #91-596 of 1970 charging all employers engaged in business affecting interstate commerce to be responsible for providing a safe working place; it is administered by the Department of Labor; the OSHA regulations are published in Title 29, Chapter XVII, Part 1910 of the CFR and the Federal Register.

osmosis: The diffusion of fluids through membranes.

ought sizes: An expression referring to conductors of sizes No. 1/0, 2/0, 3/0, and 4/0. Sometimes called "naught."

outage: The condition resulting when a component is not available to perform its intended function.

outlet: Any point on the wiring system from which current may be taken for consumption.

outlet box: A box installed in a wiring system, from which current is taken to supply some apparatus as, for instance, a lamp. See Fig. O-11.

outline lighting: An arrangement of incandescent lamps or gaseous tubes to outline and call attention to certain features, such as the shape of a building or the decoration of a window.

out of true: Inaccurate, twisted, varying from the exact.

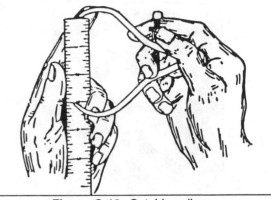

Figure O-12: Outside calipers.

output: Amount of energy delivered to an external device from the source of generation.

outside caliper: A caliper used for gauging outside measurements or sizes. Electronic calipers are shown in Fig. O-12.

outside gouge: A firmer gouge on which the bevel is ground upon the outside or convex face.

oval: An egg-shaped figure with unequal end curves.

overall: A common term meaning an outermost or total dimension.

overhang: A projecting upper part of a building, as a roof or balcony. See Fig. O-13.

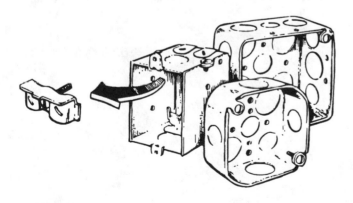

Figure O-11: Various outlet boxes used on building construction projects.

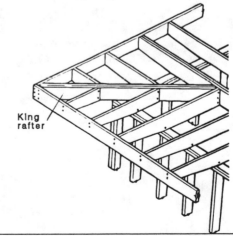

Figure O-13: Typical roof overhang.

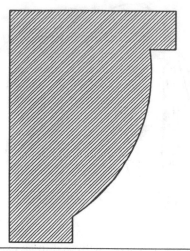

Figure O-14: Cross-section of an ovolo.

overhaul: To take apart, inspect, repair, and reassemble, as a piece of machinery.

overload: More than a normal amount of current flowing through an electrical device.

overload protection: The use of circuit breakers, relays, automatic limiters, and similar devices to protect equipment from overload damage by reducing current or voltage.

overload switch: An automatic switch for breaking a circuit in case of an overload.

overtime: Time spent working beyond specified working hours; usually at premium wages.

ovolo: A convex molding; a quarter round. See Fig. 14.

oxidation: The act of uniting, or causing a substance to unite with oxygen chemically.

oxide: Any binary compund of oxygen with an element or with an oganic radical, as iron rust.

oxidizing: Finishing mealwork by means of an acid solution.

oxidizing agent: A substance which by giving up some of its oxygen changes another substance to an oxide or some other compound.

oxyacetylene: A mixture of oxygen and acetylene gas in such proportions as to produce the hottest flame known for practical use. Oxyacetylene welding and cutting is used in almost every metalworking industry.

oxygen: A tasteless, colorless, gaseous element. It constitutes one fifth of the air by volume. It is closely related to combustion.

oystering: The using of a veneer cut from the roots and boughs of certain trees.

ozone: A triatomic form of oxygen that is formed naturally in the upper atmosphere by a photochemical reaction with solar ultraviolet radiation or generated commercially by a silent electric discharge in oxygen or air.

P

pad: A coil of tape.

padauk: Wood native to Burma and Andaman Islands. It is hard and heavy, with large open pores; characterized by reddish streaks. Used in furniture and paneling.

padding: Depositing several layers of weld beads upon a surface to restore the appearance or usability.

padlock: A detachable lock, usually used in connection with a hasp and staple. See Fig. P-1.

pad saw: A type of handsaw with narrow tapering blade.

pagoda: A tower-shaped roof or top; a feature of cabinets, cupboards, etc., designed by Chippendale and others, showing Chinese influence.

paint: 1) Pigment or color, either dry or mixed with oil or water. 2) The act of applying the same.

paint base: The body matter of paint, as latex, white lead, zinc, and others.

paint brush: Device containing a handle and bristles for applying paint. See Fig. P-2.

paint drier: An additive to paint to cause it to dry within a given period of time.

painter: One who applies paint to a surface or object.

painting: Decorating with paint.

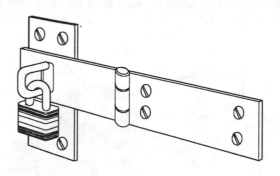

Figure P-1: Padlock with related hasp.

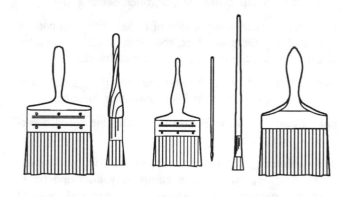

Figure P-2: Types of paint brushes in common use.

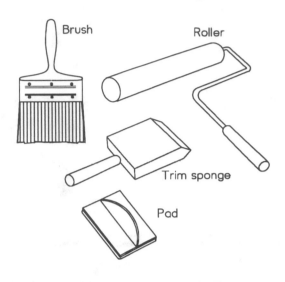

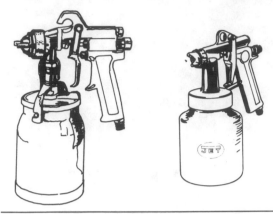

Figure P-4: Paint-spraying apparatus.

Figure P-3: Various types of paint applicators including roller and pad.

paint pad: A device used to spread paint that utilizes a foam rubber pad instead of bristles. See Fig. P-3.

paint roller: A rotating device used for applying paint. See Fig. P-3.

paint thinner: Turpentine or petroleum spirits may be used to thin heavy-bodied paints in order to make application easier. Petroleum spirits are much used on account of lower cost.

paint spraying: Applying paint with compressed air rather than with brush, pad, or roller. See Fig. p-4.

paktong: A metal composition of nickel, zinc, and copper, resembling German silver, and made in China, used for candlesticks, fire irons, and other metalwork.

palette knife: A very thin-bladed, flexible, steel knife used for applying wood fillers and other substances.

palladium: A rare, white, ductile, malleable metal occurring with platinum.

pad-mounted: A shortened expression for "pad-mount transformer," which is a completely enclosed transformer mounted outdoors on a concrete pad, without need for a surrounding chain-link fence around the metal, box-like transformer enclosure. Fig. P-5.

pan: 1) A sheet metal enclosure for a watt-hour meter, comonly called a "meter pan." 2) Pan-shaped concrete forms used to mold the underside of a concrete slab.

pane: A window glass or light.

panel: 1) A raised or sunken portion surrounded as by a frame, especially applicable to woodwork. 2) A unit for one or more sections of flat material suitable for mounting electrical devices.

panelboard: A single panel or group of panel units designed for assembly in the form of a single panel; includes buses and may come with or without switches and/or automatic overcurrent protective devices for the control of light, heat, or power circuits of individual as well as aggregate capacity. It is designed to be placed in a cabinet or cutout box that is in or against a wall or partition and is accessible only from the front. See Fig. P-6 on page 203.

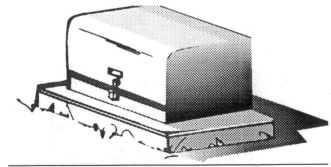

Figure P-5: Pad-mount transformer.

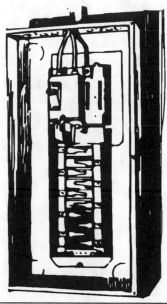

Figure P-6: Typical panelboard showing main circuit breaker and internal busbars.

panelboard schedule: Table on electrical drawings giving the characteristics of one or more panelboards used on the project, including the type and size of circuit breakers or fuse protection. See Fig. P-7.

panel clip: Metal fastener for joining edges of plywood together. See Fig. P-8.

panel door: Door with vertical and horizontal members that hold wood panels. See Fig. P-9.

Panelboard Schedule									
Panel No.	Type Cabinet	Panel Mains			Branches				
		Amps	Volts	Phase	1P	2P	3P	Prot.	Frame
A	Surface	100A	120/240	1, 3W	3	-	-	20A	70A
					-	3	-	20A	
					-	1	-	60A	
Sq. D Type NQO									
W/100A Main Brk.									

Figure P-7: Panelboard schedule used on an electrical working drawing.

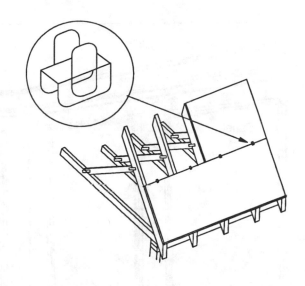

Figure P-8: Panel clip and application of panel clips on plywood panels.

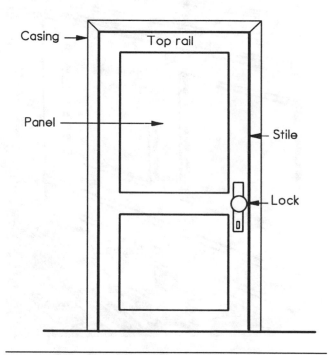

Figure P-9: Principle parts of a panel door.

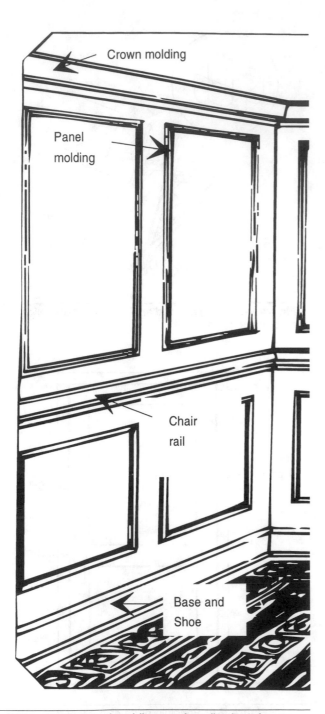

Figure P-10: Panel moldings and application of same.

panel moldings: Moldings of different shapes and materials used to finish interior panels. See Fig. P-10.

panel saw: A type of handsaw with fine teeth, used for cutting thin wood.

panel strip: A strip of molded wood or metal to cover a joint between two sheathing boards to form a panel. See Fig. P-11.

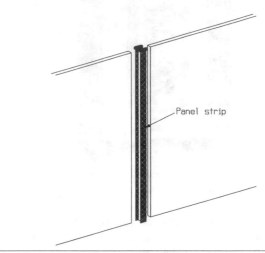

Figure P-11: Panel strip.

pantile: A roofing tile whose cross section is an ogee curve; a curved roofing tile laid alternately with con ex covering tiles. A flat paving tile, Dutch or Flemish.

panograph: An instrument for copying drawings on an enlarged or reduced scale. See Fig, P-12.

Figure P-12: Panograph.

paper birch: Sometimes called white birch. A large tree growing to a height of 50 to 75 feet. Wood is strong and hard; light brown in color. Used in the manufacture of spools and for paper pulp.

paperhanger: One who installs wallpaper to cover walls, ceilings, and partitions in a building.

papering: The act of applying wallpaper.

paper-lead cable: Cable having oil impregnated paper insulation and a lead sheath.

parabola: A conic section or curve as would be obtained by passing a plane through a cone parallel with its side. See Fig. P-13.

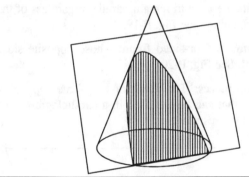

Figure P-13: Parabola.

parabolic girder: A form of bowstring girder, the outline of whose bow is that of a polygon inscribed in a parabola; used on bridge construction. See Fig. P-14.

paradox: Something seemingly incredible, yet true.

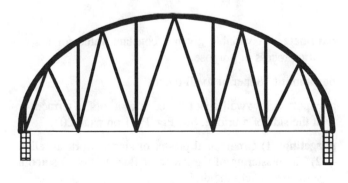

Figure P-14: Parabolic girder.

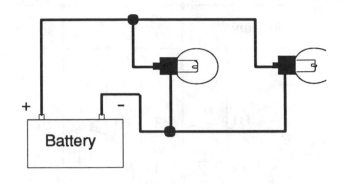

Figure P-15: Lamps connected in parallel.

paraffin: A translucent, waxy, solid substance derived principally in the distillation of petroleum.

parallax: This occurs when the scale of an electrical instrument is not in correct alignment.

parallel: Lying side-by-side; extending in the same direction and equidistant at all points; having the same direction; similar.

parallel circuit: A circuit having a common feed and a common return, between which two or more pieces of apparatus are connected, each receiving a separate portion of the current flow from the common feed. See Fig. P-15.

parallel clamps: Wood-working clamps used for holding workpieces together while being glued or worked on. Fig. P-16:

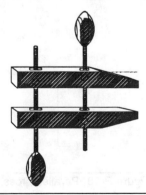

Figure P-16: Parallel wood clamp.

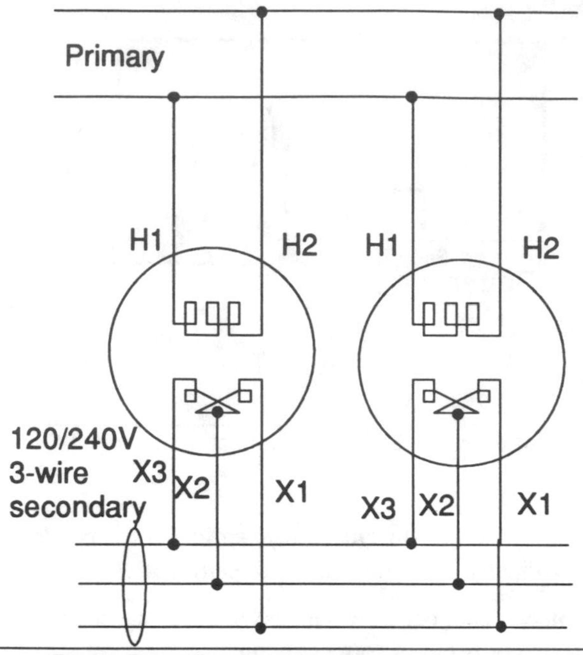
Primary
H1
H2
H1
H2
120/240V
3-wire
secondary
X3
X2
X1
X3
X2
X1

Figure P-17: Parallel-connected transformers.

parallel-connected transformer: When two or more transformers have their primary windings connected to the same source of supply in such a manner that the impressed voltage in each case is the same as that of the line. See Fig. P-17.

parallel forces: When two forces are parallel and act in the same direction but do not start from the same point, the resultant is parallel to both and equal to their sum. When the forces act in opposite directions the resultant is equal to the difference between the two. See Fig. P-18.

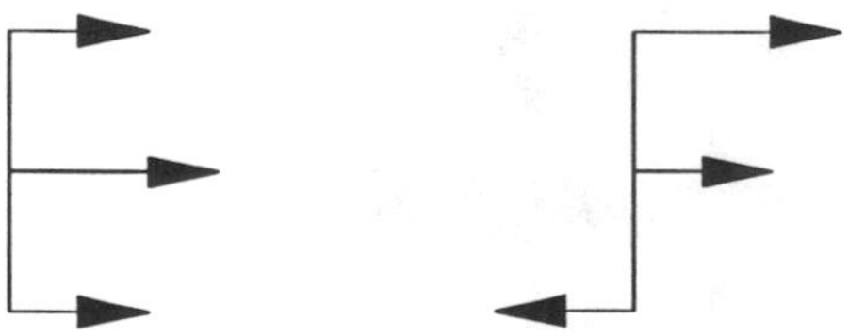
Resultant

Figure P-18: Parallel forces.

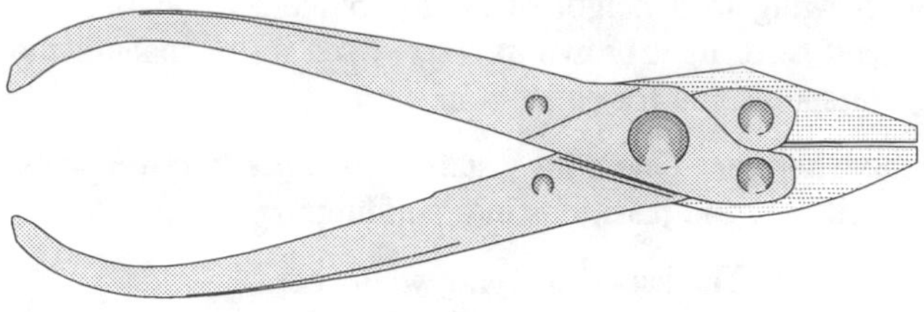

Figure P-19: Parallel-jaw pliers.

parallel jaw pliers: Pliers equipped with a toggle joint that permits the jaws to remain parallel regardless of the extent of opening. See Fig. P-19.

parallelogram: A four-sided figure whose opposite sides are parallel. See Fig. P-20.

parallelogram forces: The result of two forces operating at an angle that can be enclosed in a parallelogram. The diagonal is the resultant.

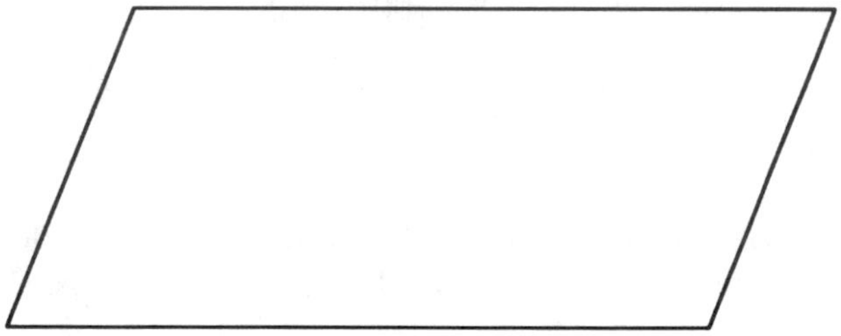

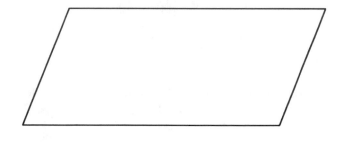

Figure P-20: Parallelogram.

parameter: A variable given a constant value for a specific process or purpose.

paramount: Superior to all others.

parapet: A low wall about the edge of a roof, a terrace, or at the side of a bridge. See Fig. P-21 on page 207.

pargeting: 1) Ornamental plaster or stucco work in relief. 2) The plastering of the inside of flues to give a smooth surface and help the draft.

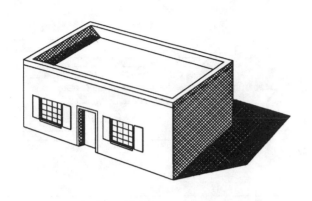

Figure P-21: Parapet wall around roof.

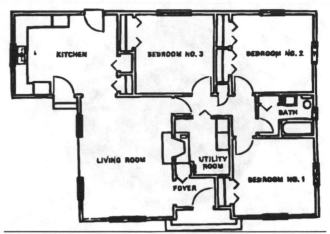

Figure P-22: Floor plan of house showing rooms divided by partitions.

parging: A rough coat of mortar applied over a masonry wall as protection or finish; may also serve as a base for an asphaltic waterproofing compound below grade.

paring: A method of wood turning.

paring chisel: A long chisel used by woodworkers in making, paring, or slicing cuts.

paring gouge: A woodworker's bench gouge that has the angle of the cutting edge ground on the inside or concave face of the blade.

paris green: A poisonous copper arsenite used as a pigment and for destroying insects.

parquetry: Wooden mosaic for floors or furniture.

particleboard: Construction panels built from reconstituted wood particles bonded with resin under heat and pressure.

parting strip: In a sash window, one of the thin strips of wood let into the pulley stile to keep the sashes apart.

parting tool: A narrow-blade turning tool used for cutting grooves, recesses, and for cutting off.

partition: 1) A permanent interior wall which serves to divide a building into rooms. See Fig. P-22. 2) A division made of real property among those who own it in undivided shares.

partition plate: The horizontal member at the top of a partition wall, serving as a cap for the studs and as a support for joists, rafters, or studs. See Fig. P-23.

party wall: A partition of brick or stone between adjoining properties.

pascal: The derived SI unit for pressure or stress: one pascal equals one newton per square meter.

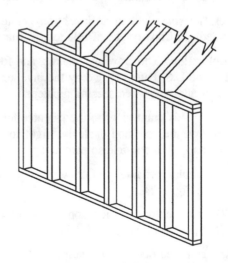

Figure P-23: House framing utilizing a top plate to cap wall studs and to offer support for rafters.

Figure P-24: Patera.

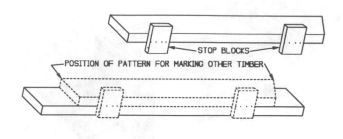

Figure P-25: Pattern for rafter cuts. *Courtesy Department of the Army.*

Pascal's law: Pressure applied to a given area of a fluid enclosed in a vessel is transmitted undiminished to every equal area of the vessel.

passive solar heating: Heating a structure by means of direct sunlight or by circulating heat through natural means.

patch: 1) To connect circuits together temporarily. 2) To repair a damaged building or piece of equipment.

patent: 1) The sole right in an invention granted by the government. 2) Conveyance of title to government land.

patent application: The drawings, claims, and forms submitted to the U.S. Patent Office by the inventor, usually through an attorney, when seeking a patent.

patent grant: This grant gives to the patentee for the term of the patent, the sole and exclusive right to manufacture, sell, and use the article so patented.

patera: A circular ornament often worked in relief on friezes. See Fig. P-24.

patina: The dark color and rich appearance of the wood in furniture, caused by age.

patio: An open area (usually a concrete slab) just outside the building; used for recreation and socializing.

pattern: A model or specimen; something made for copying, or for reproducing similar articles. See Fig. P-25.

pavement: A hard-surface covering for road or sidewalk.

pavilion: Usually a roofed structure not entirely enclosed by walls, used as a gathering place or place of amusement. See Fig. P-26.

pawl: A hinged or pivoted arm having a pointed edge or hook made to engage with ratchet teeth. See Fig. P-27. See *ratchet*.

Payne's process: A method of fireproofing wood by injecting sulphate of iron followed by the injection of a solution of sulphate of lime or soda.

payoff: The equipment to guide the feeding of wire.

pb: pushbutton; pullbox.

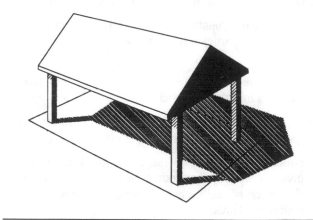

Figure P-26: Typical pavilion.

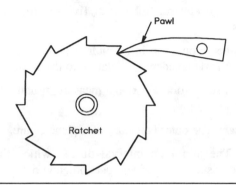

Figure P-27: Simple pawl and ratchet.

Figure P-28: Pedestal.

pe: 1)Professional engineer. 2) polyethylene.

peak load: The heaviest load that a generator or system is called on to supply at regular intervals.

peak value: The largest instantaneous value of a variable.

pearling: The carving of a series of small circles or ovals.

pearlite: The eutectoid alloy of carbon and iron containing 0.9 percent carbon, which is the iron-carbon alloy of lowest transformation point in the solid.

pearwood: A light-brown, close-grained, medium-hardwood extensively used for drafter's T-squares, triangles, etc.

peat: Partially carbonized vegetable material used as fuel. It has a very heavy water content when dug from the bogs, and must be pressed and dried before it will burn freely.

pebble dash: A finish for exterior walls made by dashing pebbles against the plaster or cement coating.

pebbles: A rough, uneven, wavy effect in a spray coat of paint, etc., can be charged to insufficient atomization or too low air pressure.

pebbling: The process of graining or imparting a surface of irregular roughness to paper to improve its attractiveness and to reduce the shiny effect.

pedestal: A base of or for a column, statue, or other object. See Fig. P-28.

pediment: A triangular member framed in by a cornice and surmounting a portico; usually of low altitude as compared with the width of its base.

peen: The small end of the head of a hammer, as the ball-peen hammer for metalworkers. See Fig. P-29.

peening: Beating over or smoothing over a metallic surface with the peen end of a hammer.

peg: A wooden pin, or spike, used for fastening together the parts of furniture in lieu of nails.

pellucid: Perfectly clear; transparent.

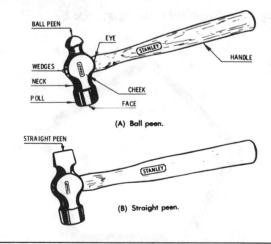

Figure P-29: Several types of peening hammers.

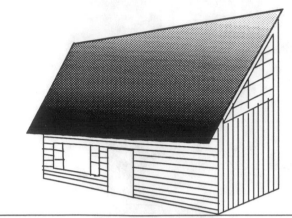

Figure P-30: Pent roof.

penciling: The tapering of insulation to relieve electrical stress at a splice or termination.

pendant: A hanging ornament.

penetrating: Having power to penetrate or permeate. In wood finishing, a penetrating stain is one that forces its way below the surface into the fibers of the wood.

penetration (nuclear): A fitting to seal pipes or cables through the containment vessel wall.

pent roof: A roof with a slope on one side only. See Fig. 30.

pentagon: A five-sided plane figure. See Fig. P-31.

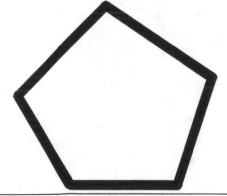

Figure P-31: Pentagon.

pentode: An electron tube with five electrodes or elements.

percentage: Proportion in a hundred parts. Rate per hundred. A part considered in relation to the whole.

perch: In stonework, a variable measure, usually about 25 cu. ft.

perimeter: The outer boundary of a plane figure.

period: The minimum interval during which the same characteristics of a periodic phenomenon recur.

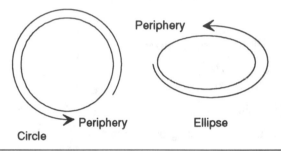

Figure P-32: Peripheries of a circle and ellipse.

periodic arrangement: An arrangement of elements in the order of their increasing atomic numbers. As a result of this arrangement the elements fall into nine natural groups, the members of each resembling each other very strongly.

peripheral speed: The speed, usually registered in feet per minute of the circumference of a part, as a wheel or a shaft.

periphery: The circumference of a circle, ellipse, or similar figure. See Fig. 32.

peristyle: A system of columns encircling a building.

permanent load: A load which is constant and unvarying; a dead load as the weight of the structure itself, or a load imposed, or both taken together.

permanent magnet: A piece of magnet steel which retains the acquired property of attracting other pieces of magnetic material after being under the influence of a magnetic field.

permanganate: A dark purple salt of permanganic acid (HMnO4).

permalloy: An alloy of nickel and iron that is easily magnetized and demagnetized.

permanent magnet moving-coil meter: The basic movement used in most measuring instruments for servicing electrical equipment. The basic movement consists of a stationary permanent magnet and a movable coil. When current flows through the coil, the resulting magnetic field reacts with the magnetic field of the permanent magnet and causes the coil to rotate.

permeability: The ability of a core of magnetic material to conduct lines of force. Some core materials have higher permeabilities than others and thus, used with a given winding, they provide inductors having greater inductance values. More precisely, the permeability of a certain magnetic material is the ratio of the flux produced with that magnetic material as the core to the flux produced with air as the core.

It is important to be aware of the fact that the permeability of a given magnetic material varies with the magnetizing force in ampere-turns per unit of core length that is applied to the core. The permeability also depends on the amount and direction of any magnetic flux that might already exist.

permissible mine equipment: Equipment that has been formally accepted by the Federal agency.

permutation: The arrangement of any determinate number of things or letters in all possible orders, one after another.

perpend: A header brick extending through the wall so that one end appears on each side of it. See Fig. 33.

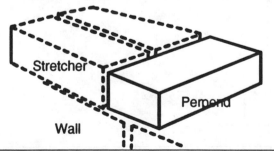
Figure P-33: Header brick extending through and exposed on each side of a wall.

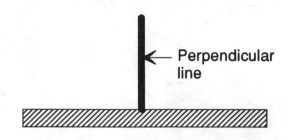
Figure P-34: Perpendicular line meeting a surface at right angles.

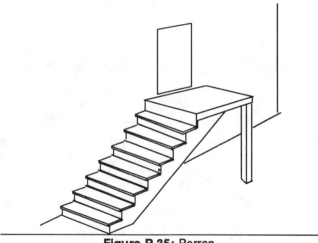
Figure P-35: Perron.

perpendicular: Meeting a given line or surface at right angles. See Fig. P-34.

perron: A staircase outside of a building, leading to the first floor. See Fig. P-35.

persimmon: Belongs to the ebony family. Docs not grow large enough to be of great commercial value. Wood is yellowish, often streaked with black; hard and strong, and takes high polish. Used for brush backs, billiard cues, and veneer.

perspective: The art or science of representing on a plane surface objects as they actually appear to the eye. See Fig. P-36 on page 212.

per-unit quantity: The ratio of the actual value to an arbitrary base value of a quantity.

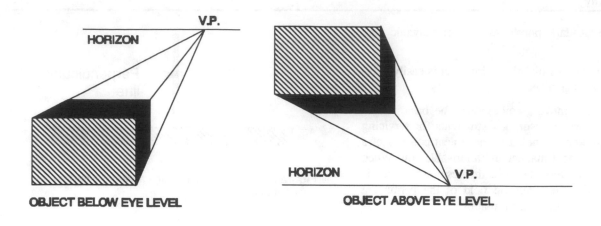

OBJECT BELOW EYE LEVEL

OBJECT ABOVE EYE LEVEL

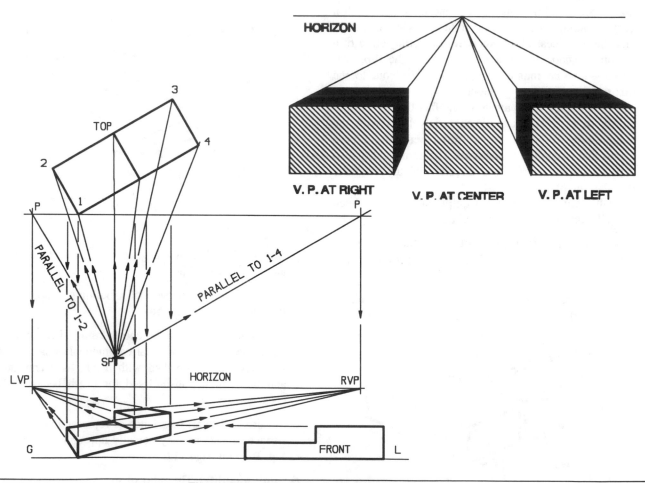

V. P. AT RIGHT   V. P. AT CENTER   V. P. AT LEFT

Figure P-36: Principles of perspective drawing.

per-unit quantity: The ratio of the actual value to an arbitrary base value of a quantity.

PES: Power Engineering Society of IEEE.

petroleum: A natural oil taken from the earth, used extensively for heating and lighting. By distillation, it yields many products valuable in the industries, such as gasoline, kerosene, paraffin, etc.

pew: One of the long benches used as seats in a church.

pewter: An alloy of tin, lead, and antimony, formerly much used for tableware.

PF: Power factor.

pH: An expression of the degree of acidity or alkalinity of a substance; on the scale of 1-10, acid is under 7, neutral is 7, alkaline is over 7.

phase: The time instant when the maximum, zero, or other relative value is attained by an electric wave.

phase angle: The angle which expresses the phase relation in an alternating-current circuit.

phase conductor: The conductors other than the neutral.

phase converter: A device that will permit the operation of a three-phase induction motor from a single-phase power source. See Fig. P-37.

phase leg: One of the phase conductors (an ungrounded or "hot" conductor) of a polyphase electrical system.

phase meter: A meter which indicates the frequency of the circuit to which it is attached. A frequency meter.

phase out: A procedure by which the individual phases of a polyphase circuit or system are identified; such as to "phase out" a 3-phase circuit for a motor in order to identify phase A, phase B, and phase C to know how to connect them to the motor to get the correct phase rotation so the motor will rotate in the desired direction.

phase protection: A means of preventing damage in an electric motor through overheating in the event a fuse blows or a wire breaks when the motor is running.

phase sequence: The order in which the successive members of a periodic wave set reach their positive maximum values; a) zero phase sequence — no phase shift; b) plus/minus phase sequence — normal phase shift.

phase shift: The absolute magnitude of the difference between two phase angles.

phasor quantity: A complex algebraic expression for sinusoidal wave.

Phillips machine screw: A metal or wood holding screw made in a variety of head forms and having a four-point star shaped recess in the head which requires a special tool for setting. See Fig. P-38.

phosphor bronze: An alloy composed of copper and tin to which a little phosphorus is added. It is largely used as a bearing metal.

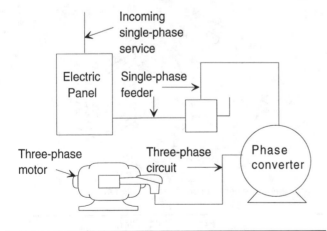

Figure P-37: Phase converter.

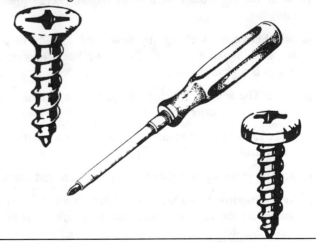

Figure P-38: Phillips head screws and Phillips screwdriver.

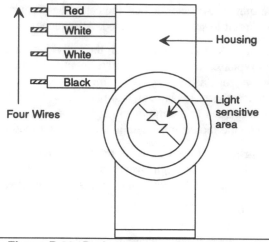

Figure P-39: Basic parts of a photoelectric cell.

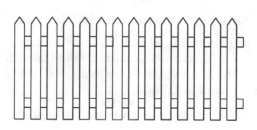

Figure P-40: Picket fence.

photocell: A device in which the current-voltage characteristic is a function of incident radiation (light).

photoelectric control: A control sensitive to incident light. See Fig. P-39 on page 214.

photoelectricity: A physical action wherein an electrical flow is generated by light waves.

photometer: An instrument for measuring the intensity of light or for comparing the relative intensity of different lights.

photon: An elementary quantity of radiant energy (quantum).

physical change: A change by which the identity of the substance is not altered; e.g., sawing a board into small pieces is a physical change.

physics: The science which treats of the constitution and properties of matter.

pi : The ratio of the circumference of a circle to its diameter, or 3.1416.

pick: The grouping or band of parallel threads in a braid.

picket: A narrow board used in making fences. It is often pointed at the top; sometimes called a pale or paling. See Fig. P-40.

pickle: A solution or process to loosen or remove corrosion products from a metal.

pickup value: The minimum input that will cause a device to complete a designated action.

picocoulomb: 10^{-12} coulombs.

picture mold: A molding attached to a wall, from which pictures are hung.

pier: A mass of masonry supporting an arch, bridge, etc. See Fig. 41.

pier glass: A large mirror placed between two windows.

pietra dura: Hard and fine stones as those used for inlay and the like.

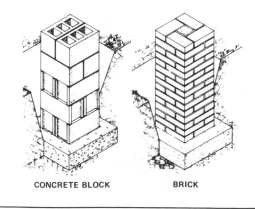

CONCRETE BLOCK   BRICK

Figure P-41: Pier.

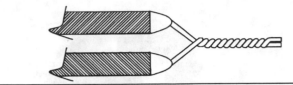
Figure P-42: Pigtail splice.

pietra dura: Hard and fine stones as those used for inlay and the like.

Piezoelectric effect: Some materials become electrically polarized when they are mechanically strained; the direction and magnitude of the polarization depends upon the nature, amount and the direction of the strain; in such materials the reverse is also true in that a strain results from the application of an electric field.

pig: Cast iron in the form of a bar ingot.

pigeon hole: Small open compartment, as in the upper portion of a roll-top desk, used as receptacle for letters, envelopes, etc.

pig iron: The casts iron of commerce as it comes from the smelting furnace; usually in bars of about 100 lb.

pigment: That substance added to paint or ink to give body and color.

pigtail: A splice made by twisting together the bared ends of parallel conductors. See Fig. P-42.

pike pole: A pole tipped with a sharp metal point; used in supporting poles in an upright position during "planting" or removal.

pilaster: A right-angled columnar projection from a pier or wall used to support a floor girder or stiffen the wall itself. See Fig. P-43.

PILC cable: Paper insulated, lead covered.

pile driver: A vertical framework provided with guides for carrying a weight which, after being elevated to the top of the framing, is allowed to fall, by force of gravity, on the head of a pile.

piling: Large timbers or poles driven into the ground or the bed of a stream to make a firm foundation.

pillar: A column to support a structure. See Fig. P-44.

pillar file: Used on narrow work, cutting grooves, etc. Same general shape as a hand file but not as wide and is obtainable in any cut.

pillow block: A bearing or support for a shaft. See Fig. P-45 on page 216.

pilot lamp: A lamp that indicates the condition of an associated circuit.

pilot wire: An auxiliary insulated conductor in a power cable used for control or data.

pin: A small peg or wooden nail.

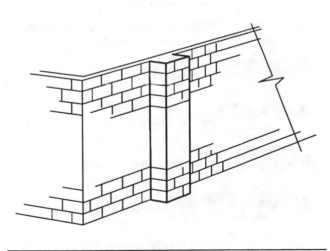
Figure P-43: Pilaster.

Figure P-44: Pillars used to support a building.

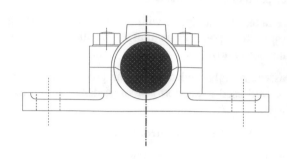

Figure P-45: Pillow block.

pincers: A jointed instrument with two handles and a pair of grasping jaws for holding an object.

pinch bar: A shop name for crowbar. See Fig. P-46.

pinch dog: A small bar of steel, with two sharpened points projecting from it at right angles, used by woodworkers for clamping pieces together when hand screws cannot be used.

pine: Includes a wide variety of woods, ranging from the soft, easily worked white pine to the hard, heavy, longleaf yellow pine which is used in heavy construction work.

pinion: The smaller gear of a pair, either bevel or spur, regardless of size. See Fig. P-48.

pin knot: This term is applied to a knot not over $1/2$ in. in diameter.

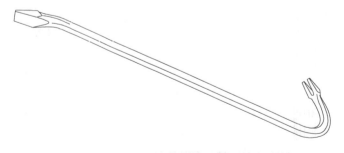

Figure P-47: Pinch bar, sometimes called *crowbar*.

Pinion

Figure P-48: Spur gears — the smaller being the pinion.

pinnacle: 1) A high or top most point. 2) A small turret or tall ornament, as on a parapet.

pin punch: A long, slender punch used for driving out tight-fitting pins. See Fig. P-49.

pin spanner: Used on round nuts having holes in the periphery to receive the spanner pin.

pint: A measure of capacity; one half a quart; one eighth of a gallon. Used in both liquid and dry measure.

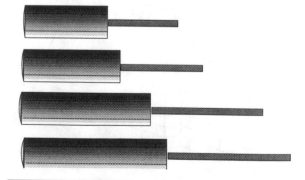

Figure P-49: Set of pin punches, used to remove drift pins.

Figure P-50: Pin vise.

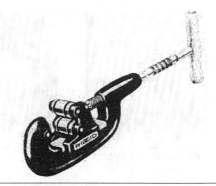

Figure P-52: Pipe cutter.

pin vise: A small hand vise with a V notch in each jaw for gripping wire or round objects. See Fig. P-50.

pin wrench: Used on round nuts which have two holes in their face to receive the pins of the wrench.

pipe coupling: A threaded sleeve used to connect two pipes. See Fig. P-51.

pipe cutter: A tool for cutting wrought iron or steel pipes. The curved end which partly encircles the pipe carries one or more cutting disks. Feed of the cutter is regulated by a screw as the tool is rotated around the pipe. See Fig. P-52.

pipe die: A screw plate used for cutting threads on pipe.

pipe fittings: A general term referring to ells, tees, various branch connectors, etc., used in connecting pipes. See Fig. P-51.

pipe hanger: Device for suspending pipe. Malleable iron hanger consists of a lag screw, a piece of pipe, a socket, and an adjustable ring.

pipe thread: The V-type thread used on pipe and tubing, cut on a taper of $\frac{3}{4}$ in. per foot, which insures a thoroughly tight joint. See Fig. P-53 on page 218.

pipe vise: Pipe vises are of two kinds; the hinged side type with V jaws for small pipes, and the chain type used for large pipes. See Fig. P-54 on page 218.

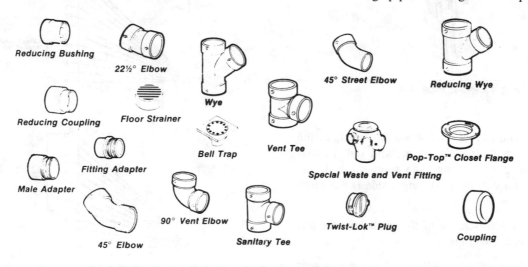

Figure P-51: Various types of pipe couplings and fittings. *Courtesy Genova, Inc.*

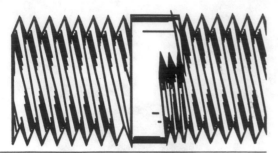

Figure P-53: Typical pipe thread.

piqua: An African tree belonging to the cedar family. It is a light brown, medium-hard wood, with grain somewhat like mahogany.

pique: French inlay.

pitch: Inclination or slope, as for roofs or stairs, or the rise divided by the span. See Fig. P-55.

pitch blende: A dark mineral with pitchlike luster found in Bohemia. A radium-bearing mineral, composed mainly of uranium oxide.

Figure P-54: Pipe vise.

pitch board: A template of thin wood or metal, in the shape of a right-angle triangle, for marking out and testing the cuts of a stair string. The shorter side is the height of the riser cut; the next longer side is the width of the tread cut.

pitch circle: The circumference of the pitch line; the circle of contact in meshed gears.

pitch diameter: The diameter of the pitch circle of a gear wheel.

pitch of a roof: The angle which the slope of a roof makes with the horizontal. See Fig. P-55.

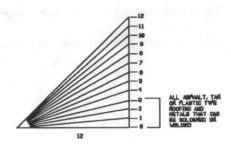

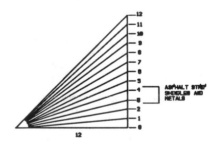

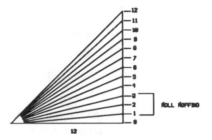

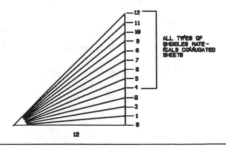

Figure P-55: Roof pitches

pitch of a screw: The distance from a point on a screw thread to a corresponding point on the next turn. In a single thread the pitch is the amount of advance in one revolution.

pith knot: A knot with a pith hole not more than ¼ in. in diameter.

pitting: 1) Small cavities in a metal surface. 2) Spraying lacquer in a room where the temperature is less than 65 deg. F., especially with high pressure.

pivoted casement: A casement window pivoted on its upper and lower edges. See Fig. P-56.

pivot pin: A point supporting something which turns.

plain sawing: Saw cuts taken parallel to squared side of log. See Fig. P-57.

plain turning: Straight or cylindrical turning.

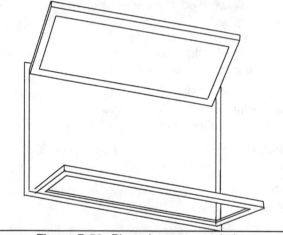

Figure P-56: Pivoted casement window.

plan: A draft or form drawn on a plane surface as a map; especially a top view or a view of a horizontal section; a diagram.

planchet: Blank piece of metal punched out of a sheet before being finished by further work, such as the blank from which coins are made.

plancier: The underside of the corona in a cornice.

plane: Level, flat, even. Also a tool for smoothing boards or other surfaces of woods. See Fig. P-58.

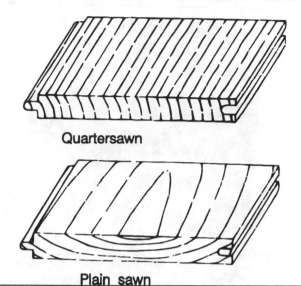

Quartersawn

Plain sawn

Figure P-57: Comparison of plain saw cuts and quartersawn cuts.

planer: A machine for producing plane surfaces on wood or metal. The work is fastened to a table which moves back and forth under the tool. See Fig. P-59 on page 220.

plane tree: Same as buttonwood or sycamore.

plane trigonometry: A branch of mathematics dealing with the measurement of triangles. Six functions are developed and the relationship existing among these six functions and their application to the solution of the right and oblique triangles are of great mathematical importance.

planimeter: An instrument for measuring the area of any plane surface, by moving a pointer around its boundary and reading the indications of a scale.

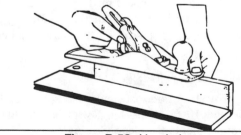

Figure P-58: Hand plane.

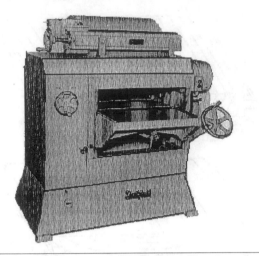

Figure P-59: Planing mill.

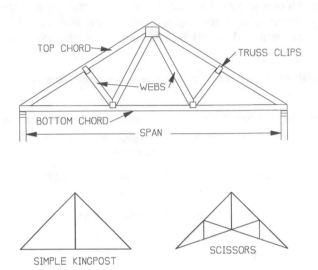

Figure P-60: Various types of trusses.

planing mill: A mill or woodworking establishment equipped with planing and other machines for surfacing, matching, and fitting boards or planks. See Fig. P-59.

plank: A wide piece of sawed timber thicker than a board.

plank truss: A roof truss or bridge truss built of planking. See Fig. P-60.

plans and specifications: Drawings and a full set of directions accompanying them.

plaque: An ornamented plate or disk, of metal, wood, ivory, porcelain, etc.

plasma: A gas made up of charged particles.

plaster: Mixed plaster is mortar to which a binder is added and is used for plastering walls and ceilings.

plasterboard: A building board made of plaster and faced with paper on both sides. Sometimes called *gypsum board*; used instead of plaster for an interior wall covering.

plastering: The act of applying plaster.

plastering trowel: A thin rectangular piece of steel 4 to 5 in. wide and 10 to 12 in. long, with handle attached, offset but parallel to the blade. See Fig. P-61.

plaster lath: The thin strips of wood nailed to studding, joists, or rafters to receive plaster.

plaster of Paris: Calcined gypsum, marketed in the form of a white powder. When mixed with water, it sets quickly, and is useful in making casts and models.

plastic: Capable of being molded or modeled. Moldable material.

plastic art: Ceramics or sculpture in which things are molded.

plastic deformation: Permanent change in dimensions of an object under load.

Figure P-61: Plastering trowel.

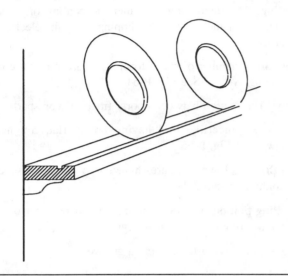

Figure P-62: Plate rail.

plastic wood: A wood compound which quickly hardens on exposure to air. Used for filling in cracks and defects.

plastics: Nonmetallic moldable compounds and the articles made from them.

plasticizers: Those substances which combine with resinoid materials to serve as a softening agent preventing permanent infusibility.

plate: The top horizontal timber of a wall on which rest, and are fastened, attic joists, roof rafters, etc.

plate cut: The cut in a rafter which rests upon the plate; sometimes called the seat cut.

plate glass: A high-grade glass cast in the form of a plate or sheet and subsequently polished; usually thicker than window glass, of smoother surface, and better quality.

platen: A flat working surface for laying out or assembling metal work, or the movable table of a planer or similar machine.

plate rail: A narrow shelf-like molding attached to an interior wall for the support of dishes, etc. See Fig. P-62.

platform: A horizontal structure usually covered with wood or metal and set on uprights to form an elevated flooring or stand.

platform framing: In this type of construction the floor platforms are framed independently; the second and third floors are supported by studs one story in height. See Fig. P-63.

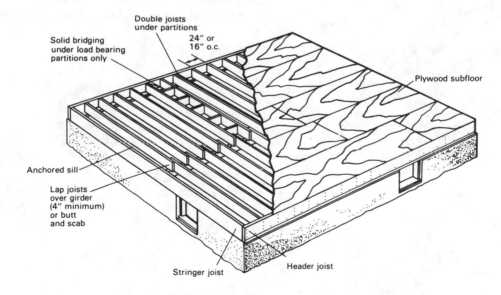

Figure P-63: Platform framing. *Courtesy U.S. Dept. of Agriculture.*

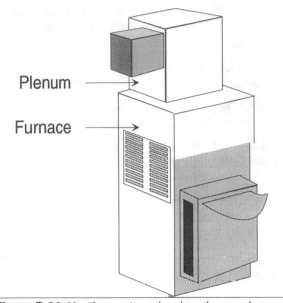

Figure P-64: Heating system showing plenum above furnace, along with related ductwork.

plating: The depositing of a metallic coating on another material or metal either by dipping or by the electrolytic process.

plenum: Chamber or space forming a part of an air conditioning system. See Fig. P-64.

play: The motion between poorly fitted or worn parts.

pliers: A pincerlike tool having broad, flat, roughened jaws. See Fig. P-65.

plinth: The lowest square-shaped part of the base of a column or pedestal.

plotting points: The determining of the position of, and the locating of points as in a graph.

plow: A grooving tool; to cut a groove.

plowing: Burying cable in a split in the earth made by a blade.

plug: A male connector for insertion into an outlet or jack.

plug fuse: A type of fuse that is held in position by a screw-thread contact instead of spring clips, as is the case with a cartridge fuse. See Fig. 66.

plug gauge: A very accurately made plug for testing the size of holes or internal diameters in machine work.

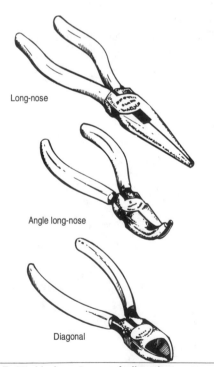

Long-nose

Angle long-nose

Diagonal

Figure P-65: Various types of pliers in common use.

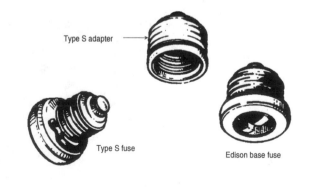

Type S adapter

Type S fuse

Edison base fuse

Figure P-66: Types of plug fuses in common use.

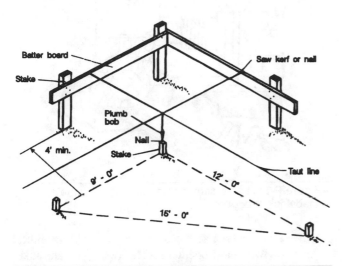

Figure P-67: Use of a line and plumb bob to accurately set batter boards.

Figure P-68: Pocket door.

plugging: Braking an induction motor by reversing the phase sequence of the power to the motor. This reversal causes the motor to develop a counter torque, which results in the exertion of a retarding force. Plugging is used to secure both rapid stop and quick reversal.

plug weld: A method of attaching plates or fixtures by welding through a hole in one or both of the parts.

plumb: To test or true up vertically, as a wall by means of a plumb line. See Fig. P-67.

plumb and level: A piece of well-finished hardwood or metal with bubble set lengthwise for testing horizontal accuracy and another bubble set crosswise for testing vertical accuracy. See Fig. P-67.

plumb bob: The weight used at the end of a plumb line. See Fig. P-67.

plumb cut: Any cut made in a vertical plane; the vertical cut at the top end of a rafter.

plumbing: Installation and repair of water pipes, tanks, bathroom fixtures, sewage lines, etc.

ply: A layer or thickness, as in something built up of several layers, each thickness being termed a ply.

pneumatic: Pertaining to, or operated by, air pressure.

pneumatic tools: tools operated by air pressure.

pocket door: Door that slides out of the way into a pocket made in the wall. It requires no floor space and, when the door is opened, it does not in any way obstruct the entire doorway. See Fig. P-68.

pointing: The finishing of joists in a brick or masonry wall to approve appearance or protect against weather.

pointing trowel: A small trowel used by bricklayers for pointing and striking up joints and for removing mortar from the face of the wall. See Fig. P-69.

Figure P-69: Pointing trowel.

224

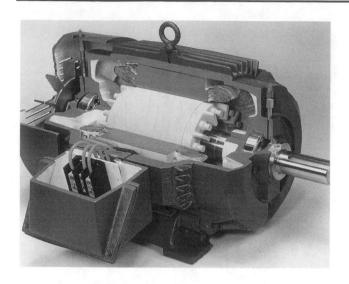

Figure P-70: Polyphase motor.

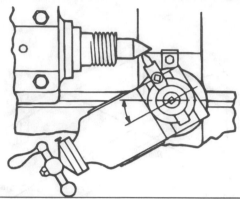

Figure P-71: Lathe center being turned in a lathe's headstock or poppet.

polarity: 1) Distinguishing one conductor or terminal from another. 2) Identifying how devices are to be connected, such as positive (+) or (-).

Polarization Index: Ratio of insulation resistance measured after 10 minutes to the measure at 1 minute with voltage continuously applied.

pole: 1) That portion of a device associated exclusively with one electrically separated conducting path of the main circuit of device. 2) A supporting circular column.

polychloroprene: Generic name for neoprene.

polycrystalline: Pertaining to a solid having many crystals.

polyethylene: A thermoplastic insulation having excellent electrical properties, good chemical resistance, and good mechanical properties with the exception of temperature rating.

polymer: A high-molecular-weight compound whose structure can usually be represented by a repeated small unit.

polyphase circuits: ac circuits having two or more interrelated voltages, usually of equal amplitudes, periods, phase differences, etc.

polyphase motor: An ac motor that is designed for either three- or two-phase operation. The two types are alike in construction, but the internal connections of the coils are different. See Fig. P70.

polypropylene: A thermoplastic insulation similar to polyethylene, but with slightly better properties.

polythtrafluoroethylene (PTFE): A thermally stable (-90° to + 250°C) insulation having good electrical and physical properties even at high frequencies.

poplar: A common tree whose wood is soft, light in weight, and is easy to work.

poppet: The headstock of a lathe. A lathe center. See Fig. P-71.

poppy heads: The ornaments which form the tops of the ends of benches or pews.

porch: A covered structure, outside of a building, and with separate roof, forming an entrance to the building. See Fig. P-72.

portable: Designed to be movable from one place to another, not necessarily while in operation.

portal: A door, also a gateway or entranceway.

portico: A space with a roof supported by columns; usually a porch before the entrance to a building. See *porch.*

positive: Connected to the positive terminal of a power supply.

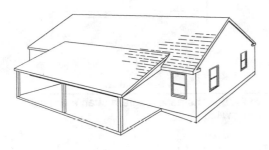

Figure P-72: Typical porch.

positive feed: When the feed motion is communicated directly by means of gears, without friction clutches or belts.

positive plate: The plate of a storage cell from which the current flows to the negative plate during the process of discharging. See Fig. P-73.

post: A timber set on end to support a wall, girder, or other member of the structure.

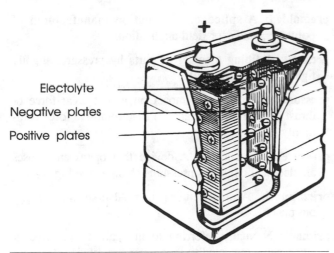

Electolyte

Negative plates

Positive plates

Figure P-73: Storage battery, showing both positive and negative plates.

post-and-beam construction: Wall construction in which beams are supported by heavy posts rather than many smaller studs.

potential: The difference in voltage between two points of a circuit. Frequently, one is assumed to be ground (zero potential).

potential energy: Energy of a body or system with respect to the position of the body or the arrangement of the particles of the system.

potentiometer: An instrument for measuring an unknown voltage or potential difference by balancing it, wholly or in part, by a known potential difference produced by the flow of known currents in a network of circuits of known electrical constants.

pothead: A terminator for high-voltage circuit conductor to keep moisture out of the insulation and to protect the cable end, along with providing a suitable stress relief cone for shielded-type conductors. See Fig. P-74.

Overhead high-voltage lines
connect to these terminals
on pothead mounted on power pole

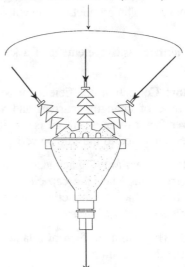

High-voltage cable down
power pole to either manhole,
padmount or submersible
transformer

Figure P-74: Pothead.

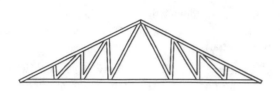

Figure P-75: Pratt truss.

power: 1) Work per unit of time. 2) The time rate of transferring energy; as an adjective, the word "power" is descriptive of the energy used to perform useful work; measurements: pound-feet per second, watts.

power, active: In a 3-phase symmetrical circuit: $p = \sqrt{3}$ VA cos θ; in a 1-phase, 2 wire circuit: $p = $ VA cos θ.

power, apparent: The product of rms volts times rms amperes.

power element: Sensitive element of a temperature-operated control.

power factor: Correction coefficient for ac power necessary because of changing current and voltage values. The power factor of an ac circuit is a ratio of the apparent power compared to the true power.

power factor correction: Capacitance is used for power factor correction. When the capacitive current is at its peak positive value, the inductive current is at its peak negative value, and vice-versa.

power feed: The automatic feed of a lathe, planer, screwcutting, or other machine.

power loss (cable): Losses due to internal cable impedance, mainly I^2r; the losses cause heating.

power pool: A group of power systems operating as an interconnected system.

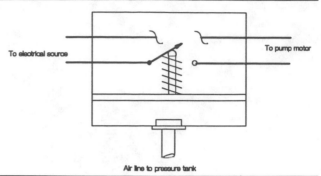

Figure P-76: Schematic diagram of an electrical pressure switch.

P-P: Peak to peak.

pratt truss: A popular form of truss for both roof and bridge construction. See Fig. P-75.

precast concrete: Concrete units (such as piles or vaults) cast away from the construction site and set in place.

precious metal: Gold, silver, or platinum.

precision grinding: Machine grinding in which the tolerances are exceedingly close.

precision lathe: A small bench lathe used for very accurate work.

prefabrication: Construction of components such as walls, trusses, or doors, before delivery to the building site.

premolded: A splice or termination manufactured of polymers, ready for field application.

press fit: A fitting together of parts by pressure; slightly tighter than a sliding fit.

pressure: 1) An energy impact on a unit area; force or thrust exerted on a surface. 2) Electromotive force commonly called *voltage*.

pressure motor control: A device that opens and closes an electrical circuit as pressures change. See Fig. P-76.

prick punch: A small center punch. Also known as a layout punch.

primary: Normally referring to the part of a device or equipment connected to the power supply circuit.

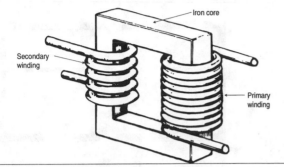

Figure P-77: Basic transformer showing primary and secondary coils.

primary cell: A device for transforming chemical energy into electrical energy. It consists essentially of a container with a solution (electrolyte) and two plates of electrodes.

primary coil: The coil into which the original energy is introduced and which sets up magnetic lines of force to link with another coil in which energy is induced. See Fig. P-77.

primary control: Device that directly controls operation of a heating system.

printed circuit: A board having interconnecting wiring printed on its surface and designed for mounting of electronic components. See Fig. P-78.

process: Path of succession of states through which a system passes.

profiling machine: A type of milling machine in which the cutter can be made to follow a profile or pattern. A very valuable machine for certain classes of work; program, computer: The ordered listing of sequence of events designed to direct the computer to accomplish a task.

projecting belt course: Usually an elaboration of a plain band course of masonry or cut-stone work projecting several inches beyond the face of the wall. See Fig. P-79.

projection: 1) A jutting out, a prominence. 2) In drawing, the method by which one or more views of an object are used as an aid in securing additional views. See Fig. P-80 on page 228.

propagation: The travel of waves through or along a medium.

property: An observable characteristic.

protected enclosure: Having all openings protected with screening, etc.

protector, circuit: An electrical device that will open an electrical circuit if excessive electrical conditions occur.

proton: The hydrogen atom nucleus; it is electrically positive; mass = 1.672×10^{-24} grams; charge = 0.16 attocoulomb.

prototype: The first full size working model.

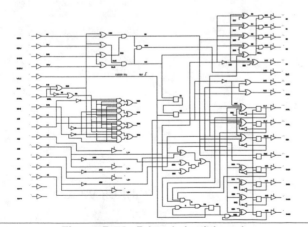

Figure P-78: Printed circuit board.

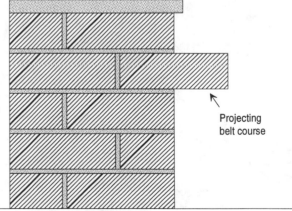

Figure P-79: Projecting belt course.

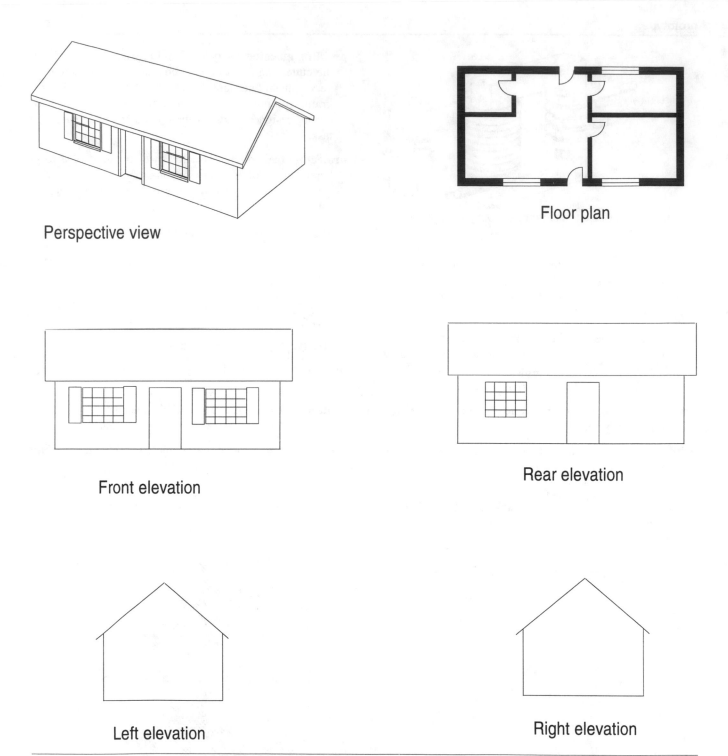

Perspective view

Floor plan

Front elevation

Rear elevation

Left elevation

Right elevation

Figure P-80: Orthographic projection of a building.

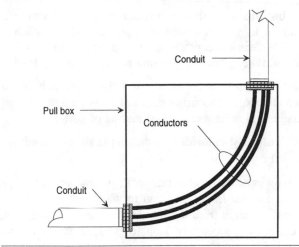

Figure P-81: Typical pull box.

proximity effect: The distortion of current density due to magnetic fields; increased by conductor diameter, close spacing, frequency, and magnetic materials such as steel conduit or beams.

PSAR (Preliminary Safety Analyses Report): Construction permit.

PSI: Pound force per square inch.

PT: Potential transformer.

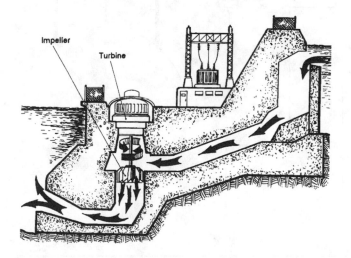

Figure P-82: Cross-section of a typical hydroelectric plant.

pull box: A sheet metal box-like enclosure used in conduit runs, either single conduits or multiple conduits, to facilitate pulling in of cables from point to point in long runs, or to provide installation of conduit support bushings needed to support the weight of long riser cables, or to provide for turns in multiple-conduit runs. See Fig. P-81.

pulley lathe: A lathe used for turning either a straight or crowned face on pulleys.

pulley stile: The vertical sides of a double-hung window frame, on which are fastened the pulleys for the sash weights.

pulley tap: A tap with a very long shank, used for tapping set-screw holes in the hubs of pulleys.

pull-down: Localized reduction of conductor diameter by longitudinal stress.

pulling compound (lubricant): A substance applied to the surface of a cable to reduce the coefficient of friction during installation.

pulling eye: A device attached to a cable to facilitate field connection of pulling ropes.

pull pin: A device for throwing mechanical parts in or out of gear, or for readily shifting in or away from a fixed relative position.

pulsating current: A direct current in which the value is not constant but the flow is in one direction.

pulsating function: A periodic function whose average value over a period is not zero.

pulse: A brief excursion of a quantity from normal.

pumice: Powdered lava used as a polishing material.

pumped storage (hydro power): The storage of power by pumping a reservoir full of water during off-peak, then depleting the water to generate when needed. See Fig. P-82.

puncture: Where breakdown occurs in an insulation.

purge: To clean.

purlins: Timbers spanning from truss to truss, and supporting the rafters of a roof. See Fig. P-83 on page 230.

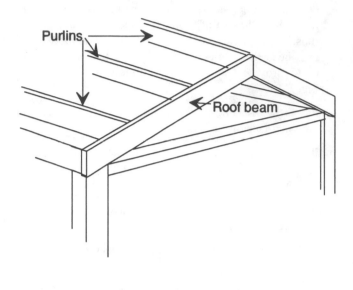

Purlins

Roof beam

Figure P-82: Roof detail showing purlins.

push button: 1) A device that completes an electric circuit as long as a button is depressed. 2) A switch in which electrical contacts are closed by pushing one button and are opened by pushing another. See Fig. P-83.

putty: A composition used for filling small holes in woodwork, and securing panes of glass in sash. Glazing compound is frequently used instead of putty.

PVC (polyvinyl chloride): A thermoplastic insulation and jacket compound.

PWR (pressurized water reactor) (nuclear power): A basic nuclear fission reaction in which water is used to transfer energy from the reactor; the water exchanges its heat with a secondary loop to produce steam for the turbine.

pyroconductivity: Electric conductivity that develops with changing temperature, and notably upon fusion, in solids that are practically non-conductive at atmospheric temperatures.

pyrometer: Thermometer that measures the radiation from a heated body.

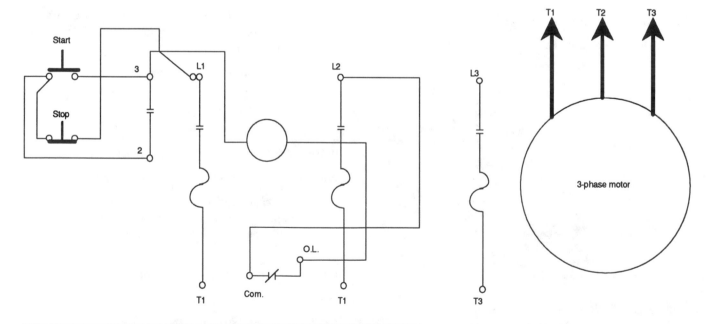

Figure P-83: Schematic diagram of a pushbutton motor-control circuit—3-wire control.

QA (Quality Assurance): All the planned and systematic actions to provide confidence that a structure, system, or component will perform satisfactorily.

QAP (nuclear): Quality assurance policy.

quadrangle: A square or quadrangular space surrounded by buildings, as on college grounds.

quadrant: The quarter of a circle or compass area, or of their circumferences. Quadrant scales are used in surveying. See Fig. Q-1.

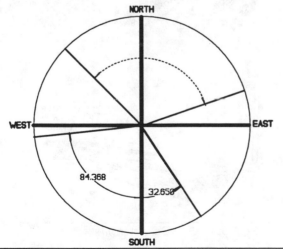

Figure Q-1: Quadrant scales used in surveying; angles are marked from the north-south line.

quadrilateral: A plane figure with four sides and four angles. See Fig. Q-2.

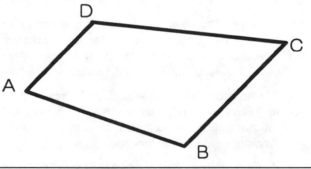

Figure Q-2: Quadrilateral

quadruplexed: Twisting of four conductors together.

quadruple thread: A thread in which there are four distinct helices, making the lead four times the pitch. A quadruple thread is usually of the square or acme type. See Fig. Q-3 on page 250.

qualified life (nuclear power): The period of time for which satisfactory performance can be demonstrated for a specific set of service conditions.

qualified person: A person familiar with construction, operation and hazards involved in a building construction project or apparatus.

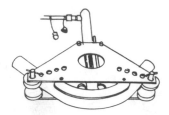

Figure Q-3: Quarter bends are frequently used in electrical and plumbing applications to make a 90° conduit or pipe turn.

qualitative analysis: The determining of how many and what elements or ingredients are present, as in a concrete or mortar mix.

quality: Degree of goodness.

quantity: A certain mass, volume, or number; that property of anything that may be increased or diminished.

quantity survey: The count of all materials necessary to construct a building project. One of the steps necessary to prepare a cost estimate for a construction project. Also see *takeoff, listing, labor units,* and *finalizing.*

quarry: 1) An excavation from which stone is removed by blasting, cutting, etc. 2) A small square or lozenge-shaped pane of glass, plain or decorated, used in forming the glass fronts of eighteenth-century bookcases.

quarry-faced masonry: That in which the face of the stone is left unfinished just as it comes from the quarry.

quarry tile: Also called "promenade" tile. A name for machine-made, unglazed tile, ¾ inch or more in thickness.

quart: Two pints; ¼ gallon. In dry measure, one eighth of a peck.

quarter: One of four equal parts of anything. One of the four principle parts of a compass.

quarter bend: A bend through an arc of 90 degrees; as in a piece of conduit. Sometimes called *elbow*. See Fig. Q-4.

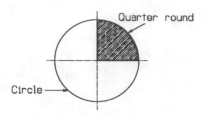

Figure Q-4: Quarter round.

quarter round: Molding that is quartered from a round piece. Use primarily at the bottom of base trim at the floor line. See Fig. Q-4.

quarter sawing: Signifies that the log is first cut lengthwise into four quarters. In sawing these quarters into boards, the cuts are made parallel with the medullary (wood) rays. Quartersawn lumber is less apt to warp than other types of cuts. See Fig. Q-5.

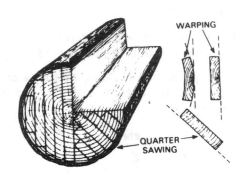

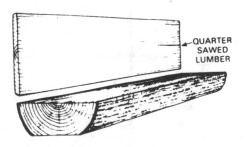

Figure Q-5: Quartersawn lumber.

quartz: A hard crystalline mineral occurring as a rock (SiO_2). It is usually colorless, but is often colored by impurities.

quaternary steel: That class of alloy steel that consists of iron, carbon, and two other special elements.

quatrefoil: A unit of decoration in the form of a four-leaved flower. See Fig. Q-6.

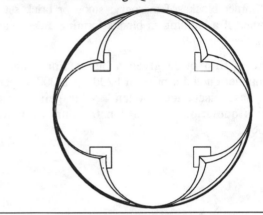

Figure Q-6: Architectural quatrefoil.

queen closer: A half brick, made by cutting the brick lengthwise.

queen truss: A truss framed with two vertical tie posts, as distinguished from the king truss which has but one. See Fig. Q-7.

quenching: The dipping of heated steel into water, oil, or other bath, to impart necessary hardness.

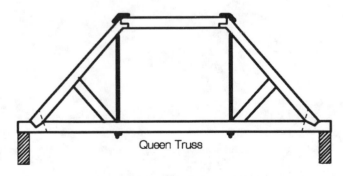

Figure Q-7: Queen truss.

quick-: A device that has a high contact speed independent of the operator; example: quick-make or quick-break. See *quick-break switch*.

quick-break switch: Usually of the knife-blade type. The blade is made of two pieces. As the switch is pulled out, the first half of the blade is withdrawn, and as the throw increases, the second half is drawn out by the action of a spring attached to the first blade.

quicklime: Unslaked lime made from nearly pure limestone.

quicksand: A mass of loose sand mixed with water to such an extent that it is not capable of supporting the weight of a heavy body.

quicksilver: The common name for mercury; also the amalgam of tin used on the backs of mirrors.

quiet enjoyment: The right of an owner to the use of property without interference of possession.

quiet title: A court action brought to establish title and to remove a cloud on the title.

quill: A hollow shaft or spindle.

quirk: A small groove in, beside, or between moldings or beads. See Fig. Q-8.

quirk bead: A bead molding separated from the surface on one side by a groove. A double quirk bead means a groove on each side of the beads.

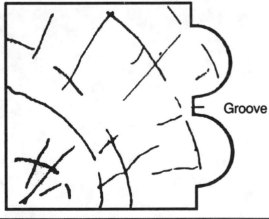

Groove

Figure Q-8: Cross-section of double quirk molding.

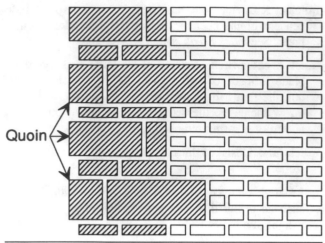

Quoin

Figure Q-9: Quoins.

quirk molding: One that has a small groove, although frequently applied to a molding having a convex and a concave curve, the two separated by a small flat.

quit claim deed: A deed given when the grantee already has, or claims, complete or partial title to the premises and the grantor has a possible interest that otherwise would constitute a cloud upon the title.

quoins: Corner blocks of masonry; stone or brick set at the corner of a building in blocks forming a decorative pattern. See Fig. Q-9.

quotation: A firm price given a buyer. For example, "the supplier quoted a price of $200 for a 200-ampere, three-phase disconnect switch." "The general contractor's quotation for the construction work was $170,500."

R

rabbet: A corruption of the term rebate. A groove or cut made in the edge of one plank, etc., so that another, similarly cut, may fit into it; rabbet joint. See Fig. R-1.

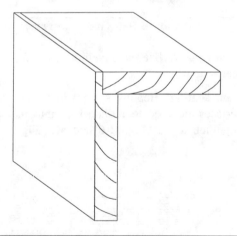

Figure R-1: Typical rabbet joint.

rabbet plane: A plane designed for cutting or planing rabbets to form rabbet joints. See Fig. R-2.

raceway: Any channel designed expressly for holding wire, cables, or bars and used solely for that purpose. See Fig. R-3.

rack (cable): A device to support cables.

Figure R-2: Rabbet plane. *Courtesy Stanley Works.*

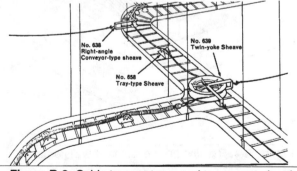

No. 638
Right-angle
Conveyor-type sheave

No. 639
Twin-yoke Sheave

No. 858
Tray-type Sheave

Figure R-3: Cable tray system used to support electrical or communication conductors.

rack and pinion: A gear assembly used to convert rotary motion to linear motion. The linear motion takes place on a slotted bar (rack) that is powered by a circular toothed gear (pinion). Used in many types of construction machinery.

racking: In approaching a corner where two walls meets, "racking" is the making of each brick or block course shorter than the course below it so workers on the walls may tie in their courses in the easiest manner. See Fig. R-4.

radar: A radio detecting and ranging system.

radial: Extending outward from a center or axis.

radial-arm saw: Circular saw that moves back and forth on an overhead arm. Both blade and arm are adjustable for cuts at various angles. See Fig. R-5.

radial bar: A wooden or metal bar equipped with a sharp steel point on one end and a pen or pencil on the other to strike large curves.

radial cut: Wood cut made at a right angle to growth rings. See *quarter sawn*.

radial feeder: A feeder connected to a single source.

radial step: Tread in winding staircase.

radian (rad): A supplementary SI unit for plane angles; the plane angle with its vertex at the center of a circle that is subtended by an arc equal in length to the radius.

Figure R-5: Radial-arm saw.

radiant energy: Energy traveling in the form of electromagnetic waves.

radiant heating: Heating system in which warm or hot surfaces are used to radiate heat into the space to be conditioned. See Fig. R-6 and also Fig. R-7 on page 237.

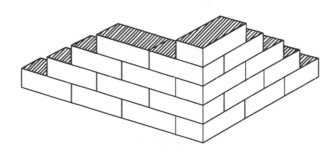

Figure R-4: Racking brick courses at wall corners.

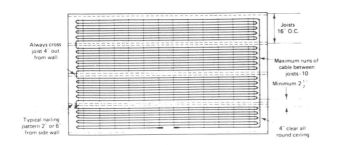

Figure R-6: Layout of radiant heating cable.

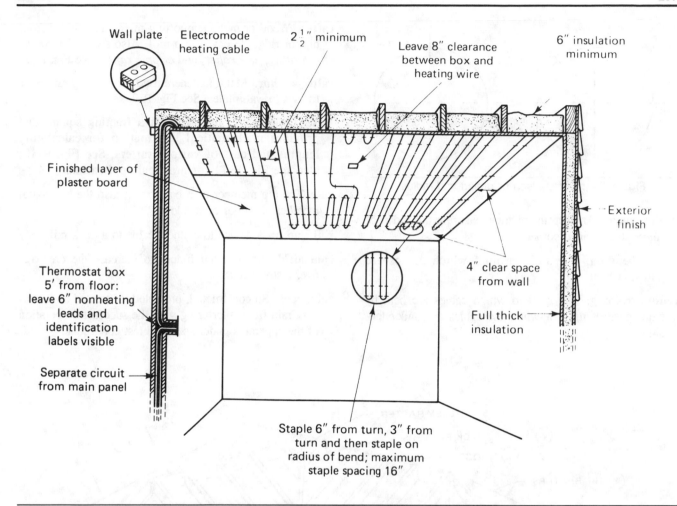

Wall plate Electromode heating cable $2\frac{1}{2}''$ minimum Leave 8" clearance between box and heating wire 6" insulation minimum

Finished layer of plaster board

Exterior finish

Thermostat box 5' from floor: leave 6" nonheating leads and identification labels visible

4" clear space from wall

Full thick insulation

Separate circuit from main panel

Staple 6" from turn, 3" from turn and then staple on radius of bend; maximum staple spacing 16"

Figure R-6: Electric radiant heating. Electrical resistance cable is buried in the ceiling plaster to form a heating panel.

radiating surface: The amount in square feet of effective heating area of a radiator.

radiation: The process of emitting radiant energy in the form of waves or particles.

radiation, blackbody: Energy given off by an ideal radiating surface at any temperature.

radiation, nuclear: The release of particles and rays during disintegration or decay of atom's nucleus; these rays cause ionization; they are: alpha particles, beta particles, gamma rays.

radiator: Heating unit for steam or hot-water space heating.

radical: Relating to the root or roots of numbers; being or containing a root.

radium: A highly radioactive metallic element obtained from the uranium mineral, pitchblende.

radius: Distance from the center of a circle, sphere or arc (curve) to the outside circumference; one-half the diameter of a circle. A comparison of radius and radian is shown in Fig. R-7 on page 238.

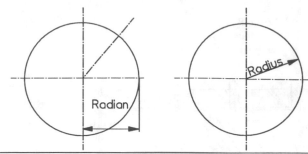

Figure R-7: Comparison or radian and radius.

radius gauge: An instrument for measuring the radaii of fillets and rounded corners.

radius, bending: The radii around which cables are pulled or bent.

radius, training: The radii to which cables are bent by hand positioning, not while the cables are under tension.

rafter: Wood or metal supporting members that run from hip, or ridge, to eaves in a roof. Also see *valley rafter*, *hip rafter*, j*ack rafter*, and *cripple rafter*. See Fig. R-8.

rafter anchor: Metal fasteners used for holding rafters or trusses to wall plates. See Fig. R-9.

rafter table: Table inscribed on a framing square that provides mathematical information for calculating lengths and cuts of various rafters. See Fig. R-10; also see Fig. R-11 for practical applications.

rail: The top member of a balustrade; also the horizontal member of a door or window.

rail clamp: A device to connect cable to a track rail.

rainshield: An inverted funnel to increase the creepage over a stress cone.

rainproof: So constructed, protected, or treated as to prevent rain from interfering with the successful operation of the apparatus under specified test conditions.

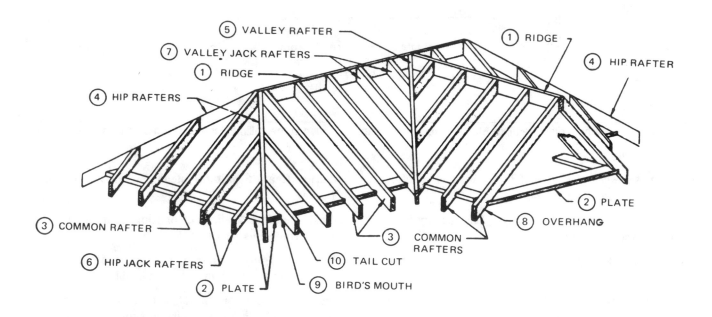

Figure R-8: Various types of wood rafters. *Courtesy U.S. Dept. of Agriculture.*

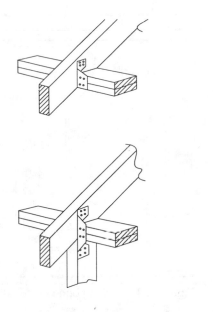

Figure R-9: Rafter anchors.

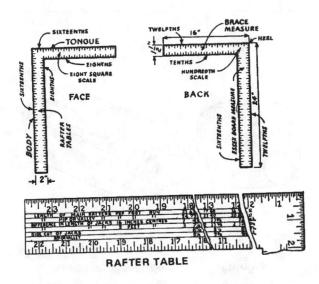

RAFTER TABLE

Figure R-10: Sample rafter tables fround on framing square. *Courtesy Stanley Works.*

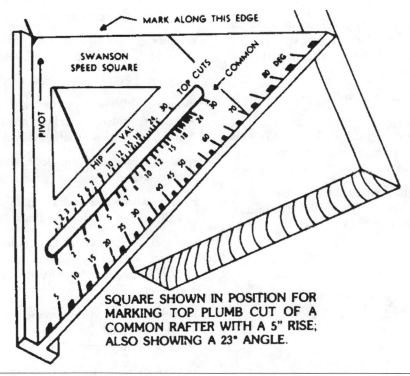

MARK ALONG THIS EDGE

SWANSON
SPEED SQUARE

PIVOT

HIP — VAL

TOP CUTS

COMMON

80 DEG

SQUARE SHOWN IN POSITION FOR
MARKING TOP PLUMB CUT OF A
COMMON RAFTER WITH A 5" RISE;
ALSO SHOWING A 23° ANGLE.

Figure R-11: Practical application of rafter tables.

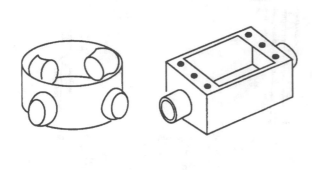

Figure R-12: Raintight electrical boxes.

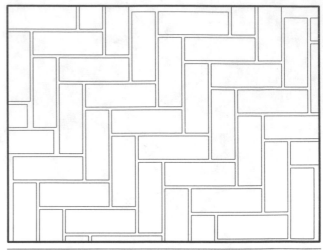

Figure R-14: Raking bond brick pattern.

raintight: So constructed or protected that exposure to a beating rain will not result in the entrance of water. See Fig. R-12.

raising: A wrinkled or blistered condition in a surface due to reaction of lacquer solvents or unoxidized oil in oil-base undercoaters, or in the application of lacquer over old paint or varnish.

rake: 1) The amount of set on a cutting tool. 2) Angle of slope or incline.

raked joint: Joint formed in brickwork by raking out some of the mortar an even distance from the face of the wall.

raker-tooth saw: Saw blade with teeth set alternately to the right and left with one straight tooth between pairs to clear away chips. See Fig. R-13.

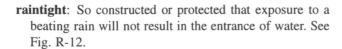

Figure R-13: Raker-tooth saw blade.

raking bond: Brick laid in an angular or zigzag fashion. See Fig. R-14.

ram: Random access memory.

rammer: A pneumatic tool designed to pack down earth before constructing a concrete foundation or slab.

ramp: A slope or inclined way used in place of steps.

random: 1) Without apparent method or system. 2) Odd sizes, considered together, are spoken of as random sizes, as shingles of different widths.

random joints: Joints made in veneer without reference to the veneer being of equal width.

random work: Stonework laid up in irregular order; as a wall built up of odd-size stones.

range: 1) A strip of land six miles wide determined by government survey, running in a north-south direction. 2) A self-contained cooking appliance — usually gas or electric.

range hood fan: Metal hood with electric exhaust fan and lights mounted directly over the kitchen range to exhaust cooking fumes.

rasp: A filelike tool having coarse projections for abrasion. See Fig. R-15.

Horse rasps – tanged

Horse rasps – tanged plater's special

Magicut® plater's special horse rasp

Pattern maker's cabinet rasps

4-in-Hand®(formerly shoe rasp)

Wood rasps – flat bastard

Wood rasps – half round bastard

Wood rasps round bastard

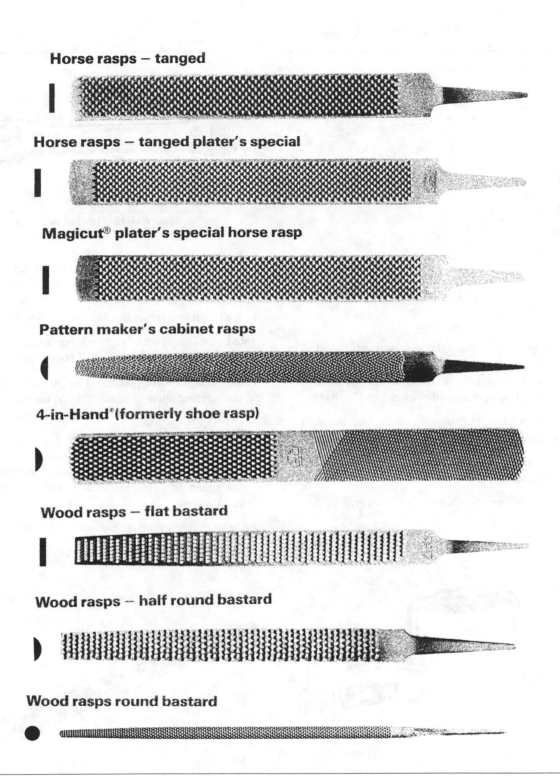

Figure R-15: Types of rasps in common use.

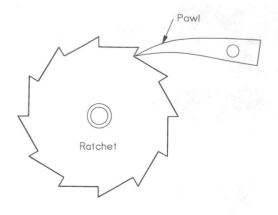

Figure R-16: Ratchet.

Figure R-18: Ratchet wrench.

ratchet: A gear with triangular-shaped teeth, adapted to be engaged by a pawl, that either imparts intermittent motion to the ratchet, or locks it against backward movement when operated otherwise. See Fig. R-16.

ratchet bar: A straight bar with teeth like those of a ratchet wheel to receive the thrust of a pawl.

ratchet bit brace: A bit brace with ratchet attachment to permit operating in close quarters. See Fig. R-17.

ratchet screwdriver: See *Yankee screwdriver*.

ratchet wrench: Type of nonslip end wrench that permits working in close quarters. Grip is released by a slight backward movement and a new hold is had automatically without removing the wrench. See Fig. R-18.

rated: Indicating the limits of operating characteristics for application under specified conditions.

rating, temperature (cable): The highest conductor temperature attained in any part of the circuit during a) normal operation b) emergency overload c) short circuit.

rating, voltage: The thickness of insulation necessary to confine voltage to a cable conductor after withstanding the rigors of cable installation and normal operating environment.

ratio: Relative value; proportion.

rattail file: Name commonly applied to round files. See Fig. R-19.

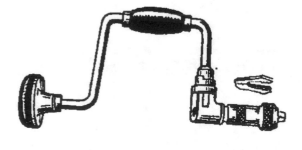

Figure R-17: Ratchet bit brace.

Figure R-19: Rattail file.

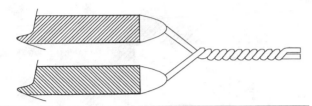

Figure R-20: Rattail splice.

rattail splice: A wire splice made by twisting the bared ends together. See Fig. R-20.

rawhide: Untanned, dressed animal skin.

rawhide mallet: Mallet with a head made of tightly rolled rawhide.

Rawlplug®: Anchor used for securing items onto masonry. The fiber, lead-lined plug is inserted into a drilled hole; a sheet metal screw is then inserted into the plug to expand the plug for a tight fit. See Fig. 21.

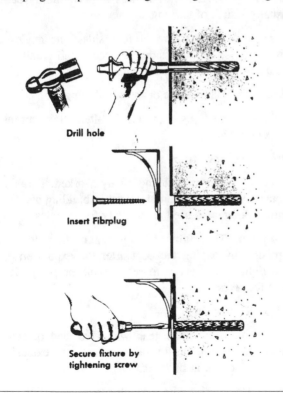

Drill hole

Insert Fibrplug

Secure fixture by
tightening screw

Figure R-21: Rawlplug.

raw material: The ingredients before being processed, that enter into a finished product.

ray: Wood tissue plane that extends outward from the pith; medullary rays.

raze: To tear down; to demolish; as to raze a building.

REA (Rural Electrification Administration): A federally supported program to provide electrical utilities in rural areas.

reactance: 1) The imaginary part of impedance. 2) The opposition to ac due to capacitance (X_C) and/or inductance (X_L).

reactor: A device to introduce capacitive or inductive reactance into an electrical circuit.

reactor, nuclear: An assembly designed for a sustained nuclear chain reaction; the chain reaction occurs when the mass of the fuel reaches a critical value having enough free neutrons or enough heat to sustain fusion.

ready-mix concrete: Concrete that is mixed at a plant and delivered to the job site in a truck equipped with a rotating drum to keep concrete mixed en route. Reversing the rotation empties the drum.

real time: The actual time during which a physical process transpires.

real estate: Land, buildings, and other appurtenances that cannot be readily removed.

realtor: A coined word used to designate an active member of a local real estate board affiliated with the National Association of Real Estate Boards.

ream: To smooth the surface of a hole and finish it to size with a reamer.

reamer: A tool with cutting edges, square or fluted, used for finishing drilled holes. A tool for removing burrs on the inside of the mouth on metal conduit.

reaming: The process of smoothing the surface of holes with a reamer.

rebar: Reinforcing bar or rod used in reinforced concrete construction. See Appendix J.

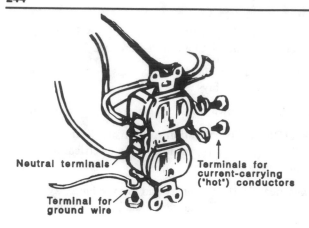

Figure R-22: Duplex receptacle with related wiring.

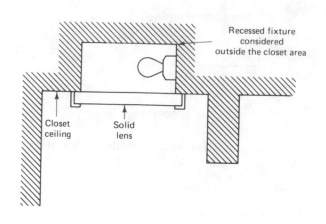

Figure R-23: Typical recessed lighting fixture.

rebate: A recess in or near the edge of one piece to receive the edge of another piece cut to fit it.

recapitulate: To summarize; to review.

receptacle: A contact device installed at an outlet for the connection of an attachment plug and flexible cord to supply portable equipment. See Fig. R-22.

recessed: Placed above, below, or behind a finished surface. See Fig. R-23.

recharging: 1) Restoring the electrolyte in a storage battery so that active electrons may flow. 2) Adding or replacing the refrigerant to a refrigeration systems, such as air conditioners and other cooling apparatus.

reciprocal: In mathematics, the reversal of a fraction or whole number. For example, the reciprocal of 2 is ½; the reciprocal of ¼ is 4.

reciprocating: Action in which the motion is back and forth in a straight line.

recognized component: An item to be used as a subcomponent and tested for safety by an independent testing laboratory.

reconnaissance: Preliminary operations prior to making an actual survey of land; that is, making notes, sketches, and a general study of the problem.

recorder: A device that makes a permanent record, usually visual, of varying signals.

rectangle: A plane figure of four sides; the angles are right angles, opposite sides are equal and parallel; the adjacent sides need not be equal.

rectangular: Having one or more right angles.

rectifiers: Devices used to change alternating current to direct current.

rectify: To change from ac to dc.

red cedar: Wood is soft and easily worked, durable in contact with soil. Has many used in building and furniture making, also used extensively for shingles.

redemption: The right of a mortgagor to redeem the property by paying the debt after the expiration date; the right of an owner to reclaim his property after a sale for taxes.

red-leg: See *high-leg*.

red oak: An oak, darker than white oak and of coarser grain, also more brittle and porous; used extensively for interior trim in buildings.

redraw: Drawing of wire through consecutive dies.

reducer: Any one of the various pipe connections so constructed as to permit the joining of pipes of different sizes, such as reducing sleeve, reducing ell, reducing tee, etc. See Fig. 24.

reducing: The use of a device or fixture for making the size of one part smaller so as to fit another, as a reducing coupling.

reducing agent: A substance that removes oxygen or some other element from a second substance.

reducing joint: A splice of two different size conductors.

reduction: The gain of electrons by a constituent of a chemical reaction.

redundancy: The use of auxiliary items to perform the same functions for the purpose of improving reliability and safety.

redwood: A giant tree of California. The wood is soft reddish in color, and light in weight.

reeding: A general term applied to half-round moldings of various kinds; also ornamentation by the use of such moldings. Reeding is the reverse of fluting. See Fig. 25.

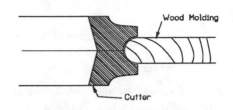

Figure R-24: Example of wood molding with half-round bead, along with the shape of the cutter used to obtain this configuration.

reel: A drum having flanges on the ends; reels are used for wire/cable storage.

reflective insulation: Thin sheets of metal or foil on paper set in the exterior walls of a building to reflect radiant energy.

reformation: An action to correct a mistake in a deed or other instrument.

refraction: The bending of a ray of light as it travels through a transparent substance. See Fig. R-25.

refrigerant: Substance used in refrigerating mechanisms to absorb heat in an evaporator coil and to release heat in a condenser as the substance goes from a gaseous state back to a liquid state.

refrigeration: A method used for cooling or lowering temperatures of objects or areas.

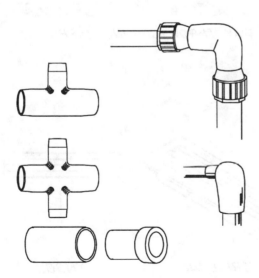

Figure R-24: Several types of pipe reducers.

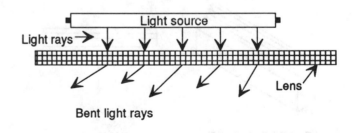

Figure R-25: Principle of refraction.

register: Combination grille and damper assembly covering on an air opening or end of an air duct. See Fig. R-26.

regulation: The maximum amount that a power supply output will change as a result of the specified change in line voltage, output load, temperature, or time.

reinforced: Strengthened by the addition of extra material. Reinforced concrete has within its mass iron or steel rods, bars or shapes to give it additional strength. See Fig. R-27.

reinforced concrete: Concrete work increased in strength by iron or steel bars imbedded in it.

reinforced jacket: A cable jacket having reinforcing fiber between layers.

reinforcing steel: Stell bars or rods used in concrete construction to give added strength.

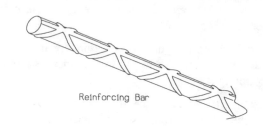

Reinforcing Bar

Figure R-27: Reinforcing rods used in reinforced concrete construction.

reinforcement symbols: Symbols used on structural working drawings to indicate placement of reinforcing steel in a reinforced concrete structure. See Fig. R-28.

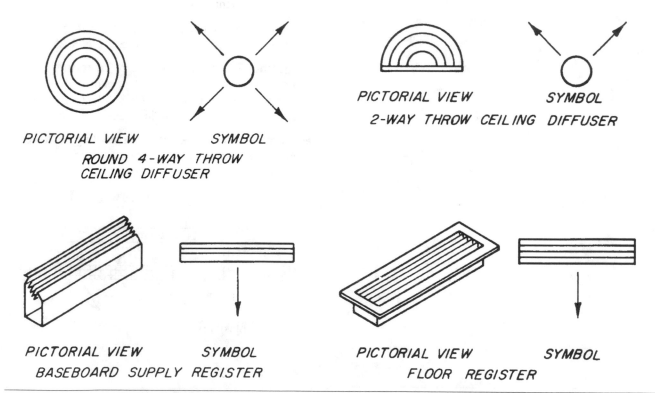

PICTORIAL VIEW   SYMBOL
ROUND 4-WAY THROW
CEILING DIFFUSER

PICTORIAL VIEW   SYMBOL
2-WAY THROW CEILING DIFFUSER

PICTORIAL VIEW   SYMBOL
BASEBOARD SUPPLY REGISTER

PICTORIAL VIEW   SYMBOL
FLOOR REGISTER

Figure R-26: Several types of registers and diffusers used with HVAC systems.

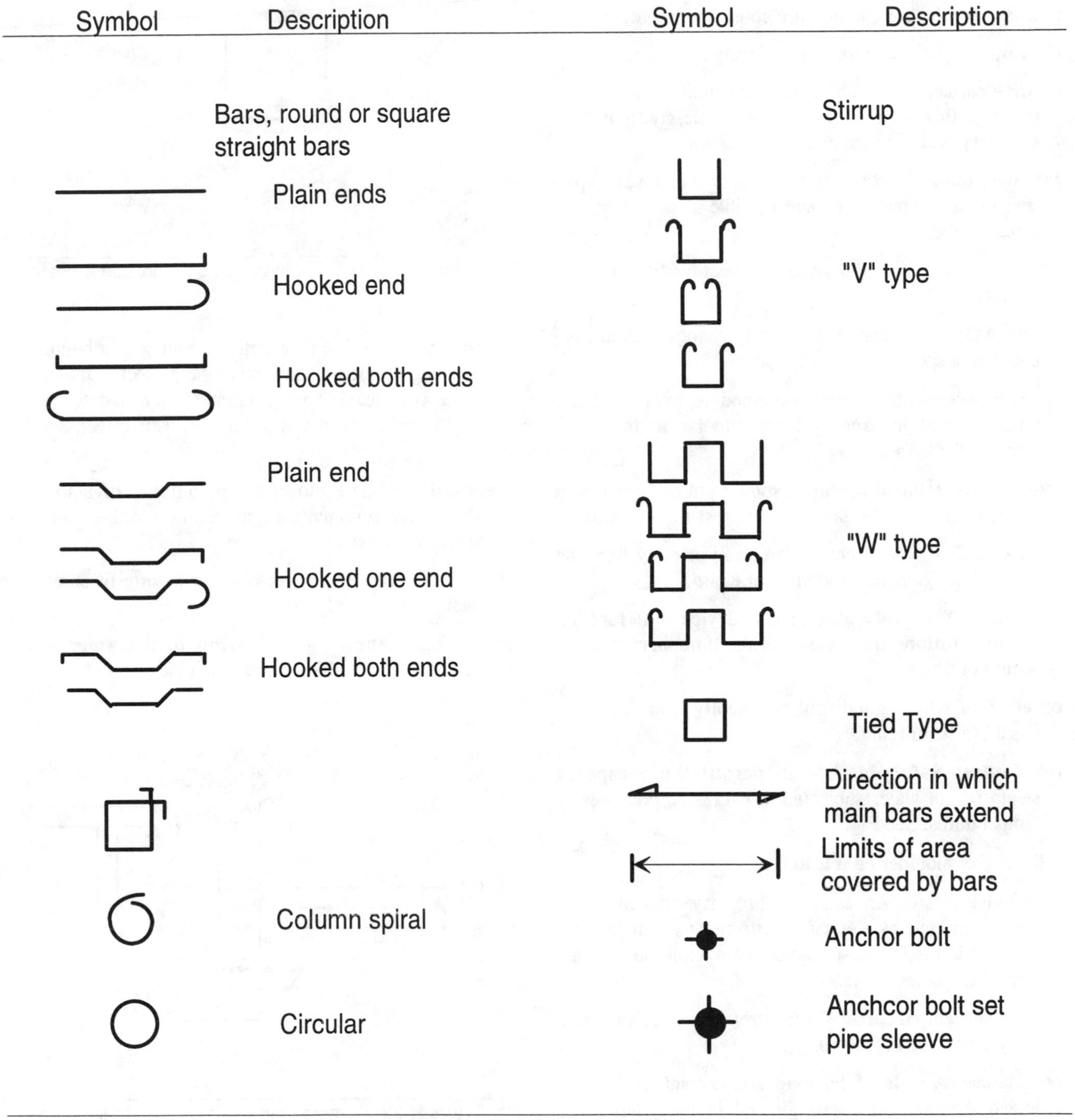

Symbol	Description	Symbol	Description
	Bars, round or square straight bars		Stirrup
	Plain ends		"V" type
	Hooked end		
	Hooked both ends		
	Plain end		"W" type
	Hooked one end		
	Hooked both ends		Tied Type
			Direction in which main bars extend
			Limits of area covered by bars
	Column spiral		Anchor bolt
	Circular		Anchcor bolt set pipe sleeve

Figure R-28: Reinforcement symbols.

related trades: The various trades whose work is necessary for the completing of a construction project.

relating: Having reference to; referring.

relative capacitance: The ratio of a material's capacitance to that of a vacuum of the same configuration; will vary with frequency and temperature.

relative humidity: Ratio of amount of water vapor present in air to greatest amount possible at same temperature.

relative motion: Motion of one object with respect to another.

relay: A device designed to abruptly change a circuit because of a specified control input.

relay, overcurrent: A relay designed to open a circuit when current in excess of a particular setting flows through the sensor.

release: The relinquishment of some right or benefit to a person who already has some interest in the property.

release of lien: The discharge of certain property from the lien of a judgement, mortgage, or claim.

reliability: The probability that a device will function without failure over a specified time period or amount of usage.

relief: That which stands out prominently from a surface is said to be in relief.

relief valve: Safety device to permit the escape of steam or hot water subjected to excessive pressures or temperatures. See Fig. R-29.

relish: The shoulder on a tenon.

remainder estate: An estate in property created at the same time and by the same instrument as another estate and limited to arise immediately upon the termination of the other estate.

remote-control circuits: The control of a circuit through relays, etc. See Fig. R-30.

renaisance: A style of building and decoration that followed the medieval. It originated in Italy during the fifteenth century.

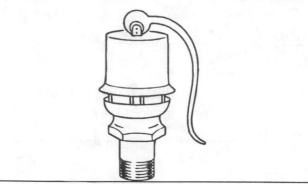

Figure R-29: Relief value used on a hot-water heater.

rendering: Life-like perspective drawing of a building or building project to show what the project will look like when completed. Tree, cars, and other objects are usually added to the drawing for a realistic effect. See Fig. R-31.

repeatability: The closeness of agreement among repeated measurements of the same variable under the same conditions.

repetition: The doing, making, or saying of something again or repeatedly.

replacing: Renewing; restoring to a former place, condition, or the like; taking the place of.

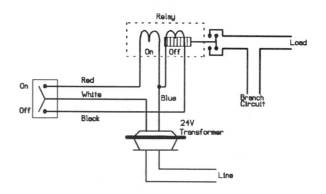

Figure R-30: Remote-control circuit used to control a residential lighting system.

Figure R-31: Rendering of a residence. *Courtesy Davis Publications, Inc.*

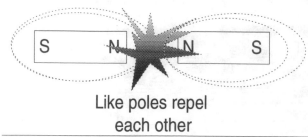

Like poles repel
each other

Figure R-32: Principles of repulsion.

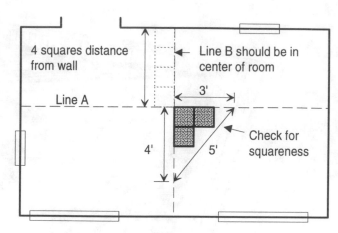

*Note: Start laying tiles in center
of room as shown.*

Figure R-33: Working lines for laying resilient tile in a
room or area.

replica: An exact copy or reproduction.

representative: Typical; being of the best style or type.

reproducibility: The ability of a system or element to maintain its output or input over a relatively long period of time.

reproduction cost: Normal cost of exact duplication of a property, as of a certain date.

repulsion: In electrical circuits, the action of a force by which two similarly charged bodies tend to repel each other. See Fig. R-32.

requisition: A formal request, as for materials or supplies.

reservoir, thermal: A body to which and from which heat can be transferred indefinitely without a change in the temperature of the reservoir.

residual elements: Elements present in an alloy in small quantities, but not added intentionally.

residual magnetism: The small amount of magnetism left in a piece of iron or steel after the magnetizing force has been removed.

residual stress: Stress present in a body that is free of external forces or thermal gradients.

resilience: The act or power of springing back; capability of a strained body to recover its size and shape after deformation.

resilient tile: Floor tile made from either asphalt, rubber, vinyl, or vinyle asbestos, and set into a mastic; usually comes in 12″ × 12″ squares. See Fig. R-33.

resin: The polymeric base of all jacketing, insulating, etc. compounds, both rubber and plastic. Any of various oily gummy substances obtained from certain trees, soluble in alcohol, ether, etc., but not in water.

resinoid: A general term applid to synthetic resinous substances as distinguished from natural resins.

resistance: The opposition in a conductor to current; the real part of impedance.

resistance furnace: A furnace that heats by the flow current against ohmic resistance internal to the furnace.

resistance, thermal: The opposition to heat flow; for cables it is expressed by degrees centigrade per watt per foot of cable.

resistance welding: Welding by pressure and heat when the work piece's resistance in an electric circuit produces the heat.

resistivity: A material characteristic opposing the flow of energy through the material; expressed as a constant for each material; it is affected by temper, temperature, contamination, alloying, coating, etc.

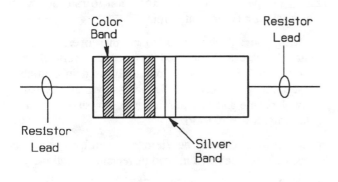

Figure R-34: Parts of a resistor.

resistor: A device whose primary purpose is to introduce resistance. Small resistors are used in practically all electronic projects and are identified by colored bands encircling the device. When viewing a resistor, the gold or silver band should be on the right-hand side. Each color to the left of this gold or silver band represents a number, as follows:

Black	0
Brown	1
Red	2
Orange	3
Yellow	4
Green	5
Blue	6
Violet	7
Gray	8
White	9

With these figures in mind, the resistor in Fig. R-34 with the colors (from left to right) yellow (4), violet (7), and brown (1) is read as follows:

- The first yellow band represents the number 4, and the second band represents the number 7; the two combined colors give a value of 47. The third color (brown) gives the number of zeros in the value. Since brown represents 1, the third band indicates that the resistor has 1 zero in its value. So the value of the resistor is 470 ohms.

resistor, bleeder: 1) Used to drain current after a device is de-energized. 2) To improve voltage regulation. 3) To protect against voltage surges.

resolution: The degree to which nearly equal values of a quantity can be discriminated.

resolver: A device whose input and output is a vector quantity.

resonance: A condition reached in an electrical circuit when the inductive reactance neutralizes the capacitance reactance leaving ohmic resistance as the only opposition tot he flow of current.

resonating: The maximizing or minimizing of the amplitude or other characteristics provided the maximum or minimum is of interest.

response: A quantitative expression of the output as a function of the input under conditions that must be explicitly stated.

restoration: The bringing back of an article or building as nearly as possible to its original state.

restriction: A device in a deed for controlling the use of land for the benefit of the land.

restriction covenant: A clause in a deed limiting the use of the property conveyed for a certain period of time.

restrike: A resumption of current between contacts during an opening operation after an interval of zero current of ¼ cycle at normal frequency or longer.

retaining wall: A wall of masonry or other substance to prevent the sliding of earth or other material. See Fig. R-35 on page 252.

retard: To hinder, delay, to prevent from acting.

return: The end railings of a fire escape balcony.

return bend: A pipe bend or a fitting shaped like the letter U.

return grille: A vented frame, attached to ductwork, in which return air, in an airconditioning or hot-air heating system, is taken in for return to the cooling or heating apparatus for recirculation. Return grilles are normally placed at opposite levels from outlets. See *Register.*

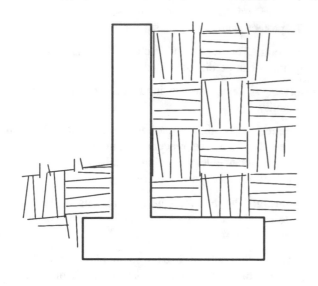

Figure R-35: Details of a retaining wall.

Return nosing: In stair construction the mitered, over-hanging end of a tread, outside the balusters.

reveal: Space between window or door frame and the outside edge of the wall. The vertical side of an opening in a wall, especially that portion of the side of a door or window between the line where the window frame or doorframe stops and the outer edge of the opening.

reverse lay: Reversing the direction of lay about every five feet during cabling of aerial cable to facilitate field connections.

reversible process: Can be reversed and leaves no change in system or surroundings.

reversion: The residue of an estate left to the grantor, to commence after the determination of some particular estate granted out by him.

revolution: The act of revolving.

revolution counter: Also called "speed indicator." It is a device for counting the revolutions of a shaft. By pressing a pointer against the end of a shaft, the revolutions are registered on a dial.

revolutions per minute: An expression to rate of speed of machines; abbreaviated rpm.

revolving door: A door with four vanes operating in a curved frame and mounted on a central vertical axis about which it revolves. Commonly used in commercial buildings to permit easier access by a greater number of people and to retain an even temperature within the store. See Fig. R-36.

revolving field: When the field coils and poles in an electric motor revolve instead of remaining stationary.

RF: Radio frequency 10kGz to GHz.

RFI: Radio frequency interference.

rheology: The science of the flow and deformation of matter.

rheostat: A variable resistor, which can be varied while energized, normally one used in a power circuit.

rib: A skeleton arch that is part of the frame work that supports a vault.

ribbon strip: A board attached to studding to carry floor joists.

ridge: The top of the roof where two slopes meet.

ridge capping: The covering which runs along the ridge of a roof.

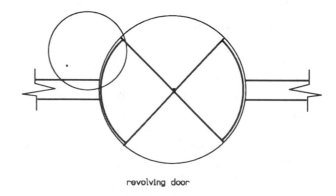

revolving door

Figure R-36: Plan view of a revolving door.

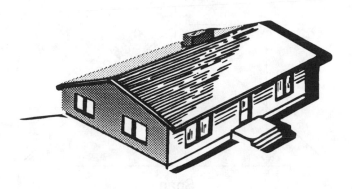

Figure R-37: Typical ridge roof. Also called gable roof.

ridge course: Final course of shingles or tiles that is installed during a roofing job; is applied to the roof ridge.

ridge pole: The highest horizontal timber in a roof to which the rafters are fastened. Sometimes called *ridgeboard*.

ridge roof: A roof whose rafters meet in an apex; its end view is that of a gable. See Fig. R-37.

ridge tiles: Those roofing tiles used to cap the ridge of a roof.

ridge ventilator: An attic ventilating device that runs continuously along the roof ridge. See Fig. R-38.

Ridge vent

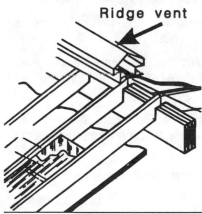

Figure R-38: Ridge ventilator.

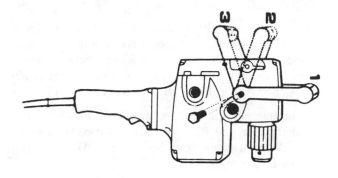

Figure R-39: Right-angle drills are handy when working in tight places.

riffler: A small rasp or file, usually curved, used for filing surfaces or for enlarging holes.

right angle: An angle of 90° formed by one straight line standing perpendicular to another.

right-angle drill: An electric drill that is geared to bore at a right angle to the housing containing the drill motor. This tool is very handy for use where tight working conditions exist. See Fig. R-39.

right line: A straight line; the shortest distance between two points.

right of way: An easement over another's land — also used to describe strip of land used as a roadbed by a railroad or other public utility for a public purpose.

right triangle: A plane object with three adjoining sides and containing one 90° angle and two 45° angles. See Fig. R-40.

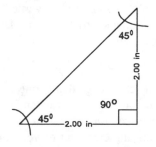

Figure R-40: Right triangle.

rigidity: Stiffness, resistance to change of form.

rigid metallic conduit: A metal enclosed pipe designed expressly for holding electrical conductors. The use of rigid metal conduit is permitted under nearly all atmospheric conditions and occupancies subject to see conditions. See Article 336 of the National Electrical Code.

rigid nonmetallic conduit: A pipe made of nonmetallic materials, such as PVC (plastic), and used for containing electrical conductors. See Article 347 of the National Electrical Code.

ring: In mathematics, a plane figure included between two circumferences having the same center.

ring-out: 1) A circular section of insulation or jacket. 2) The continuity testing of a conductor.

ring shake: A separation of the wood between the annual rings.

ripping: Sawing wood in the direction of, or parallel to the grain.

ripping bar: See *wrecking bar*.

riparian: Pertaining to the banks of a river, stream, waterway, and so forth.

riparian owner: One who owns lands bounding upon a river or water course.

ripple: The ac component from a dc power supply arising from sources within the power supply.

ripsaw: Hand saw with teeth designed for cutting along the length of a board (parallel to the wood grain).

rise: 1) The vertical distance between the springing of an arch and the center or highest point of the intrados. 2) The height of a stair step. 3) The vertical distance from the top of a building's structure wall plate to the upper end of the roof ridge. See Fig. R-44.

rise and run: Term used to indicate the degree of incline.

riser: 1) A board set on edge that connects the treads of a stair; upright member of a stair extending from trend to tread. 2) A vertical run of conductors in conduit or busway, for carrying electrical power from one level to another in a building. 3) A vertical run of pipes — either water, waste, or vent — in a plumbing system.

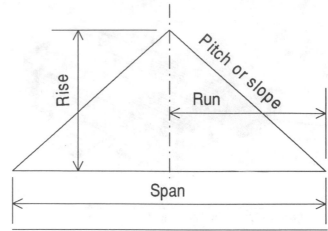

Figure R-44: Rise, run, span, and pitch of a roof.

riser diagram: Electrical, plumbing, or similar drawing that provides an instant overview of the service and related components. A riser diagram of an electric service is shown in Fig. R-45,

riser valve: Device used to manually control flow of refrigerant in vertical piping.

rivet: A short, metal, bolt-like fastening, without threads, that is clinched by hammering. Rivets are designated by the shape of the beads as flat or panhead, buttonhead, countersink, mushroom, soap, or swollen neck. See Fig. 46.

riveting: The heading over or clinching of rivets. Small rivets are headed when cold; large ones when hot.

rivet set: A steel punch with a hollow or cupped face, used in the setting of rivets.

rivet symbols: Symbols that are used on working drawings to designate the type of rivets and how they are to be installed. See Fig. 47.

RMS (Root-mean-square): The square root of the average of the square of the function taken throughout the period.

rock crystal: Colorless, transparent quartz.

rock duster: A machine to distribute rock dust over coal to prevent dust explosions.

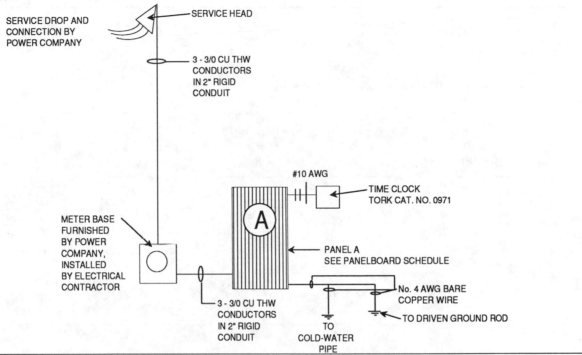

Figure R-45: Typical riser diagram for an electric service.

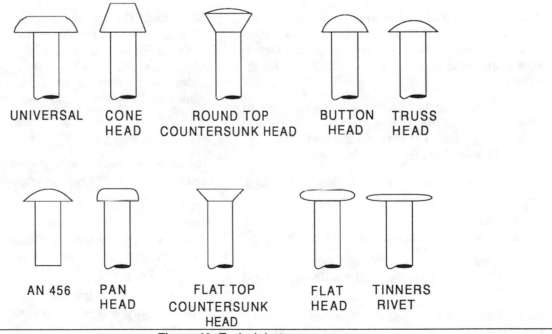

Figure 46: Typical rivets.

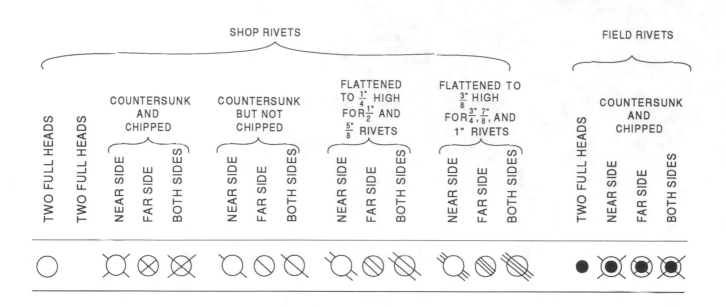

Figure 46: Rivet drawing symbols.

Rockwell test: A method for testing the hardness of metal. A hardened steel ball or a diamond cone is pressed into the item to be tested and the depth of the impression is measured with a dial micrometer or other instrument.

rod: 1) A measuring device, consisting of a piece of wood, used for determining the exact height of risers in a flight of stairs. 2) The shape of solidified metal convenient for wire drawing, usually $\frac{5}{16}$ inch or larger.

rodman: One who uses or carries a surveyor's leveling rod.

roller bearing: A bearing made of hardened-steel rollers instead of the round steel balls used in ball bearings.

roller: A piece of heavy powered equipment designed to compact and level the earth or asphalt during road and site work.

rolling: Reducing the cross-sectional area of metal stock or otherwise shaping metal products using rotating rolls.

roll roofing: This roofing material is used for low roof pitches of 1 to 12 and up to 4 to 12. It is manufactured in rolls 36 inches wide and 36 to 48 feet long. Of the 36-inch width, 17 inches — the exposed part — is usually covered with a colored mineral material. The remaining 19-inch wide plain surface is overlapped by the next roll.

ROM: Read only memory.

Roman brick: Brick that is thinner and longer than the common brick. The Roman brick is usually $1\frac{1}{2}'' \times 4'' \times 12''$ in dimensions.

Romanesque: The architectural style that grew out of the Roman era. It preceded the Gothic. It was used chiefly in ecclesiastical buildings between the seventh and twelfth centuries.

Romex: General Cable's trade name for type NM cable; but it is used generically by electrical workers to refer to any nonmetallic sheathed cable. Fig. R-47.

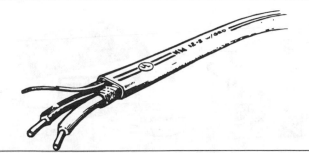

Figure R-47: Nonmetallic sheathed cable, often called *Romex.*

roof: Top exterior covering of a building — used to help make structures waterproof and weather tight. In general, the roof consists of a supporting structure composed of rafters and/or trusses, some type of sheathing on top of the structure, and a roof covering.

roof boards or roofers: The sheathing or undercovering of a roof which serves as a foundation for shingles, slate, or other roofing material.

roof deck: Roof sheathing attached over rafters or trusses. See Fig. R-48.

roof framing: The supporting structure of a roof.

roofing nails: Galvanized nails with extra-wide heads for attaching asphalt and shingles made of similar materials. See *Shingle nail.*

roof sheathing: See *roof deck.*

roof truss: Timbers or structural steel fastened together for the support of a roof.

roof types: Building roofs have been given various names based on their design. See Fig. R-49.

Rosendale cement: The name given to a natural cement made from rock found near Rosendale, New York.

rope: Twisted hemp fibers that are made into a strong, flexible cord. Nylon and other synthetic materials are also used for strong ropes.

rope knots: Various ways of tying a rope for various applications. See *knots.*

rosewood: A dark colored wood, heavy, hard, and brittle, used as a veneer.

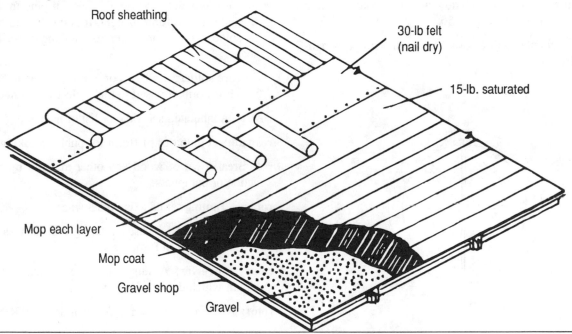

Figure 48. Roof sheathing details.

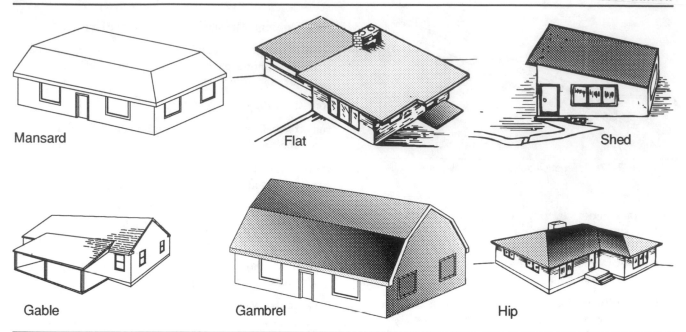

Figure R-49: Popular types of roofs.

rose window: A circular window with divisions radiating from the center. See Fig. R-50.

rosette: A circular unit of ornamentation with parts radiating from the center.

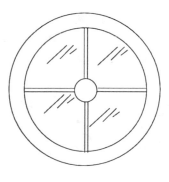

Figure R-50: Rose widow.

rosin: The product of several species of pine from which it exudes in the form of gum. It is much used as a flux in soldering tinwork, and for making varnishes and soaps.

resin-core solder: Solder containing a rosin flux and which is normally used to make electrical connections.

roster: A tabulated schedule or list of names.

rostrum: A pulpit or platform for public speaking.

rot: Breakdown of wood and other material by bacterial action or dampness.

rotary: Turning on its axis, like a wheel.

rotary blower: An incased rotating fan such as is used for forced draft in furnaces.

rotary converter: A single machine connected to an ac circuit which delivers dc or vice versa.

rotor: Rotating part of a mechanism. See Fig. R-51.

rottenstone: Decomposed limestone marketed in the form of a fine powder and used in the polishing of varnished surfaces.

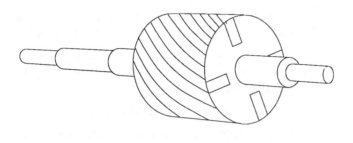

Figure R-51: Motor rotor.

Figure R-52: Router. *Courtesy Sears.*

rotunda: The circular room under a dome.

roughcast: Stucco thrown against a wall to form a rough finish.

rough inspection: The first inspection made of an electrical installation after the conductors, boxes and other equipment have been installed in a building under construction.

roughing in: The first stage of an electrical installation, when the raceway, cable, wires, boxes and other equipment are installed; that electrical work which must be done before any finishing or cover-up phases of building construction can be undertaken.

rough lumber: Undress lumber as it comes from the mill. Standard dimensions are usually maintained; that is, a 2″ × 4″ board usually measured that dimension, whereas a finished board will be somewhat less.

roundel: A term used to refer to any circular ornament, such as a medallion, patera, rosette, etc.

round off: To delete the least significant digits of a numeral and adjust the remaining by given rules.

rout: To cut or gouge out with a router.

router: A two-handled tool for smoothing the face of depressed surfaces in woodwork. See Fig. R-52.

router bits: Cutting bits used in a router, jointer, or other power tools to cut various patterns in wood.

rowlock: Pattern of brickwork consisting of a course of brick laid on edge with ends exposed. See Fig. R-53.

rowlock-back wall: A wall made with the bricks of the exterior face laid flat, and the bricks of the backing laid on edge.

RPM: Revolutions per minute.

rubber, chlorosulfonated polyethylene (CP): A synthetic rubber insulation and jacket compound developed by DuPont as Hypalon®.

rubber, ethylene propylene: A synthetic rubber insulation having excellent electrical properties.

rubble: Roughly broken quarry stone.

Figure R-53: Rowlock brick pattern.

rule: Measuring tool graduated in various increments such as inches and fractions thereof; millimeters, or other elements.

rule joint: A type of knuckle joint with projecting shoulders that abut when the joined pieces are fully opened, as in the ordinary six-foot folding rule.

run: The run of roof is the shortest horizontal distance measured from a plumb line through the center of the ridge to the outer edge of the plate. In equally pitched roofs, the run is always equal to half of the span or generally half the width of the building.

rung: Rounded cross strip as on a ladder or chair.

run of stairs: Horizontal part of a step without nosing; horizontal distance of face of one riser to another.

run of work: A number of jobs following one another in steady succession.

running board: 1) A device to permit stringing more than one conductor simultaneously. 2) Board used as protection for Romex cable in a residential attic or basement.

running bond: Pattern of brickwork consisting of a course of brick laid on edge with ends exposed.

running fit: Refers to the fitting together of parts with just sufficient clearance to permit freedom of motion.

runs: In painting, runs are caused by applying too much material in one spot, or using too much thinner, resulting in sags or curtains.

rupture stress: The unit stress at the time of failure.

rush: The stems of a marsh-growing plant, used for chair seats since early times.

rust: Hydrated ferric oxide. $2Fe_2O_3 3H_2O$.

rusticated work: Squared stones with edges beveled or grooved to make the joints stand out.

rust joint: A joint in which some oxidizing agent is employed; either to cure a leak in plumbing pipes or to help the joint withstand higher pressures.

R value: Means used to measure the insulating value of materials. The higher the R value, the greater the insulating value.

S

sabre saw: Portable electric saw using a thin blade that cuts by a reciprocating (up-and-down) motion. See Fig. S-1.

sabre-saw blade: Blade used in sabre saw. Both wood-cutting and metal-cutting varieties are available. To handle a wide range of materials, the proper selection of blades is required. See chart on page 262.

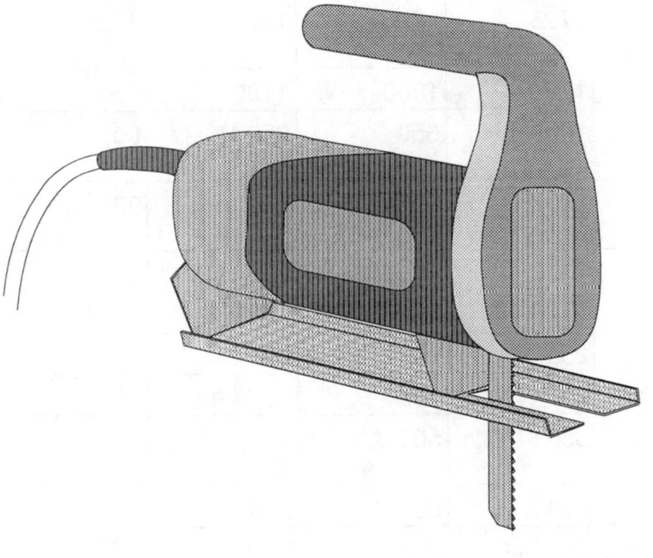

sacrificial protection: Prevention of corrosion by coupling a metal to an electrochemically more active metal which is sacrificed.

saddle: Short horizontal member set on top of a post to spread the load of the girder over it; piece of wood, stone, or metal placed under a door. See Fig. S-2.

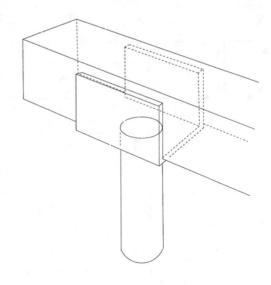

Figure S-1: Sabre saw. The blade cuts on the up stroke.

Figure S-2: A saddle used on top of a post to spread the load of the girder over it.

Guide for Selecting Sabre Saw Blades

Material	Saw Pitch Work Thickness				Strokes per Minute Work Thickness			
	1/8"	1/4"	1"	More	1/8"	1/4"	1"	More
Carbon Steel	24	14	10	6	500	500	250	150
Free Machining	18	14	10	6	500	500	250	150
Chrome Vanadium	18	14	12	8	250	175	100	90
Silicon Manganese	18	14	10	6	250	100	50	
Armor Plate	18	12	10	8	175	90		
High Speed	21	14	10	8	100			
Stainless Steel	24	14	10	8	90			
Thin-Wall Tubing	32	14			250	100		
Rolled Shapes	24	14	10	8	175	100	90	
Cast Iron	18	14	10	6	500	250	100	90
Aluminum	18	10	6	6	1725	1725	1300	650
Babbit-Lead	14	8	6	6	1725	1725	1725	1725
Soft Brass	18	14	8	6	1725	1300	1100	650
Phosphor Bronze	18	14	10	6	1100	650	250	90
Tobin Bronze	18	12	10	6	650	500	250	90
Copper	18	10	6	6	1300	1100	650	500
Magnesium	14	10	6	4	1725	1725	1725	1725
Manel	18	12	10	6	250	200	90	90
Nickel	14	8	6	6	200	100	90	90
Nickel Silver	18	14	10	6	650	500	250	90
Silver	24	18	14	8	650	500	250	90

Guide for Selecting Sabre Saw Blades
(Continued)

Material	Saw Pitch Work Thickness				Strokes per Minute Work Thickness			
	1/8"	1/4"	1"	More	1/8"	1/4"	1"	More
Asbestos	24	14	6	4	1725	1725	1725	1350
Bakelite	10	8	6	4	1725	1725	1725	1725
Wall Board	14	10	6	4	1725	1300	1100	650
Cork	10	8	6	6	1725	1725	1725	1300
Fiber	14	10	6	6	650	500	250	100
Pressed Wood	24	18	6	4	1725	1725	1300	650
Plastics	14	10	6	4	1725	1725	1725	1725
Porcelain	24	18		500	100			
Hard Rubber	32	18	14	8	1300	100	90	90
Transite	32	18	14	8	1300	1100	90	90
Paper	10	8	6	4	1725	1300	1100	675
Hard Wood	18	10	6	4	1725	1725	1725	1725
Soft Wood	14	6	4	3	1725	1725	1300	1100

saddle valve: Valve body shaped so it may be silver brazed to a refrigerant tubing surface.

SAE steels: Numerical system used to designate various types of steels. The first figure of the number indicates the general class to which the steel belongs; that is, carbon steel, nickel steel, nickel chromium, etc. The second figure usually indicates (in cases of alloys) the percentages of the main alloying element. The last two or three figures indicate the carbon content in hundredths of 1 percent. Therefore, specification 2345 calls for a nickel steel having 3 percent carbon.

SAE: Society of Automotive Engineers.

safe carrying capacity: Anything so constructed as to carry a certain weight without a breakdown.

safe edge: The uncut edge of a file that makes possible the protection of an adjacent surface when operating in a groove or corner. See Fig. S-3.

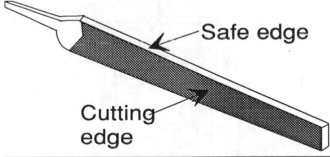

Figure S-3: File with one cutting edge and with remaining edges being "safe".

safety color codes: Various colors designated to symbolize certain safety conditions. The most common ones follow:

- Red. Fire equipment; flammable-liquid container; stop button on machinery or electrical equipment.

- Yellow. Caution; power source; waste container.

- Orange. Exposed cutting or moving parts; danger area on a machine; starter button.

- Violet. Radiation hazard.

- Green. Safety; first-aid equipment.

safety conductor: A safety sling used during overhead line construction.

safety control: A device that will stop the refrigerating unit if unsafe pressures and/or temperatures are reached.

safety factor: The ratio of the maximum stress which something can withstand to the estimated stress which it can withstand.

safety motor control: Electrical device used to open a circuit if the temperature, pressure, and/or the current flow exceed safe conditions.

safety plug: Device that will release the contents of a container above normal pressure conditions and before rupture pressures are reached.

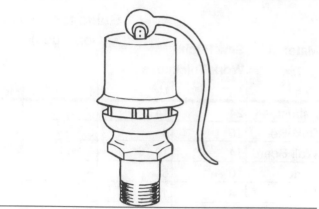

Figure S-5: Saftey value used on hot-water heater.

safety switch: A knife-blade switch enclosed in a metal box and operated externally. See Fig. S-4.

safety valve: A safety device for allowing steam or water to escape from a boiler when the pressure may become dangerous. Usually adjusted to permit not more than 5 pounds of pressure above maximum allowable working pressure of the boiler. See Fig. S-5.

sag: To dip, bend, or cause to bend downward; a depression, especially in the middle. A departure from original shape by its own weight, as the sag of a door.

sag, apparent: Sag between two points at 60°F and no wind.

sag, final: Sag under specified conditions after the conductor has been externally loaded, then the load removed.

sag, initial: Sag prior to external loading.

sag, maximum: Sag at midpoint between two supports.

sag section: Conductor between two snubs.

sag snub: Where a conductor is held fixed and the other end moved to adjust sag.

sag, total: Under ice loading.

salamander: Portable kerosene heater with high heat output in comparison to size. Used on building sites where heat is needed for workers or to keep materials from freezing.

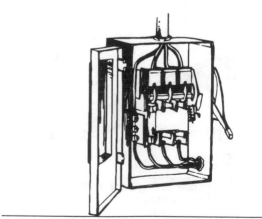

Figure S-4: Electrical safety switch.

salon: A large and magnificent room; usually a room for holding receptions, or displaying exhibits, as distinguished from an assembly room.

saltbox: Building in the early years of the English colonies reflected the traditions of humble rural English buildings. The New England house of the seventeenth century is characterized by a natural use of materials in a straight forward manner. The box-like appearance is relieved by a prominent chimney, a sparse distribution of small casement-type windows. The one-room house often was expanded by adding a room against the chimney end, forming a large house with a centrally-located chimney. The well know "saltbox" shape house also provided rooms by extending the rear roof slope. Other useable space was made by placing windows in the gable end forming a half story. In larger houses the upper floor projected beyond the lower floors creating an overhang known as a jetty.

sampling: A small quantity taken as a sample for inspection or analysis.

sand: Small grain of mineral, largely quartz, that is the result of disintegration of rock.

sander: Various types of power tools used to finish woodwork and wood floors. See Fig. S-6.

sand finish: Refers to sandpaper finish on wood surfaces and to the sand-blast method of renovating the exterior of brick or stone buildings.

sandalwood: A heavy close-grained wood, of fragrant odor, native to the East Indies.

sandblasting: The driving of sand against an object or surface by air pressure. Used for renovating the exterior of stone and brick buildings.

sanding: The finishing of a surface on a sanding machine or by hand.

sandpaper: Paper coated with sharp sand; used as an abrasive, particularly for finishing surfaces of woodwork. Also called "garnet paper."

sandstone: Composed of grains of fine sand cemented together by silica, oxide or iron, or carbonate of lime; used as a building stone.

Figure S-6: Electric belt sander. *Courtesy Sears.*

sanitary: Pertaining to, or tending to promote health.

sanitation: The neutralization or removal of conditions injurious to health. Purification of water supply; disposal of sewage, etc.

sanitary engineer: One who supervises the planning and construction of water supplies, sewage systems, etc.

sanitary sewer: Underground pipe or tunnel for carrying off domestic sanitary wastes.

sap: The juice of plants that is necessary to growth.

sapwood: The new wood next to the bark of a tree.

sash balance: A device — dispensing with weights, pulleys, or cord — operated with a spring to counterbalance a double-hung window sash.

sash: A frame in which window glass is set or retained. See Fig. S-7.

sash bars: The strips in a sash which separate the narrow panes of glass.

sash chain: The chain used to carry the weights in double hung window sash.

sash lock: Metal slip-lock used to secure double-hung windows. See Fig. S-8.

sash pulley: The small pulley set in a window frame, over which the sash cord or chain runs.

S beam: A standard (S) steel beam designation that has recently replaced the I beam.

scab: 1) Supporting members of wood structures. 2) Nickname given to workers who are nonmembers of organized labor groups.

scabble: To dress off rough stones for rubble work.

scaffold: A temporary structure for the support of workers and materials. A folding scaffold is shown in Fig. S-10.

scaffold height: The distance between various stages of scaffolding, usually about 4 or 5 feet, representing the height within which a bricklayer can carry on his work efficiently.

scaffold nails: Double-headed nails for use in building temporary wood structures like scaffolding and concrete forms. The extra head facilitates removing the nails after the structure has served its purpose. See Fig. S-11.

scagliola: An imitation of colored marble obtained in plastering, used for floors, columns, and other ornamental interior work.

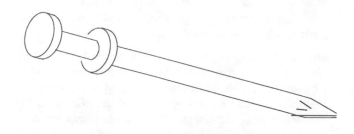

Figure S-11: Double-headed scaffold nail.

Scalar: A quantity (as mass or time) that has a magnitude described by a real number and no direction.

scale: 1) A measuring device graduated into divisions. 2) The size of a drawing in relation to the size of the object represented. 3) In architecture or building, drawings are usually made to scales of either ⅛ inch or ¼ inch equals a foot. See *architect's scale*.

scaled drawing: A drawing made smaller than the work that it represents, but to a definite proportion that should be specified on the drawing.

scalene: A triangle in which no two sides are equal; also a cone or cylinder in which the axis is inclined to the base.

scaling: Drawing to scale, or measuring a construction working drawing with an appropriate scale — like an architect's scale — to determine building dimensions. See Fig. S-12.

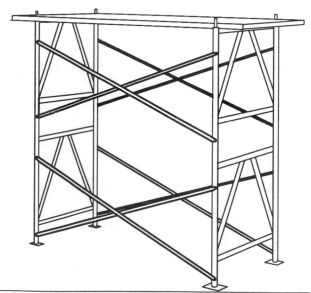

Figure S-10: Folding scaffold.

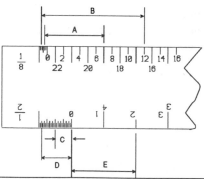

Figure 12: Typical architect's scale.

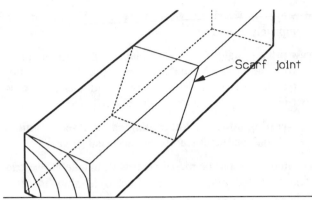

Figure S-13: Typical scarf joint.

GRILLE SCHEDULE

MARK	MFGR.	MODEL	SIZE	CFM	REMARKS
1	Krueger	SHI 0	9x9	146	OBD Alum.
2		SH3	6x6	50	
3				60	
4				75	
5				80	
6				100	
7			9x9	144	
8				152	
9				170	
10				175	
11				181	
12				226	
13		SH4	6x6	65	

Figure S-14: Sample grille and diffuser schedule used on HVAC working drawing.

scamillus: The small groove that separates the necking of the Greek Doric column from the shaft.

scan: To examine sequentially, part by part.

scantling: Small timber as 2 by 3, 2 by 4, etc., used for studding.

scarf joint: A joint made by notching and lapping the ends of two timbers, fastening them together with bolts, nails, straps, or glue. See Fig. S-13.

scarify: To roughen up, as a road, for repairs.

scavenger pump: Mechanism used to remove fluid from sump or containers.

schedule: Systematic method of presenting notes or lists of materials or equipment on a working drawing in tabular form. See Fig. S-14.

scientific: According to exact and accurate rules; systematic.

scintillation: The optical photons emitted as a result of the incidence of ionization radiation.

scissors truss: A type of roof truss, so named from its resemblance to a pair of scissors. It has been frequently used for supporting roofs over halls and churches.

sconce: An ornamental bracket for holding candles or a lantern. Wall sconces are also available with electric lamps.

scope: Slang for oscilloscope.

score: A cut, scratch, or note cut in any type of building material such as plaster, concrete, or wood.

scoring: Scratches made in the scratch or brown coat of plaster to give better adhesion to the finish coat.

scotia: A concave molding often found in the base of a column. See Fig. S-15.

scram (nuclear): The rapid shutdown of a nuclear reactor.

SCR brick: A type of large brick normally used in 6-inch masonry walls.

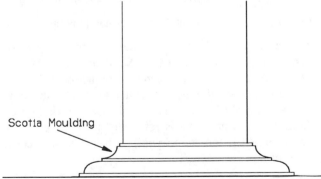

Figure S-15: Scotia shown at the base of a column.

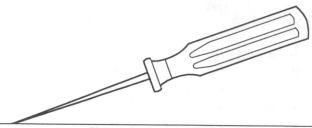

Figure S-16: Scratch awl.

scraper: A flat plate of steel used by woodworkers to smooth wood surfaces.

scratch awl: A sharp-pointed piece of steel used for marking on metal or wood. See Fig. 16.

scratch brush: A wire brush used for removing rust, dirt, and foreign matter from metal surfaces.

scratch coat: The first coat in masonry or plastering that is scratched to give a better hold for succeeding coats.

screeds: Strips of plaster about 8 inches wide and the depth of the first two coats, used as thickness guides in applying the remaining plaster. A metal depth guide that is used when stuccoing surfaces.

screed finish: Rough finish on a concrete slab that is made with a straightedge.

screeding: Leveling of a concrete slab with a straight-edge.

screen pack: A metal screen used for straining.

screw: A threaded fastener used in securing wood joints. See *lag screw, machine screw,* and *sheet metal screw.*

screw clamp: Same as hand screw or hand clamp.

screw extractor: Self-threading device used to remove broken screws, bolts, and taps. Sometimes called "easy out." In use, the broken fastener must first be drilled for a short distance through its center to allow the screw extractor to allow the counterclockwise threads to grab hold. Once gripped, the screw extractor must continue to be turned (counterclockwise) to back out the screw, bolt, or tap.

screw eye: A wood screw with the head formed into a completely closed ring or circle.

screw pitch gauge: A gauge usually made up of many leaves, the edge of each being cut to a thread of indicated size. Used for determining the number of threads per inch on a given screw, bolt, or nut.

screw-cutting lathe: A lathe adapted to thread cutting, being equipped with lead screw and change gears.

screwdriver: A bar or rod of steel with handle at one end and flattened blade at the other to fit slots in screwheads.

scribe awl or scriber: A pointed steel instrument for making fine lines on wood or metal for layout work. See *scratch awl.* See Fig. S-17.

scribing: Using a scriber or compass to transfer one irregular shape on one object to another.

scroll saw: A thin-bladed saw for cutting curved designs. A jig saw is a power-driven scroll saw. Also see *sabre saw.* See Fig. S-18.

scupper: Outlet at the end(s) of flat roofs to allow water runoff.

scutch: A tool resembling a pick on a small scale, with flat cutting edges, for trimming bricks for particular uses.

Escutcheon: A metal plate, as around a keyhole.

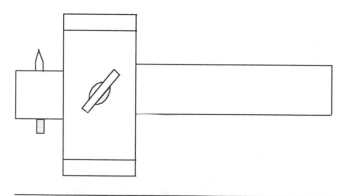

Figure S-17: Typical scriber.

scuttle: Access openings to either attic or roof.

SE: Service entrance.

sealed motor compressor: A mechanical compressor consisting of a compressor and a motor, both of which are en-closed in the same sealed housing, with no external shaft or shaft seals, and with the motor operating in the refrigerant atmosphere.

sealed motor compressor: A mechanical compressor consisting of a compressor and a motor, both of which are enclosed in the same sealed housing, with no external shaft or shaft seals, and with the motor operating in the refrigerant atmosphere.

sealed: Preventing entrance.

sealer: Compounds applied to wood, metal, or masonry surfaces to fill pores to help prevent moisture from entering.

sealing compound: The material poured into an electrical fitting to seal and minimize the passage of vapors.

seam: The joining of two edges. On sheet metal joints, the edge of one piece is folded or turned over the edge of another. See Fig. S-18.

seaming iron: Also called *grooving tool*. Used for setting seams in sheet metal work.

seasoning of lumber: By the kiln-drying process, boards are placed in a drying room or kiln to hasten the seasoning. By the air-drying process, lumber is allowed to dry naturally under sheds.

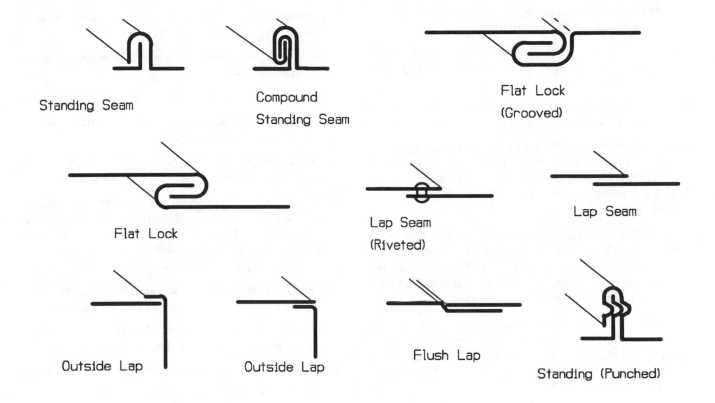

Figure S-18: Various types of seams used in sheet metal work.

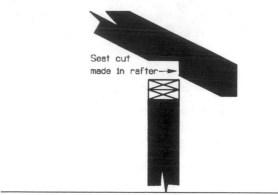

Seat cut
made in rafter→

Figure S-19: Seat cut made in roof rafter.

seat: A stationary part in a valve that provides a leakproof contact.

seat angle: Small steel angle riveted to one member to support the end of a beam or girder.

seat cut: Cut made in rafters to fit on top of the wall plate and to keep the rafter at the correct angle or slope. See Fig. S-19.

secant: The secant of the angle is the quotient of the hypotenuse divided by the adjacent side.

secondary: The second circuit of a device or equipment which is not normally connected to the supply circuit. In a step-down transformer, the high-voltage side is called *primary*, while the low-voltage side is called *secondary*.

section: 1) A drawing representing the internal parts of an object as if it had been cut straight through vertically or horizontally. Also partial sections may be taken through particular parts. A section of land established by government survey and containing 640 acres.

sectional: 1) Of, or pertaining to, a section or district; local; consisting of sections; divisible into sections. 2) A drawing that shows some portion cut away, in order that the drawing may be more easily understood, is called a "sectional view."

sector: (1) A part of a circle bounded by two radii and the arc subtended by them. (2) Any mechanical part of similar shape.

sediment: The matter which settles at the bottom of a liquid.

sedimentary rock: Rock formed under water by pressure or by cementation.

Seebeck effect: The generation of a voltage by a temperature difference between the junctions in a circuit composed of two homogeneous electrical conductors of dissimilar composition; or in a nonhomogeneous conductor the voltage produced by a temperature gradient in a nonhomogeneous region.

segment: The part of a circle included between the chord and its arc.

segmental rack or segmental wheel: An arc or portion of a gear wheel, used for imparting reversible motion to a spindle.

segregation: The separating of parts as from a main body and bringing these parts together as a unit.

seismograph: An instrument which automatically records an earthquake shock.

seizin: Possession of real estate by one entitled thereto.

self-acting: Operating automatically; performing an operation without outside assistance.

self-centering punch; A punch designed to mark the center of an object for drilling.

self-excitation: Direct current obtained from the brushes of a dc generator to provide current for its electromagnetic field. In alternators the term refers to a dc generator built on the alternator shaft. Used to provide direct current for the alternator field.

self-excited alternator: An ac generator that also produces a direct current for magnetizing its main fields.

self-feed bit: Large circular bit with replaceable cutters for cutting holes for pipe and other circular objects.

self inductance: Magnetic field induced in the conductor carrying the current.

self-tapping screw: See *sheet metal screw.*

semicircle: A half circle, bounded by the circumference line and diameter.

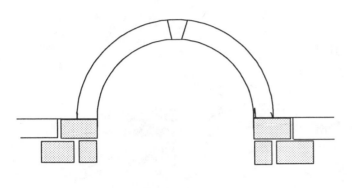

Figure S-20: Semicircular arch.

semicircular arch: An arch whose intrados is a half circle. See Fig. S-20.

semi-transparent: Translucent, allowing the passage of light, but not permitting a clear view of any object.

semiconductor: A material that has electrical properties of current flow between a conductor and an insulator.

sensible heat: Heat that causes a change in temperature of a substance; measured with a thermometer; as opposed to latent heat.

sensor: A material or device that goes through a physical change or an electronic characteristic change as conditions change.

separable insulated connector: An insulated device to facilitate power cable connections and separations.

separate property: Property owned by a husband or wife which is not community property; acquired by either spouse prior to marriage or by gift or devise after marriage.

separately excited: A machine that gets the current needed for the excitation of its field from some outside source.

separator: 1) Material used to maintain physical spacing between elements in cables, such as a layer of tape to prevent jacket sticking to individual conductors. 2) Sections of steel pipe forming spacers between I beams bolted together serving as a structural unit.

sepia: Dark brown drawing paper with reddish tint frequently used by architects.

septic tank: A plumbing unit used for decomposing solid sewage matter. It is designed to dispose of these wastes in a completely sanitary and odorless manner by natural bacterial action which dissolves most of the solids into liquids and gases. See Fig. S-21.

septic tank system: Private sewage disposal section for an individual home.

sequence controls: Devices that act in series or in time order.

sequence: In regular order, or systematic arrangement.

series: When two or more electrical components are connected so that the current feeding one must pass through the others, they are said to be connected in series (one after another, as in a string). The following four rules state the condition that exists in a series circuit.

- The current is the same in all parts of a series circuit.

- The sum of the voltage crops across all the resistors in a series resistive circuit is equal to the applied (source) voltage.

- The total resistance in a series circuit is equal to the sum of the individual resistances.

- The total voltage applied to a series circuit divides between the resistors in direct proportion to their resistance.

series parallel circuit: A circuit made up of two or more simple parallel circuits all joined in series.

series-wound motor: A dc motor with the armature and field connected in series. Used on machines where variable loads occur. The speed varies with the load.

serrated: Having a notched edge or saw-like teeth.

serration: Like the toothed edge of a saw.

service: The conductors and equipment used for delivering energy from the electrical supply system to the wiring system of the premises served. See Fig. S-22.

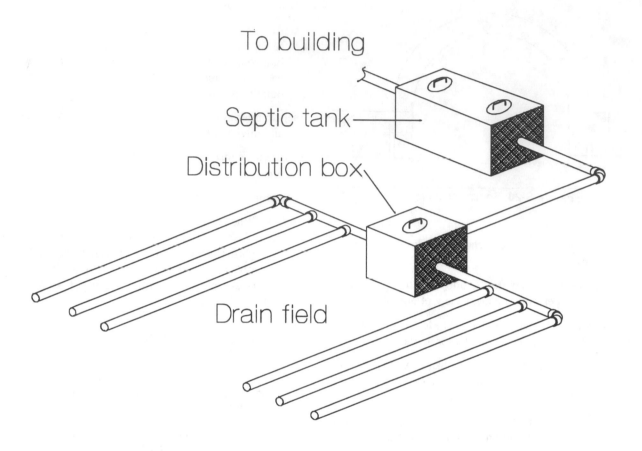

Figure S-21: Septic tank and drain fields.

service cable: The service conductors made up in the form of a cable.

service conductors: The supply conductors that extend from the street main or transformers to the service equipment of the premises being supplied.

service drop: Run of cables from the power company's aerial power lines to the point of connection to a customer's premises.

service ell: An elbow having an outside thread on one end.

service entrance: The point at which power is supplied to a building, including the equipment used for this purpose (service main switch or panel or switchboard, metering devices, overcurrent protective devices, conductors for connecting to the power company's conductors and raceways for such conductors).

service equipment: The necessary equipment, usually consisting of a circuit breaker or switch and fuses and their accessories, located near the point of entrance of supply conductors to a building and intended to constitute the main control and cutoff means for the supply to the building.

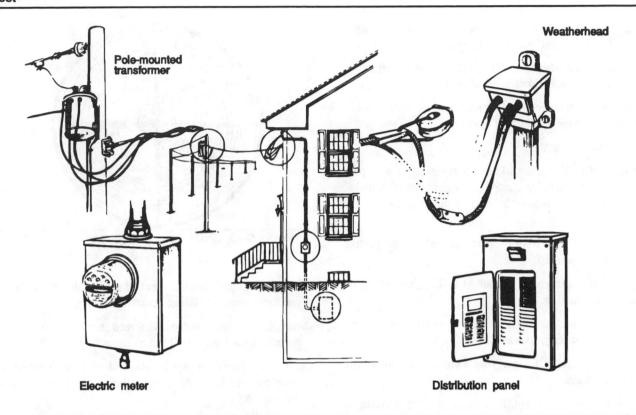

Figure S-22: Parts of a typical electrical service.

service lateral: The underground service conductors between the street main, including any risers at a pole or other structure or from transformers, and the first point of connection to the service-entrance conductors in a terminal box, meter, or other enclosure with adequate space, inside or outside the building wall. Where there is no terminal box, meter, or other enclosure with adequate space, the point of connection is the entrance point of the service conductors into the building.

service mast: A pipe or conduit projecting through the roof of a building. The top end contains a weatherhead to accept the service conductors, while the lower end usually connects to a meter base or CT (current transformer) cabinet.

service panel: See *panelboard*.

service pipe: The small pipe which conveys liquid or gas from a main pipe to its places of use.

service raceway: The rigid metal conduit, electrical metallic tubing, or other raceway that encloses the service entrance conductors.

service tee: A tee having inside thread on one end and on the branch but outside threads on the other end of the run.

service valve: A device, attached to a refrigeration system, that provides an opening for gauges and/or charging lines.

serving: A layer of helically applied material.

servomechanism: A feedback control system in which at least one of the system signals represents mechanical motion.

set: (1) A small tool for sinking nail heads below the surface. (2) To adjust a tool, as to set a plane bit.

setback: The distance from curb or other established line, within which no building may be erected.

setting (of circuit breaker): The value of the current at which the circuit breaker is set to trip.

setting-down machine: A machine used to close down the seams left by the burring machine.

settle: A bench or seat.

settlement: The unequal sinking or lowering of any part of a structure, usually caused by weakness of foundation, skimping of materials used in the structure, or by unseasoned lumber.

severalty ownership: Real property owned by one person only; sole ownership.

severy: A compartment in a vaulted ceiling, especially in Gothic construction.

sewage: Waste material that is carried out of an area or structure via sewer pipes or tunnels.

sewer: Pipe or tunnel for carrying away sewage or storm water for sanitary purposes.

shaded-pole motor: A small ac motor that utilizes a shaded coil; used for light-start loads; has no brushes or commutator.

shade line: The shadowed area of a building wall caused by the roof projection at the eaves.

shaft furnace: A furnace used for pouring wire bars from continuous melting of cathodes.

shake: A split or check in timber which usually causes a separation of the wood between annual rings. See Fig. S-23.

shakes: Handmade shingles. See Fig. S-24.

shall: Mandatory requirement of the National Electrical Code.

shapes: A general term applied to rolled structural metal, as I beams, channels, Z bars, angles, etc.

sharp sand: A clean sand containing coarse angular grains.

shaving: Removing about 0.001 inch of metal surface.

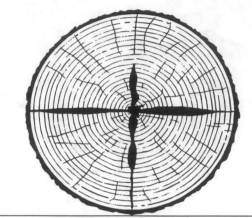

Figure S-23: Shakes or checks in the end of a log.

shear: The lateral displacement in a body due to an external force causing sliding action.

shear plate: Metal plates inserted between structural members for load distribution.

shear wall: Building wall designed to offer protection from adverse elements such as hurricanes, earthquakes, and the like.

shears: A tool with two blades for cutting metals. See Fig. S-25.

sheath: A metallic close fitting protective covering.

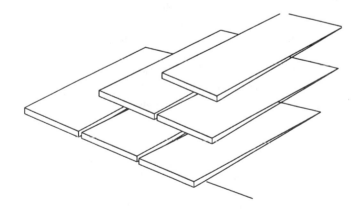

Figure S-24: Typical shakes or handmade shingles.

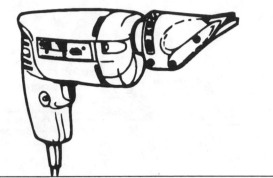

Figure S-25: Power shears.

sheathing: Usually wide boards nailed to studding or roofing rafters, to form a foundation for the outer surface covering of the side walls or roof in a dwelling. See Fig. S-26.

sheathing paper: Insulating paper used over roof and wall sheathing on a structure to limit the air infiltration and moisture penetration.

sheave beams: Steel beams forming the overhead support for an elevator.

shed: A one-story structure open on at least one side, either attached to or detached from a building. See Fig. S-27.

shed roof: A building roof with only one low-angle slope. See Fig. S-27.

sheet metal screw: Threaded, self-feeding fastener used to secure pieces of sheet metal together. See Fig. S-28.

sheet metal worker: Workers who fabricate and install ductwork, vents, and other HVAC components.

sheet rock: Another name for gypsum board. Used as an interior wall sheathing in place of plaster.

sheet steel: Thin sheets of steel used by sheet-metal workers. Thickness is gauged by number. Heavy sheets are called plates.

sheet tin: Thin sheets of iron or steel, coated with tin which serves to prevent corrosion.

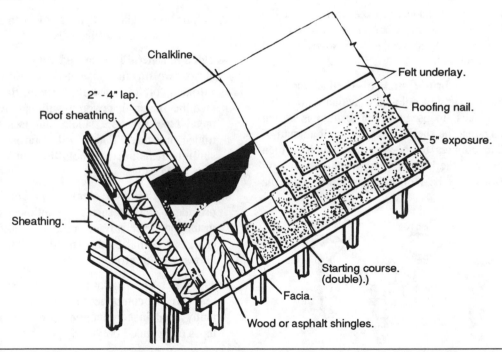

Figure S-26: Components of a roof structure, showing sheathing, sheathing paper, and similar details.

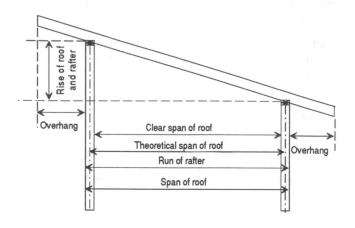

Figure S-27: Typical shed with details of shed roof.

sheet-metal working: Work performed on metal in sheet form.

sheeting: Same as sheathing.

shellac: A flake material made from the secretion of the lac insect. This flake shellac, when cut with alcohol, is known both as shellac and shellac varnish; white and orange are the usual colors.

shellac stick: A hard, brittle stick of shellac used for wood repairs. A variety of colors are available to match most wood finishes. In use, a metal spatula is heated and touched against the shellac stick. A small amount of melted shellac adheres to the spatular which is then quickly wiped onto the hole or crack in the wood. Several applications may be required, depending upon the size of the defect.

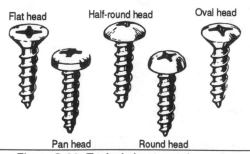

Figure S-28: Typical sheet metal screws.

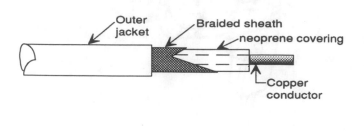

Figure S-29: Braid shield used in coaxial cable.

shellac varnish: A varnish very generally used by pattern-makers. It is made by dissolving flake shellac in alcohol.

shield: The conducting barrier against electromagnetic fields.

shield, braid: A shield of interwoven small wires, used in electronic work. See Fig. S-29.

shielded nonmetallic sheathed cable: A factory assembly of two or more insulated electric conductors in an extruded core of moisture-resistant, flame-resistant nonmetallic material, covered with an overlapping spiral metal tape and wire shield and jacketed with an extruded moisture-, flame-, oil-, corrosion-, fungus-, and sunlight-resistant nonmetallic material. Abbreviated type SNM cable.

shield, insulation: An electrically conducting layer to provide a smooth surface in intimate contact with the insulation outer surface; used to eliminate electrostatic charges external to the shield, and to provide a fixed known path to ground.

shield, tape: The insulation shielding system whose current carrying component is thin metallic tapes, now normally used in conjunction with a conducting layer of tapes or extruded polymer.

shim: Thin piece of material used to bring members to an even or level bearing.

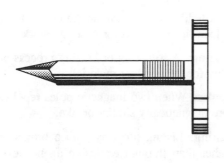

Figure S-30: Shingle nail.

shingle nail: Wide-headed nail used by roofers to secure shingles to building roof. See Fig. S-30.

shingles: Small pieces of wood or other material used for covering roofs and side walls.

ship auger: Spiral-shanked drill bit for boring wood. A feed screw on the tip of the bit draws the bit into the wood, while the outside cutting lip starts the cut.

shiplap: Boards which are rebated on both edges, the two rebates being cut on the opposite side. Shiplap is used as sheathing, siding, and sometimes for flooring.

shoe mold: Quarter-round molding used at the intersection of the baseboard and finished floor.

shore: To support as by a stout timber, usually as a prop.

shore feeder: From ship to shore feeder.

shore hardness: A measure of the hardness of a plastic.

shoring: Timbers braced against a wall as a temporary support. Also the timbering used to prevent a sliding of earth adjoining an excavation.

short-circuit: An often unintended low-resistance path through which current flows around, rather than through, a component or circuit.

short cycling: Refrigerating or water-pump system that starts and stops more frequently than it should.

short length: Refers to lumber less than eight feet in length.

short time (nuclear power): Operation for 2 hours out of a 24-hour period.

short-time overload rating (nuclear): The limiting overload current that one third (must be at least three) of the conductors in an assembly through a penetration can carry, with all other conductors fully loaded.

shoulder nipple: A nipple of any length which has an unthreaded portion of pipe between the two threaded ends.

shoved joint: Mortar joint produced by laying brick in a thick bed or mortar and forming a vertical joint by pushing the brick against he brick already laid in the same course.

shower head: Plumbing fitting used in showers to spray water.

show rafter: A short rafter, often ornamented where it may be seen below the cornice.

shrine: (1) A receptacle of sacred relics. (2) A tomb or chapel.

shrinkable tubing: A tubing which may be reduced in size by applying heat or solvents.

shroud: Housing over a condenser or evaporator.

shunt: A device having appreciable resistance or impedance connected in parallel across other devices or apparatus to divert some of the current; appreciable voltage exists across the shunt and appreciable current may exist in it.

shunt-wound motor: Used when the motor speed must be constant, irrespective of variation in load.

shutter: Louvered or slatted decorations attached to the outsides of windows. Originally used to cover and protect the window, but on most modern homes, they are used as decoration only.

side-cutting pliers: The type of pliers used most by electricians. This type of pliers have both gripping jaws and cutting edges on the side for cutting wires.

sidewall load: The normal force exerted on a cable under tension at a bend; quite often called sidewall pressure.

siding: Finishing material nailed to the sheathing of wood frame buildings and forming the exposed surface.

sieve: A screen used for removing stones and large particles from sand.

sight glass: Glass tube or glass window in a refrigerating mechanism that shows the amount of refrigerant or the oil in the system, or the pressure of gas bubbles in the liquid line.

signal: A detectable physical quantity or impulse (as a voltage, current, or magnetic field strength) by which messages or information can be transmitted.

signal circuit: Any electrical circuit supplying energy to an appliance that gives a recognizable signal.

silica gel: Chemical compound used as a drier.

silicon controlled rectifier (SCR): Electronic semiconductor that contains silicon.

sill: Horizontal timber forming the lowest member of a wood frame house; lowest member of a window frame.

sill anchor: Threaded bolt, used in conjunction with washers and nuts, embedded in the foundation in a vertical position to secure the sill to the foundation walls.

sill high: The height from floor to sill.

sill plate: A support member laid on top the foundation wall in wood-frame construction. See *sill*.

silt: A finely divided earthy material deposited from running water.

silt test: A field test to determine the concentration of fine silt or loam in sand used for concrete work.

silver brazing: A soldering or brazing process using a silver alloy for the bonding material.

similar poles: When two magnetic poles repel each other, they are magnetically similar or like.

simple listing: Listing property with a broker for sale or rent other than through exclusive agency or an exclusive right-to-sell contract; an open listing, usually verbal.

sine wave, ac: Wave form of single frequency alternating current; wave whose displacement is the sine of the angle proportional to time or distance.

single-phase circuit: An ac circuit having one source voltage supplied over two conductors.

single-phase motor: Electric motor that operates on single-phase alternating current.

single-phasing: The abnormal operation of a three-phase machine when its supply is changed by accidental opening of one conductor.

single-pole switch: A switch that opens and closes only one side of a circuit. See Fig. S-31.

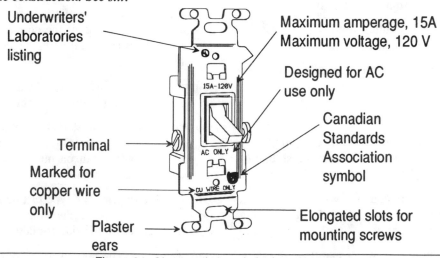

Figure 31. Characteristics of single-pole switch.

sinking: When the color sinks in, resulting in flat or semigloss spots, it is due to a porous undercoating.

sintering: Forming articles from fusible powders by pressing the powder just under its melting point.

siphoning: Loss of water in a trap or other vessel due to unequal pressure in the plumbing system.

site: The location of a building or the place on which a building is to be erected.

site plan: A plan view (as if viewed from an airplane) that shows the property boundaries and the building(s) drawn to scale and in its (their) proper location(s) on the lot. These plans will also include sidewalks, drives, streets, and similar details. Utilities such as water lines, sanitary sewer lines, telephone lines, and electrically power lines may also appear on site plans.

skeleton construction: Building constructed of steel framing with the enclosure walls supported at each story.

sketch: A suggestive presentation, either graphic or literary.

sketched: Outlined; rough drafted, a slight preliminary draft; generally used to refer to freehand drawing.

skew: Oblique; not at a right angle. Work is said to be askew when it is out of square.

skew chisel: A chisel with a straight cutting edge made at an angle other than a right angle with the center line of the tool; used in turning.

skew nailing: The driving of nails obliquely.

skewback saw: A handsaw whose back is curved in order to lighten its weight without lessening its stiffness.

skewback: The surface at each end of an arch upon which the first bricks are laid, and from which an arch springs.

skin effect: The tendency of current to crowd toward the outer surface of a conductor; increases with conductor diameter and frequency.

skinning: Removing insulation from electrical conductors before making splices or connections.

skintled brickwork: An irregular arrangement of bricks with respect to the normal face of the wall, the bricks being set in and out to produce an uneven effect; also the rough effect caused by mortar squeezed out of the joints.

skirt: The horizontal band which connects the legs of a chair beneath the seat, or the legs of a table beneath the top, sometimes called the "apron."

skirting: The finishing board which covers the plastered wall where it meets the floor. Same as baseboard.

sky lease: Lease for a long period of time of space above a piece of real estate; upper stories of a building to be erected by the tenant; upon the termination of lease, the improvement belongs to the lessor.

skylight: A glassed area in a ceiling or roof to provide for light. See Fig. S-32.

skyscraper: A term applied to high and lofty buildings of many stories, as the modern office building.

slab: (1) A thin piece of stone, marble, concrete, or the like, having a flat surface. (2) The outside pieces cut from a log when sawing it into boards.

slab foundation: Reinforced concrete slab poured directly onto the ground. Used for patios, stoops, garage floors, and the like.

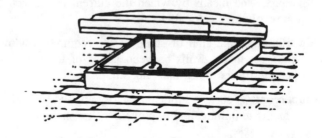

Figure S-32: Skylight.

slag: Aggregate used in lightweight concrete; a by-product of ore wmelting.

slake: A quicklime/putty mixture used to finish plaster.

slamming stile: The vertical strip against which a door abuts when closed, and into which the bolt of the lock engages.

slamming stile: The vertical strip against which a door abuts when closed, and into which the bolt of the lock engages.

slat: A thin piece of wood used in the seat or back of a chair.

slate: Hard natural material that easily splits into thin sheets for use as flooring or roofing.

sledge: A long-handled heavy hammer uscd with both hands.

sleeper: Wood strips embedded in concrete to provide a nailing base for the underflooring. Heavy beam or joist.

sleeve: Pipe or other hollow object into which other objects are placed. Frequently used in building foundations during the pour to later accept electrical or communication conductors as well as plumbing pipes.

sliding doors: Doors running on hortizontal tracts.

slip: The difference between the speed of a rotating magnetic field and the speed of its rotor.

slip-joint pliers: Pliers with adjustable jaws; also called combination pliers.

slip rings: The means by which the current is conducted to a revolving electrical circuit.

slip sill: A simple slab of stone of the required window width, set in the walls between jambs of the masonry opening. Easier to set than "lug sill."

slippercoat: A surface lubricant factory applied to a cable to facilitate pulling, and to prevent jacket sticking.

sliver: A defect consisting of a very thin elongated piece of metal attached by only one end to the parent metal into whose surface it has been rolled.

slope: The ratio of roof rise to its width.

slot: A long, narrow groove, particularly one cut to receive some corresponding part of a mechanism.

smoke detector: An electronic device designed to detect the presence of abnormal smoke in a confined area, and to sound an alarm if so detected. Most "smoke detectors" are also designed to detect abnormal heat (above a certain degree), and will also sound an alarm if this condition is met.

smoke shelf: An offset at the bottom of a chimney or top of a fireplace chamber that is designed to break up air currents that may cause fireplace smoke to enter unwanted areas. See Fig. S-33.

smoothing plane: A small plane usually not over 9 inches long with an iron width varying from 1¾ in. to 2¼ in.

smoothing trowel: Used by plasterers and cement workers for finishing surfaces. See *trowel*.

snake: A tape usually made of metal for cleaning out plumbing pipe or else used to "fish" wires through electrical conduits.

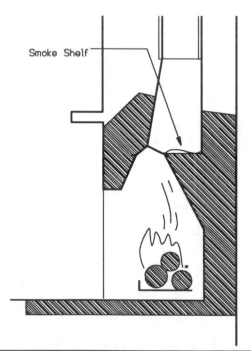

Figure S-33: Fireplace smoke shelf.

snips: Plier-like shears used for cutting sheet metal. See *shears*.

soap: Slang for electrical wire-pulling compound.

socket chisel: The strongest kind of woodworker's chisel. The upper end of the shank terminates in a socket into which the handle is driven.

socket: The receptacle into which the threaded portion of an incandescent lamp or plug is fitted.

socketing: A method of joining by means of wedging one piece of wood into the cavity of another.

socket wrench: A straight or ratchet handle used in conjunction with steel jackets that fit over nuts or bolts. The combination is used to loosen or tighten the fasteners.

socle: A projecting member at the foot of a wall or pier, or beneath the base of a column.

sod: Mats of growing grass embedded in soil. Strips are cut and rolled up for transportation to the building site where they are placed on bare earth to form a lawn.

soffit: The underside of an arch, staircase, cornice, or the like.

soft drawn: 1) A relative measure of the tensile strength of a conductor. 2) Wire which has been annealed to remove the effects of cold working. 3) Drawn to a low tensile.

softwood: An everygree or conifer tree such as fir, hemlock, pine, spruce, and the like.

soil boring: Test hole bored in soil to determine its characteristics.

soil pipe: A pipe for carrying off human waste from a building.

soil stack: Vertical cast-iron pipe conveying sewage from branch pipes to house sewer.

soil test: Testing soil samples to determine its physical properties prior to construction work.

solar azimuth: Angle at which the sun is located — east or west — from due south.

solar cell: The direct conversion of electromagnetic radiation into electricity; certain combinations of transparent conducting films separated by thin layers of semiconducting materials.

solar heat: Heat from visible and invisible energy waves from the sun.

solder: To braze with tin alloy.

solder gun: Hand-held electric heating tool used to melt solder.

soldering: Joining two metals by adhering another metal to them at a low melting temperature.

solderless wire connector: A metal or plastic device designed for connecting two or more electrical wires without the use of solder. See *terminal*.

soldier: A brick that is set in a wall in a vertical, rather than horizontal, position.

soldier course: Course of brick consisting of brick set on end with the narrow side exposed.

sole: A foot piece or rest such as would be laid on top of a subfloor to carry studding.

sole plate: A foundation plate to which a piece of machinery is bolted. See *sill plate*.

solenoid: Electric conductor wound as a helix with a small pitch; an electromagnet having a movable iron core.

solid bridging: Braces between floor joists consisting of short pieces with the same cross section as the joists. See Fig. S-34.

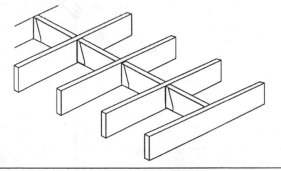

Figure S-34: Solid bridging.

solid state: A device, circuit, or system which does not depend upon physical movement of solids, liquids, gases or plasma.

solid type PI cable: A pressure cable without constant pressure controls.

solidly grounded: No intentional impedance in the grounding circuit.

soluble oil: Specially prepared oil whose water emulsion is used as a curing grinding or drawing lubricant.

solution: 1) Homogenous mixture of two or more components. 2) Solving a problem.

southern pine: Long-leaf, yellow pine used principally in heavy construction work. Is often spoken of as southern pine although there are a number of varieties native to the southern states.

solvent: A substance used to dissolve another substance such as paint thinner.

soundproofing: Use of techniques or materials to lessen the transmission of sounds in buildings.

SP: Single pole.

space heater: A heater for occupied spaces.

spackle: A compound used for patching holes or cracks in plaster.

spall: Bad or broken brick or chips of stone.

span: The distance between abutments or supports. The horizontal spread of a roof between exterior walls. See Fig. 35

spandrel: 1) The irregular triangular space between an arch and the beam above the same, or the space between the shoulders of two adjoining arches. 2) The angle of rise of a stairway. 3) In steel skeleton construction, the outside wall from the top of a window to he sill of the window above.

spark: A brilliantly luminous flow of electricity of short duration that characterizes an electrical breakdown.

spark arrester: A covering installed on chimney tops to cut down on sparks flaying out of the flue.

spark gap: Any short air space between two conductors.

spark test: A voltage withstand test on a cable while in production with the cable moving; it is a simple way to test long lengths of cable.

SPDT: Single-pole double-throw.

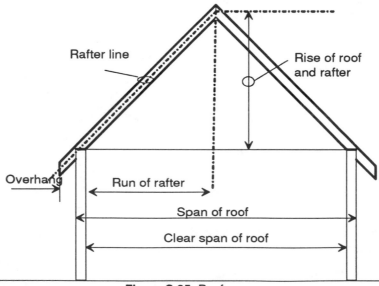

Figure S-35: Roof span.

special warranty deed: A deed wherein the grantor limits his liability to the grantee to anyone claiming, by, from, through or under him, the grantor.

specific gravity: Ratio of weight of a certain volume of material to the weight of an equal volume of water.

specification: A detailed statement of particulars.

specific heat: Ratio of the quantity of heat required to raise the temperature of a body 1 degree to that required to raise the temperature of an equal mass of water 1 degree.

specific performance: A remedy in a court of equity compelling the defendant to carry out the terms of the agreement or contract which was executed.

specify: To designate so as to distinguish from other things. An architect may specify a particular light fixture, for example.

specs: Abbreviation for the word "specifications", which is the written precise description of the scope and details of an electrical installation and the equipment to be used in the system.

spectrum: The distribution of the amplitude (and sometimes phase) of the components of the wave as a function of frequency.

spigot: The end of a pipe which fits into a bell. Also a word used synonymously with faucet.

spike: 1) A large nail, but thicker in proportion. See *nail*.

spike knot: A know sawed lengthwise.

spiral: A curved formed by a fixed point moving about a center, and continually increasing the distance from it, as in a spiral staircase.

spiral stair: Stair that turns in a circular direction as it rises. See Fig. S-36 .

spire: A tapering tower; a steeple. See Fig. S-37.

spirit level: An instrument for testing the horizontal and vertical accuracy of work. It consists of a glass tube or bulb nearly full of spirit and enclosed in a wood or metal case. When the bubble in the tube is in a central position, it indicates the accuracy of the work being tested.

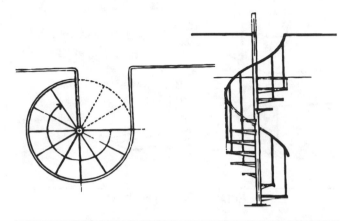

Figure S-36: Plan and elevation view of spiral staircase.

splash block: Concrete trough at the base of downspouts to receive and carry water away from the building. See Fig. S-38.

splay: To make with a bevel, to spread out, to broaden; a slanted or beveled surface.

splice: The electrical and mechanical connection between two pieces of cable. See *Western Union splice*, *fixture splice*, etc.

splice tube: The movable section of vulcanizing tube at the extruder.

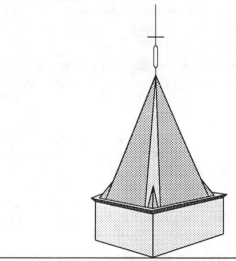

Figure S-37: Spire.

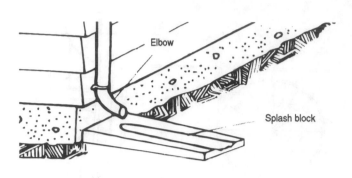

Figure S-38: Splash block.

spline: Narrow strips of wood used for strengthening wood joints.

split: Lengthwise separation of wood fibers. Crosswise separation is called *break*.

split bolt connector: A type of U bolt with matching nut used to splice two or more electrical conductors.

split fitting: A conduit fitting which may be installed after the wires have been installed.

split foyer: Home that divides into two or more levels at the entryway.

split-phase motor: Motor with two stator windings. Winding in use while starting is disconnected by a centrifugal switch after the motor attains speed, then the motor operates on the other winding.

split system: Refrigeration or air conditioning installation that places the condensing unit outside or remote from the evaporator. It is also applicable to heat pump installations.

split-wire: A way of wiring a duplex receptacle outlet with a 3-wire, 120/240 volt single-phase circuit so that one hot leg and the grounded conductor (neutral) feed one of the receptacle outlets and the other hot leg and grounded conductor (neutral) feed the other receptacle outlet. This gives the capacity of two separate circuits to one duplex receptacle.

spokeshave: A kind of double-handled plane for dressing curved woodwork.

spot-face: The machining of a flat surface around the end of a hole to allow a bolt head or nut to seat squarely on the surface.

spray cooling: Method of refrigerating by spraying refrigerant inside the evaporator or by spraying refrigerated water.

spread footing: Footing designed for wider bearing on weak soils, often with reinforcing steel and of shallow depth in proportion to width.

sprig: A small, headless wire nail or brad.

spring hinge: A hinge with a spring built into it. Used for self-closing of screen doors, etc.

springer: The stone from which an arch springs.

sprinkler system: An arrangement of overhead piping equipped with sprinkler heads or nozzles which automatically release sprays of water in case of fire.

spruce: An evergreen which grows abundantly throughout the eastern states. An inexpensive wood used extensively in the building trades.

sprung molding: A curved molding; also a thin molded piece of wood used for cornices, and attached to blocks or brackets fixed to the cornice frame.

SPST: Single-pole single-throw.

spur: A sharp-pointed tool used for cutting various lengths of rotary veneer as it is cut from a log of longer length.

spurious response: Any response other than the desired response of an electric transducer or device.

square: A right-angle measuring tool used for checking and marking right angles and measurements. See *steel square*.

square measure: An area measured by multiplying the length by the width (height by the width) to obtain square inches, square feet, square yards, or other squared measure; area.

squinch: A small arch or corbeled set-off, running diagonally, and cutting off a corner of the interior of a room or tower, to bring it from the square to the octagon, to carry the spire.

squirrel cage: Fan that has blades parallel to the fan axis and moves air at right angles or perpendicular to the fan axis.

squirrel-cage motor: An induction motor having the primary winding (usually the stator) connected to the power and a current is induced in the secondary cage winding (usually the rotor).

SSR: Solid state relay.

stability factor: Percent change in dissipation factor with respect to time.

stack: A large chimney usually of brick, stone, or sheet metal for carrying off smoke or fumes. Any vertical line of soil, waste, or vent piping.

staff bead: The molding strip used between the masonry and window frame to shut out the weather.

staging: A temporary structure of boards and posts.

stain: A wood finish which usually does not obscure the grain as would be the case with paint.

stair: A stair is a single step in a flight of stairs.

staircase: A flight of stairs with landings, newel posts, handrails, and balustrade.

stairs: Means the complete set of steps between two floors of a building and may consist of one of more flights. Simplest type is the "straight stair" or straight run stair so called because it leads from one floor to another without a turn or landing. "Close string" stairs are built with a wall on each side. "Open string" stairs have one side open to a room, so that a handrail is necessary. "Doglegged" stairs or "platform" stairs have landings near the bottom or top, usually introduced to change direction.

stair treads: The horizontal boards of a stair.

staking out: The driving of stakes into the earth to indicate the foundation location of the structure to be built. The stakes are often connected by a cord in order to secure a clean edge in the excavation.

stall: A small booth or compartment.

stanchion: A support or post of iron or wood.

standard conditions: Used as a basis for air conditioning calculations; temperature of 68 degrees Fahrenheit, pressure of 29.92 inches of mercury, and relative humidity of 30 percent.

standard deviation: 1) A measure of data from the average. 2) The root mean square of the individual deviations from the average.

standard pressure: Term applied to valves and fittings suitable for a working steam pressure of 125 pounds per square inch.

standard reference position: The nonoperated or de-energized condition.

standing wave: A wave in which, for any component of the field, the ratio of its instantaneous value at one point to that at any other point, does not vary with time.

standoff: An insulated support.

staple: A U-shaped piece of wire or iron with sharpened points for driving into wood. Used mainly to secure electrical and communication cables.

stapler: Manual or power-driven tool for driving staples.

star connection: Three-phase generators and transformers have three coils that may be connected in star, Y, or delta. When one terminal of each coil is connected together and the other three terminals are brought out separately, the connection is called *star* or *Y*.

star drill: A tool with a star-shaped point used for drilling in stone or masonry.

star shake: A radial split or crack in a log or timber as a result of being cut green and drying too rapidly. The cracks or splits which may be seen radiating from the center in an end view of a timber.

starling: An enclosure made by driving piles close together as for protection about a bridge or pier.

starter: 1) An electric controller for accelerating a motor from rest to normal speed and to stop the motor. 2) A device used to start an electric discharge lamp.

starting newel: The post at the bottom of a stair, supporting the balustrade.

starting relay: An electrical device that connects and/or disconnects the starting winding of an electric motor.

starting torque: The turning effort produced by a motor upon its shaft through the electro-magnetic effect at the initial flow of current.

starting step: The lowest step at the bottom of a stair.

starting winding: Winding of an electric motor used only during the brief period when the motor is starting.

static: Interference caused by electrical disturbances in the atmosphere.

stator: The portion of a rotating machine that includes and supports the stationary active parts. See Fig. S-39.

statute of frauds: Requires certain contracts relating to real estate, such as agreements of sale, to be in writing, in order to be enforceable.

stay: A prop or a guy for supporting canopies, steel chimneys, etc. A bar for holding parts together.

steady state: When a characteristic exhibits only negligible change over a long period of time.

steam: Water in vapor state.

steam heating: Heating system in which steam from a boiler is conducted to radiators in a space to be heated.

steel beam: A standard (S) or a wide-flange (W) steel structural member used in building construction.

steel girder: A built-up steel beam, receiving a vertical load and bearing vertically on its supports.

steel square: A large square used by carpenters. See *framing square*.

steel symbols: Symbols used on construction working drawings to describe structural steel shapes.

steel tape: Steel measuring tape in lengths of 6′ to 100′. Tape is wound in a roll in its container, pulled out by hand for use, and rewound either manually or by spring winder.

steel wool: A pad constructed of fine steel threads in various sizes for cleaning wood and metal surfaces.

steeple: A spire; a tall structure rising above the body of a building.

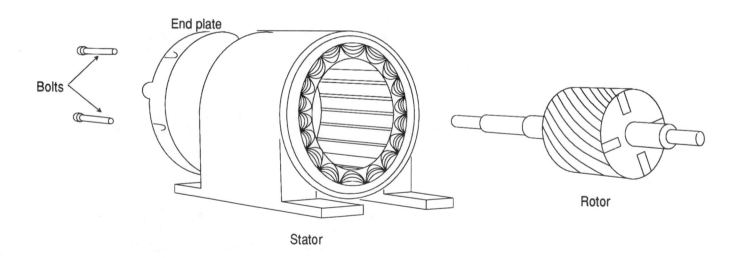

Figure 39: Basic parts of an induction motor.

stencil: A thin plate of metal or other material, with letters or pattern cut out, used for marking. When placed on a surface and color is laid on, a certain figure or design is made.

stepladder: A self-supporting folding ladder, usually made in heights of from 2 to 8 feet, with 6 feet being the most common.

stepped flashing: Metal flashing that follows the slope of a roof, as the side of a chimney.

stepped footings: Footings that remain level and plumb, but drop to different levels to remain beneath the frostline (or other required depth) on a hillside or sloping terrain.

steradian: The supplemental SI unit for solid angle; the three dimensional angle having its vertex at the center of a sphere and including the area of the spherical surface equal to that of a square with sides equal in length to the radius of the sphere.

stick shellac: A solid stick of shellac used for filling and repairing defects in wood surfaces. In used, a metal pallet is heated, rubbed against the shellac stick to melt the shellac against the pallet, and then the pallet is quickly rubbed over the defect to fill a crack, hole, etc.

stile: An upright piece in framing or paneling.

Stillson wrench: The pipe wrench of common use, named for its inventor.

stilts: Devices worn by plasterers or drywall installers to give them extra height so they can reach upper walls and ceilings without using a ladder or scaffold.

stipple: To form a rough or textured finish on a smooth surface.

stippling: Using a brush to form a texture in wet plaster.

stirrup: Metal strap in a U-form supporting one end of a wood beam.

stirrup or hanger: A drop support attached to a wall or girder to carry the end of a joist or beam.

stoa: A covered cloister, portico, or colonnade.

stone mason: One who builds foundations, walls, etc., of stone.

stone masonry: Using cut or natural stone for building structures or veneer.

stool: The wood shelf inside of and across the bottom of a window. See *sill*.

stoop: A platform or step at the door of a house.

stop: A piece attached to some part to prevent motion beyond a certain point as a doorstop in a building.

storage cell: One of the sections of a storage battery.

storm door: An extra outside door used in winter to avoid chilling the interior of a building and to lessen the effects of rain and wind at an entrance.

storm sash: An extra or outer sash used as a protection from severe winter weather.

storm sewer: Pipe, culvert, or other channel for carrying off rain or flood waters.

storm window: Extra outside window that fits over the regular window to cut down on air infiltration and heat loss in a building.

story: A floor of a building. For example, a 16-story building has 16 floors or levels.

stove bolt: Flat- or round-head bolts used to secure both metal and wooden objects.

straightedge: (1) A parallel, straight strip of wood or metal used for gauging the linear accuracy of work. (2) Tool for screeding concrete to a smooth finish.

strain: 1) A change in characteristic resulting from external forces. 2) To screen foreign materials from a substance.

strand: A group of wires, usually twisted or braided.

strand, annular: A concentric conductor over a core; used for large conductors (1000 kcmil @ 60 Hertz) to make use of skin effect; core may be of rope, or twisted I-beam.

strand, bunch: A substrand for a rope-lay conductor; the wires in the substrand are stranded simultaneously with the same direction; bunched conductors flex easily and with little stress.

strand, class: A system to indicate the type of stranding; the postscripts are alpha.

strand, combination: A concentric strand having the outlet layer of different size; done to provide smoother outer surface; wires are sized with +5% tolerance from nominal.

strand, compact: A concentric stranding made to a specified diameter of 8%-10% less than standard by using smaller than normal closing die, and for larger sizes, preshaping the strands for the outer layer(s).

strand, compressed: The making of a tight stranded conductor by using a small closing die.

strand, concentric: Having a core surrounded by one or more layers of helically laid wires each of one size, each layer increased by six.

strand, herringbone lay: When adjacent bunches have opposite direction of lay in a layer of a rope-lay cable.

strand, nonspecular: One having a treated surface to reduce light reflection.

strand, regular lay: Rope stranding having left-hand lay within the substrands and right-hand lay for the conductor.

strand, reverse-lay: A stranding having alternate direction of lay for each layer.

strand, rope-lay: A conductor having a lay-up of substrands; substrand groups are bunched or concentric.

strand, sector: A stranded conductor formed into sectors of a circle to reduce the overall diameter of a cable.

strand, segmental: One having sectors of the stranded conductor formed and insulated one from the other, operated in parallel; used to reduce ac resistance in single conductor cables.

strand, unilay (unidirectional): Stranding having the same direction of lay for all layers; used to reduce diameter, but is more prone to birdcaging.

strap hinge: Surface-mounted hinge normally used on rough finishes such as gates, outbuildings, and the like.

strap wrench: Pipe wrench utilizes a strap instead of jaws to grip pipe and other circular objects.

stratification of air: Condition in which there is little or no air movement in the room; air lies in temperature layers.

stress: 1) An internal force set up within a body to resist or hold it in equilibrium. 2) The externally applied forces.

stress-relief cone: Mechanical element to relieve the electrical stress at a shield cable termination; used above 2kV.

stretcher: Brick laid with its length parallel to the wall and side exposed.

strike: Part of a door latch.

striking distance: The effective distance between two conductors separated by an insulating fluid such as air.

striking: The process of establishing an arc or a spark.

string: One of the inclined sides of a stair supporting the treads and risers.

string course or sailing course: It consists of a course of brick or stone, projecting from a wall horizontally, for decorative purposes, or to break the plainness of a large expanse of wall.

stringer: A heavy plank or timber generally in horizontal position in a structure.

stringers: Members supporting the treads and risers of a stair.

strip cooler: A device to cool strips of compound.

strip: 1) A long narrow piece of wood. 2) To remove insulation or jacket from electrical conductors.

struck joint: Tooled mortar joint that slopes from the outside of the top brick to the a short distance from the edge of the lower brick. A recessed mortar joint.

structural bond: Tying brick veneer to an enter supporting wall so the entire assembly acts as a single supporting unit.

structural drawings: Construction working drawings that detail the supporting members of the building.

structural load: The load due to the structure itself as distinguished from the imposed load.

structural steel: The various shapes used by engineers in the erection of bridges, buildings, etc., as I beams, H beams, Z bars, channels, etc.

strut: A compression member other than a column or pedestal.

strut girder: A lattice girder whose top and bottom members are connected by vertical struts and braced by diagonal braces or by counter-bracing.

strut tenon: A tenon, such as is used on a diagonal piece or strut, usually on heavy timbers.

stub tenon: A short tenon.

stucco: Plaster or cement used for external surfacing of walls.

stuck molding: A molding that is built to a form, on the floor, or on a table, and "stuck" in position when finished.

stud bolt: A bolt threaded at both ends with blank space between to permit gripping with a pipe wrench.

stud: An upright beam or scantling as in the framework of a house.

studs: Vertically set skeleton members of a partition or wall to which the lath is nailed.

Styrofoam: A type of polystyrene insulation.

subbase: The lowest part of a base.

subcontractor: One who contracts to do a portion of a job, receiving directions from and being responsible to the general contractor of the whole.

subdivision: A tract of land divided into lots suitable for home-building purposes.

subfloor: A wood floor that is laid over the floor joists and on which the finished floor is laid.

sublet: To engage a subordinate contractor to handle a piece of work.

subletting: A leasing by a tenant to another, who holds under the tenant.

subordination clause: A clause in a mortgage or lease, stating that rights of the holder shall be secondary or subordinate to a subsequent encumbrance.

sub-panel: A panelboard in a residential system which is fed from the service panel; or any panel in any system which is fed from another, or main, panel supplied by a circuit from another panel.

subrail: A molded member or shoe planted on the top edge of a stair string to carry the lower end of the baluster.

substation: An assembly of devices and apparatus to monitor, control, transform or modify electrical power.

substrate: Base or foundation. The sheathing used on a roof to support the roofing is the *substrate*.

substructure: The lower portion of a structure upon which something else is built up.

suction: The manner in which certain kinds of plaster "pull" when worked with the trowel. The adhesion.

sugar pine: A very large species of pine native in California and Oregon. Its diameter may be as great as 15 ft. and it height 200 ft. The wood is clear and light in color and easily worked. Much used for building trim, interior and exterior.

sump: A depression in a roof to receive rain water and deliver it to the downspout. Also a depression in a floor, like in a basement, to catch and hold water for pumping and draining out.

sump pump: An automatic water pump used in basements to raise water to the sewer level.

sunk panels: Panels recessed below the surrounding surface.

superconductors: Materials whose resistance and magnetic permeability are infinitesimal at absolute zero (273°C).

superstructure: Anything built up or founded on something else.

supervised circuit: A closed circuit having a current responsive device to indicate a break or ground.

supplement of an angle: Its difference from 180 degrees.

supplemented: Additions made to; something added.

supply air: Air coming directly from the furnace in a forced hot-air heating system. The air that goes back to the furnace to be reheated is called *return air*.

surbase: A molding or border above a base, as above a baseboard in a room.

surface action: Any kind of action that affects a surface; that is, action of smoke fumes, moisture, etc. on a painted surface.

surface blowoff: A valve or plugged connection in a boiler, located just above the water line and used to draw off grease, oil, and dirt.

surface plate: A large plate of cast iron whose surface is made perfectly flat by very careful workmanship. It is used for testing flat surfaces.

surge: 1) A sudden increase in voltage and current. 2) Transient condition.

surge arrestor: A device used on electrical circuits to detect and check sudden increases in voltage and current.

surrender: The cancellation of a lease by mutual consent of lessor and lessee.

survey: The process by which a parcel of land is measured and its area ascertained.

surveying: The science of measuring land.

surveyor: A person who locates property boundaries and gives a deed description for recording. Other duty of a surveyor include making topography maps, site plans, and the like on a building site.

surveyor's compass: An instrument for determining the difference in direction between any horizontal line and a magnetic needle.

suspended ceiling: A ceiling installed beneath the structural ceiling, normally hung on wires. Such ceilings are handy in commercial structures where surfaced electrical wiring and plumbing pipes, air ducts, etc., may be installed between the structural ceiling and the suspended ceiling. See Fig. S-40.

swale: Drainage flow on a building lot.

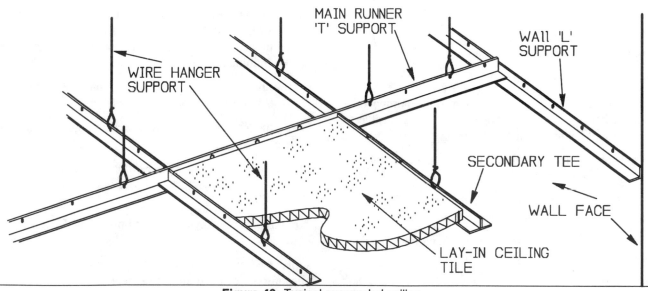

Figure 40: Typical suspended ceiling.

swarf: Metal bits hones off of cutting tools such as wood chisels, knives, etc.

sweat: To coat surfaces to be joined with solder, then causing the surfaces to adhere by the application of heat.

sweat joint: Joint my in copper plumbing pipe with solder.

sweet gum: A tree of large growth whose wood is soft but tough. It takes a very beautiful finish, but has a great tendency to warp; also know as "red gum."

sweep: The curve of the cutting edge on a woodcutting gouge.

swinging door: Door that swings both in and out on the same set of side hinges.

switch: A device for opening and closing or for changing the connection of a circuit.

switch, ac-dc general-use snap: A type of general-use snap switch suitable for use on either direct or alternating current circuits and for controlling the following:

- Resistive loads not exceeding the ampere rating at the voltage involved.

- Tungsten-filament lamp loads not exceeding the ampere rating of the switches at the rated voltage.

- Motor loads not exceeding 80 percent of the ampere rating of the switches at the rated voltage.

switch, ac general-use snap: A general-use snap switch suitable only for use on alternating current circuits and for controlling the following:

- Inductive loads not exceeding one-half the ampere rating at the voltage involved, except that switches having a marked horsepower rating are suitable for controlling motors not exceeding the horsepower rating of the switch at the voltage involved.

- Resistive and inductive loads (including electric discharge lamps) not exceeding the ampere rating at the voltage involved.

- Tungsten-filament lamp loads not exceeding the ampere rating at 125 volts, when marked with the letter T.

switch box: Metal or fiber outlet box for containing wall switches for lighting control.

switch, general-use snap: A type of general-use switch so constructed that it can be installed in flush device boxes or on outlet covers, or otherwise used in conjunction with wiring systems recognized by the National Electrical Code.

switch, general-use: A switch intended for use in general distribution and branch circuits. It is rated in amperes and is capable of interrupting its rated voltage.

switch, isolating: A switch intended for isolating an electrical circuit from the source of power. It has no interrupting rating and is intended to be operated only after the circuit has been opened by some other means.

switch, knife: A switch in which the circuit is closed by a moving blade engaging contact clips.

switch-leg: That part of a circuit run from a lighting outlet box where a luminaire or lampholder is installed down to an outlet box which contains the wall switch that turns the light or other load on or off; it is a control leg of the branch circuit.

switch, motor-circuit: A switch, rated in horsepower, capable of interrupting the maximum operating overload current of a motor having the same horsepower rating as the switch at the rated voltage.

switchboard: A large single panel, frame, or assembly of panels having switches, overcurrent, and other protective devices, buses, and usually instruments mounted on the face or back or both. Switchboards are generally accessible from the rear and from the front and are not intended to be installed in cabinets.

S wrench: A wrench shaped like the letter S having either fixed or adjustable openings.

swivel: Coupling that allows part(s) to rotate freely.

swivel vise: A bench vise that may be rotated on its base to bring the work that it holds into better position.

sycamore: The button wood tree. A very large tree attaining height as great as 150 feet. Wood moderately hard, very difficult to split; weight 38 pounds per cu. ft.; light to brownish color, often beautifully marked; takes fine finish; used extensively in furniture manufacture, interior trim, and in the form of plywood for certain types of construction.

symbol: A mark or character used as an abbreviation. See appendices for architectural, mechanical and electrical symbols.

synchronize: To cause to agree in time; happen simultaneously.

symmetrical: Exhibiting symmetry.

symmetry: The correspondence in size, form and arrangement of parts on opposite sides of a plane or line or point.

synchronism: When connected ac systems, machines or a combination operate at the same frequency and when the phase angle displacements between voltages in them are constant, or vary about a steady and stable average value.

synchronous machine: A machine in which the average speed of normal operation is exactly proportional to the frequency of the system to which it is connected.

synchronous motor: A motor that maintains a constant speed as long as the speed of the generator supplying it remains constant.

synchronous speed: The speed of rotation of the magnetic flux produced by linking the primary winding.

synchronous: Simultaneous in action and in time (in phase).

synchrotron: A device for accelerating charged particles to high energies in a vacuum; the particles are guided by a changing magnetic field while they are accelerated in a closed path.

system: A region of space or quantity of matter undergoing study.

T

tabernacle: 1) A church or place of worship. A niche, or recess, in which an image may be placed.

table saw: Power sawing tool that is used to cut wood members. The blade may be positioned for various cuts including angles, thickness, etc. See Fig. T-1.

tabulate: To arrange items or data in a table or list.

tachometer: An instrument for measuring revolutions per minute.

tack: Small, short-shanked fastener.

tack weld: A series of small welds used to secure two pieces of metal together.

tackle: The chain, rope, and pulleys, or blocks, used for hoisting purposes in the erection of heavy work.

tail: End of roof rafter that protrudes over end of top plate. See Fig. T-2.

tail beam: A relatively short beam or joist supported by a wall at one end and by a header at the other.

tailing: The part of a projecting brick or stone inserted in a wall.

Figure T-1: Table saw.

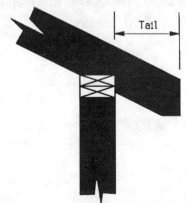

Figure T-2: The part of a rafter that protrudes over the end of the top plate is called *tail*.

tail joist: A joist that has one end terminating against a header joist.

take-off: The procedure by which a listing is made of the numbers and types of electrical components and devices for an installation, taken from the electrical plans, drawings and specs for the job.

take-up: Any device for taking up slack or removing the looseness of parts due to wear or other cause.

taking up: Relates to the making of adjustment for wear, as in machinery.

talc: Soapstone; used in making lubricants.

tallow: Made from animal fat. A large quantity is used in the manufacture of lubricating greases and leather dressings.

Tamo, Japanese Ash: A wood that varies greatly both in color and marking. Used for paneling. Very unusual grain effects are obtained with Tamo veneer.

tamping: Ramming down, such as the earth before pouring a concrete slab.

tandem: Two parts grouped together.

tandem extrusion: Extruding two materials, the second being applied over the first, with the two extruders being just a few inches or feet apart in the process.

tang: The shank of a cutting tool, or that portion that is driven into the handle. For example, the tang of a file. See Fig. T-3.

tangent of an angle: The quotient of the opposite side divided by the adjacent side.

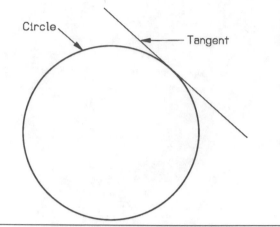

Figure T-4: Tangent of a circle.

tangent (geometry): A line that touches a curve at a point so that it is closer to the curve in the vicinity of the point than any other line drawn through the point; See Fig. T-4. Touching a line or surface at a point without intersecting; (trigonometry) in a right triangle it is the ratio of the opposite to the adjacent sides for a given angle.

tangible: Evident; real.

tank test: The immersion of a cable in water while making electrical tests; the water is used as a conducting element surrounding the cable.

tap: A splice connection of a wire to another wire (such as a feeder conductor in an auxiliary gutter) where the smaller conductor runs a short distance (usually only a few feet, but could be several feet) to supply a panelboard or motor controller or switch. Also called a "tap-off" indicating that energy is being taken from one circuit or piece of equipment to supply another circuit or load; a tool that cuts or machines threads in the side of a round hole. See Fig. T-5.

tap drill: Drill used to form hole prior to placing threads in hole. The drill is the size of the root diameter of tap threads.

tap drill sizes: Dimensions used to designate size of drill bit to fit various screw threads and diameters. See Appendices.

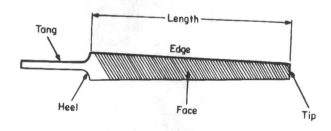

Figure T-3: Parts of a metal file, including the tang.

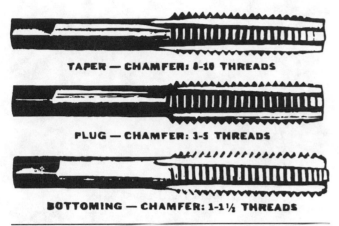

TAPER — CHAMFER: 8-10 THREADS

PLUG — CHAMFER: 3-5 THREADS

BOTTOMING — CHAMFER: 1-1½ THREADS

Figure T-5: The three common types of taps.

tap wrench: The double-armed lever with which a tap is gripped and operated during the process of tapping holes. See Fig. T-6.

tap bolt: A bolt usually threaded for its entire length. It is finished only on the point and the underside of the head. Tap bolts are made with both square and hexagon heads.

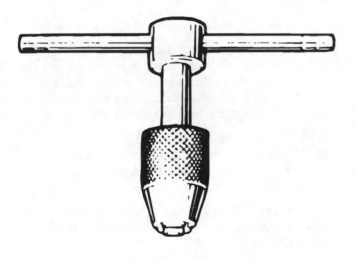

Figure T-6: Tap wrench is used to hold taps while tapping a hole.

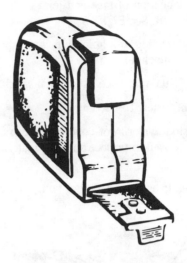

Figure T-7: Type measures have become the favoirite measuring tool for construction work.

tape: 1) A flexible, measuring scale made of thin steel, or of linen or cotton, usually contained in a circular case into which it may be rewound after use. See Fig. T-7. 2) A relatively narrow, long, thin, flexible fabric, film or mat or combination thereof; helically applied tapes are used for cable insulation, especially at splices; for the first century the primary insulation for cables above 2kV was oil saturated paper tapes.

taper: A gradual and uniform decrease in size, as a tapered socket, a tapered shaft, a tapered shank.

taper reamer: Reamer of the ordinary fluted type for reaming tapered holes; as a reamer used to prepare a hole for a tapered pin.

taper turning: Lathe tuning accomplished by setting over the tailstock or by the use of a taper attachment.

taper tap: A tap tapered in the direction of its length, in order to afford ease in cutting when commencing a screw thread in a drilled hole.

taper per foot: The manner of expressing the amount of taper.

taper gauge: A gauge for testing the accuracy of tapers, either inside or outside.

taping: The application of tape and joint compound when installing drywall. See Fig. T-8.

tapping: The threading of a hole by means of a tap, either by hand or by machine.

target, sag: A visual reference used when sagging.

tarnish: Loss of luster. To become dull.

tarpaulin: A stout waterproof covering of canvas or other material often used to protect building materials that are placed outside.

taut: Tense, tight; as a rope pulled taut, thus eliminating sag.

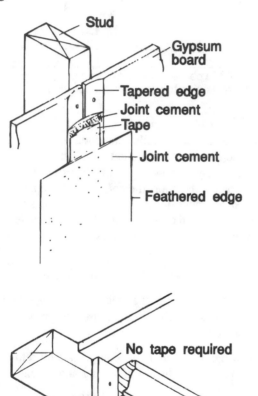

Figure T-8: Principles of drywall taping.

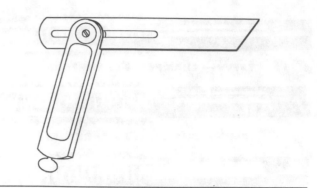

Figure T-9: T bevel.

tax deed: A deed for property sold at public sale by a political subdivision, such as a city, for nonpayment of taxes by the owner.

tax: A charge assessed against persons or property for public purposes.

T bevel: Tool used to check angles in layout work. See Fig. T-9.

T bolt: A bolt shaped like the letter T, the head being a transverse piece, that fits into the recessed undercut T slots, as on the table of a milling machine or planing machine.

TC: 1) Thermocouple 2) Time constant 3) Timed closing.

TDR: 1) Time delay relay 2) Time domain reflectometer, pulse-echo testing of cables; signal travels through cable until impedance discontinuity is encountered, then part of signal is reflected back; distance to fault can be estimated. Useful for finding faults, broken shields or conductor.

teak: An East Indian tree of large size. The wood is very durable, and is highly prized for shipbuilding and interior trim in buildings.

Technical Appeal Board (TAB) of UL: A group to recommend solutions to technical differences between UL and a UL client.

technical: Pertains to some particular art, science, trade, occupation, or the like, as technical school, technical term, etc.

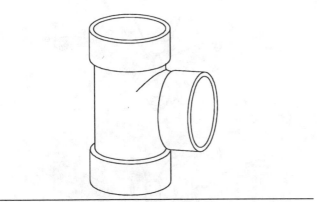

Figure T-10: Plumbing Tee.

tee: A fitting for connecting pipes of unequal sizes, or for changing direction of pipe runs. See Fig. T-10.

Teflon®: A DuPont trade name for polytetrafluoroethylene which is used as high temperature insulation and has low dissipation factor and low relative capacitance.

telegraphy: Telecommunication by the use of a signal code.

telemetering: Measurement with the aid of intermediate means that permits interpretation at a distance from the primary detector.

telephone: The transmission and reception of sound by electronics.

temper: A measure of the tensile strength of a conductor; indicative of the amount of annealing or cold working done to the conductor.

temperature, operating: The temperature at which a device is designed or rated for normal operating conditions; for cables: the maximum temperature for the conductor during normal operation.

temperature humidity index: Actual temperature and humidity of a sample of air compared to air at standard conditions.

temperature, coefficient of resistance: The unit change in resistance per degree temperature change.

temperature, ambient: The temperature of the surrounding medium, such as air around a cable.

temperature, emergency: The temperature to which a cable can be operated for a short length of time, with some loss of useful life.

template: Guide or pattern used to layout repetition work.

temporary: This single word is used to mean either "temporary service" (which a power company will give to provide electric power in a building under construction) or "temporary inspection" (the inspection that a code-enforcing agency will make of a temporary service prior to inspection of the electrical work in a building under construction).

tenancy in common: Form of cstate held by two or more persons, each of whom is considered as being possessed of the whole of an undivided part.

tenancy at will: A license to use or occupy lands and tenements at the will of the owner.

tenant at sufferance: One who comes into possession of lands by lawful title and keeps it afterwards without any title at all.

tenant: A person who holds real estate under a lease (lessee).

tenement: Everything of a permanent nature which may be holden.

tenon: A tongue projecting from the end of a piece of timber, which, with the mortise into which it fits, constitutes a mortise-and-tenon joint. See Fig. T-11.

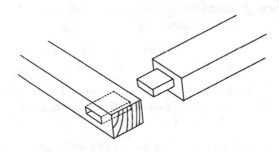

Figure T-11: A mortise-and-tenon joint.

tenon saw: An ordinary backsaw.

tensile strength: The greatest longitudinal stress a material—such as a conductor—can withstand before rupture or failure while in service.

tensile stress: A pulling or stretching force subjected to buildings and building materials.

tension: A pulling force; the opposite of compression.

tension, working: The tension that should be used for a portable cable on a power reel; it should not exceed 10% of the cable breaking strength.

tension spring: Any spring designed to be operated under a pulling strain.

tension, initial conductor: The tension prior to any external load.

tension, final unloaded conductor: The tension after the conductor has been stretched for an appreciable time by loads simulating ice and wind.

terminal: A device used for connecting cables. The finish to a newel or standard.

termination: 1) The connection of a cable. 2) The preparation of shielded cable for connection.

termite: An winged ant-like insect that weakens building structures by consuming wood. See Fig. T-12.

termite shield: Metal shield placed between foundation and wood members to deter termites from entering (and consuming) the wood structure. See Fig. T-13.

terrace: A raised level space as a lawn, having at least one vertical or sloping side.

terra cotta: A burned-clay product widely used for ornamental work on the exterior of buildings.

terrazzo flooring: A flooring of the granolithic type with polished surface; the body consists of fragments of colored stone imbedded in neat cement.

terre tenant: One who has the actual possession of land.

tessera: A small square stone or tile used in making mosaic pavements, walks, etc.

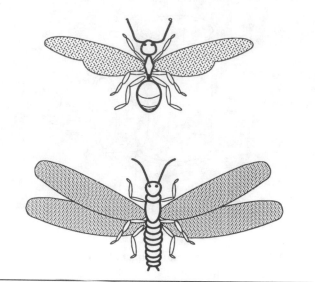

Figure T-12: Flying ant above and termite below—showing the difference between the two..

testlight: Light provided with test leads that is used to test or probe electrical circuits to determine if they are energized.

test, proof: Made to demonstrate that the item is in satisfactory condition.

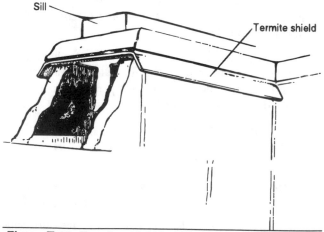

Figure T-13: Termite shield installed atop foundation wall.

test, voltage breakdown: a) Step method—applying a multiple of rated voltage to a cable for several minutes, then increasing the applied voltage by 20% for the same period until breakdown. b) Applying a voltage at a specified rate until breakdown.

test, voltage life: Applying a multiple of rated voltages over a long time period until breakdown: time to failure is the parameter measured.

test, volume resistivity: Measuring the resistance of a material such as the conducting jacket or conductor stress control.

test, water absorption: Determination of how much water a given volume of material will absorb in a given time period; this test is being superseded by the EMA test.

theorem: A truth capable of demonstration. A proposition to be proved.

theory: An attempt to account for a number of closely related observations or phenomena.

therm: Quantity of heat equivalent to 100,000 Btu.

thermal insulation: Material used in building construction to impede the progress of heat transmission. See Fig. T-14.

Figure T-14: Thermal insulation used in attic space.

thermal shock: Subjecting something to a rapid, large temperature change.

thermal relay (hot wire relay): Electrical control used to actuate a refrigeration system. This system uses a wire to convert electrical energy into heat energy.

thermal unit: A unit chosen for the comparison or calculation of quantities of heat; used as a standard for comparison of other quantities.

thermal conductivity: The ability of a metal to transmit heat through its mass.

thermal cutout: An overcurrent protective device containing a heater element in addition to, and affecting, a renewable fusible member that opens the circuit. It is not designed to interrupt short-circuit currents.

thermal endurance: The relationship between temperature and time of degrading insulation until failure, under specified conditions.

thermal protector (applied to motors): A protective device that is assembled as an integral part of a motor or motor-compressor and that, when properly applied, protects the motor against dangerous overheating due to overload and failure to start.

thermally protected (as applied to motors): When the words thermally protected appear on the nameplate of a motor or motor-compressor, it means that the motor is provided with a thermal protector designed to protect the motor from overloads.

thermionic emission: The liberation of electrons or ions from a solid or liquid as a result of its thermal energy.

thermistor: An electronic device that makes use of the change of resistivity of a semiconductor with change in temperature.

thermocouple: A device using the Seebeck effect to measure temperature.

thermodisk defrost control: Electrical switch with bimetal disk that is controlled by electrical energy.

thermodynamics: Science that deals with the relationships between heat and mechanical energy and their interconversion.

thermodynamics: The science that treats heat as a form of energy.

thermoelectric heat pump: A device that transfers thermal energy from one body to another by the direct interaction of an electrical current and the heat flow.

thermoelectric generator: A device interaction of a heat flow and the charge carriers in an electric circuit, and that requires, for this process, the existence of a temperature difference in the electric circuit.

thermometer: An instrument for measuring variations in temperature.

thermoplastic: Materials which when reheated, will become pliable with no change of physical properties.

thermoset: Materials which may be molded, but when cured, undergo an irreversible chemical and physical property change.

thermostat: A device for automatic regulation of temperature.

thermostatic valve: Valve controlled by thermostatic elements.

thermostatic expansion valve: A control valve operated by temperature and pressure within an evaporator coil, which controls the flow of refrigerant.

thickness gauge: A gauged shaped like a pocketknife, and has blades varying in thickness by thousandths of an inch. It is used in adjusting parts with a desired amount of clearance.

threading: The cutting of screw threads, either internal or external.

thimble: The section of a vitreous clay flue that passes through a wall.

T hinge: Consists of a strap with a "butt" at right angles to it. Used mainly for outside work on gates, doors, etc.

thinner: Substance used to thin paint or varnish.

thinwall conduit: Thin metal pipe used for containing electrical conductors in an electrical system.

threads: Grooves cut in metal to secure a threaded bolt, machine screw, or similar fastener.

three-phase current: Alternating current with three different cycles or phases displaced 120° apart.

three-way switch: A switch that is used to control a light, or set of lights, from two different points.

three phase circuit: A polyphase circuit of three interrelated voltages for which the phase difference is 120°: the common form of generated power.

three-square file: A term commonly applied to a three-cornered file; used for saw sharpening.

threshold: 1) The entrance to a building. 2) The plank, stone, or piece of timber under a door. A strip of wood or metal with beveled edges, used over the finish floor and the sill of exterior doors. See Fig. T-15

throat: The opening from the fireplace into the smoke chamber.

thumb plane: A name occasionally applied to a small plane 4 or 5 inches long, having a bit about 1 inch in width.

thumb screw: A screw designed to be turned with the thumb and finger.

thurm: To work with saw and chisel across the grain.

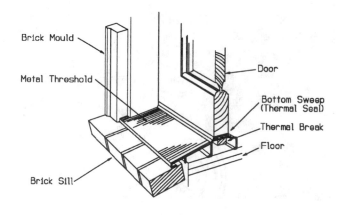

Figure T-15: Sections of a door base, including threshold.

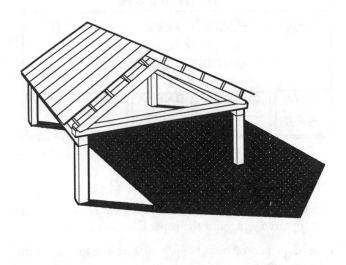

Figure 16: Tie beam.

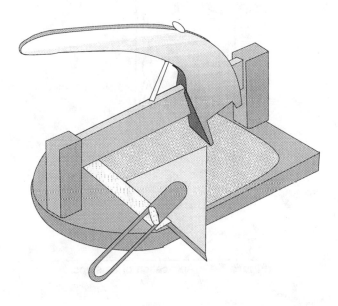

Figure T-17: Tile cutter.

thumper: A device used to locate faults in a cable by the release of power surges from a capacitor, characterized by the audible noise when the cable breaks down.

thyratron: A gas-filled triode tube that is used in electronic control circuits.

tie: A piece inserted or attached to other pieces to hold them in position.

tieback member: A timber, oriented perpendicular to a retaining wall, that ties the wall to a deadman buried behind the wall.

tie beam: A beam that ties together or prevents the spreading out of the lower ends of the rafters of roof trusses. See Fig. T-16.

tie rod: A steel rod used in concrete forms to hold the forms in place during pouring of the concrete.

tie wire: Metal wires used to secure concrete forms in place during pouring.

tig welding: Arc-welding process that uses an inert-gas shield over the weld puddle to prevent contamination and oxidation.

tile: Terra cotta, cement, or glass pieces used for roofing; also made in artistic designs and finishes for floor and wall covering.

tile cutter: Portable PVC cutter for cutting floor tile. Measurement settings are in both standard and metric. In use, the tile is placed in the cutter for the proper cut, scribed two or three times with a tungsten-tipped scribing tool, and the hinged portion of the tool is bent downward at the edge of the worktop. A perfect cut is made where scribed. See Fig. T-17.

timber: Heavy lumber greater in dimensions that 4″ x 4″ stock.

timer-thermostat: Thermostat control that includes a clock mechanism. Unit automatically controls room temperature and changes it according to the time of day.

timers: Mechanism used to control on and off times of an electrical circuit.

tin snips: The ordinary hand shears used by sheet-metal workers. See Fig. T-18 on page 302.

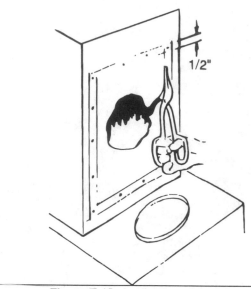

Figure T-18: Application of tin snips.

tin: A silvery-white metal with specific gravity of 7.3. It is of very great value in the industries, chiefly in the making of alloys. Used for building roofs and roof flashing.

tinned: Having a thin coating of pure tin, or tin alloy; the coating may keep rubber from sticking or be used to enhance connection; coatings increase the resistance of the conductor, and may contribute to corrosion by electrolysis.

tints: Light colorings, especially colors containing some white.

title block: A small portion of a construction working drawing used to identify the contents of the drawing. The outlined space usually in the lower right corner, or in strip form across the bottom of a drawing, containing name of company, title of drawing, scale, date, and other pertinent information.See Fig. T-19.

title sheet: First page of a set of working drawings that usually identifies the contents of the set.

title insurance: A policy of insurance which indemnifies the holder for any loss sustained by reason of defects in the title.

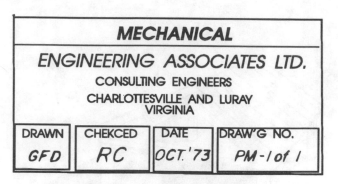

Figure T-19: Typical title block.

title by adverse possession: Acquired by occupation and recognized as against the paper title owner.

title: Evidence of ownership, which refers to the quality of the estate.

T joint: The ordinary three-way pipe fitting, in which one branch is at right angles to the other two and midway between them.

toeing: The driving of nails or brads obliquely, near the end of one piece, to attach it to another. See Fig. T-20.

toenailing: The driving of nails slantwise as in floor laying to avoid having nailheads show on the surface.

toggle: A device having two stable states: i.e.—toggle switch which is used to turn a circuit ON or OFF. A device may also be toggled from slow to fast, etc.

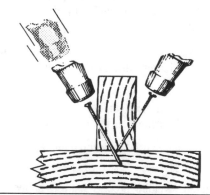

Figure T-20: Toeing wall stud to floor plate.

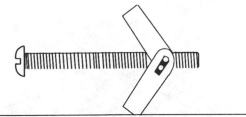

Figure T-21: Typical toggle bolt.

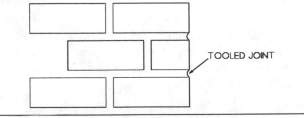

TOOLED JOINT

Figure T-23: Tooled joint.

toggle bolt: Used for attaching articles to a hollow wall. It consists of a screw with a swivel piece attached near the end. This piece may be swung into a length-wise position while the bolt is being inserted, after which it swings to a right-angle position, thus permitting "pulled up" on the bolt. See Fig. T-21.

tolerance: The permissible variation from rated or assigned value.

tongue: A projecting bead so cut on the edge of a board that it may fit into a corresponding groove on the edge of another piece.

toolroom: The shop room in which tools and parts are stored and from which they are issued to the workmen. Also that part of a shop where jigs, fixtures, etc., are made, stored, and repaired.

tongue and groove: Wood joint made with one board having a protrusion that fits into a groove of an adjoining board. See Fig. T-22.

tooled joint: Masonry joint (brick or concrete blocks) whereas the joint is scribed with a tool that pushes the mortar back into the joint. See Fig. T-23.

tooth ornament: One of the peculiar marks of the early English period of Gothic architecture, generally inserted in the hollow moldings of doorways, windows, etc.

toothing: Leaving a section of brickwork toothed so that the brickwork to follow can be bonded into it. It consists of allowing alternate courses to project a sufficient distance to assure a good bond with the portion to be built later. See Fig. T-24.

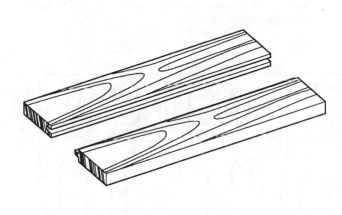

Figure T-22: Tongue and groove joint.

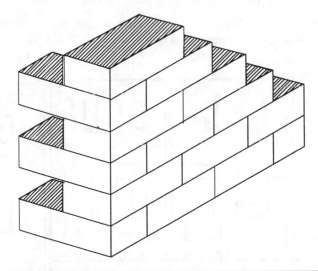

Figure T-24: Toothing a brick wall so the brickwork to follow can be bonded to it.

top plates: Wooden members running horizontally over a stud wall.

topping-off: The finishing touches put to an electrical installation; mounting plates on wall switches, receptacles and other wiring devices; receptacles and other wiring devices; installing fixtures, etc.

toroid: A coil wound in the form of a doughnut; i.e.—current transformers.

torque: The turning effort of a motor.

torquing: Applying a rotating force and measuring or limiting its value.

torque wrench: Wrench having a measuring device built into the tool to indicate the amount of pressure exerted during a tightening process.

Torrens system: A system of title records provided by state law.

torso: A term applied to columns with twisted shafts.

tort: An actionable wrong.

torus: A large convex molding of nearly semicircular section, largely used as a base molding.

tower: A structure larger than a pinnacle and less tapering than a steeple; frequently a part of a large building.

townhouse: Multi-family dwellings sharing common walls. Sometimes called *row houses*. See Fig. T-25.

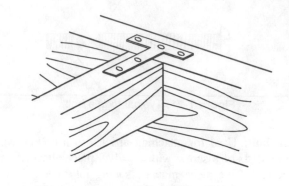

Figure T-26: T-plate used to strengthen wood joint.

township: A territorial subdivision, six miles long, six miles wide, and containing 36 sections, each one mile square.

TPE: Thermal plastic elastomer.

T plate: A metal plate shaped like a letter T used for strengthening a joint where the end of one and the side of another meet. See Fig. T-26.

TPR: Thermal plastic rubber.

tracer: A means of identifying cable.

tracery: The ornamentation of panels, circular windows, window heads, etc.

Figure T-25: Typical townhouses.

tracing paper: Thin, semi-transparent paper on which drawings are made for reproduction.

tracing: Sketching; designing; drawing. Usually transparent material on which a drawing has been made for quantity production.

tract: Parcel of land.

tract development: A planned subdivision on a parcel of land. See Fig. T-27.

trade: A craft such as carpenter, electrician, plumber, and the like.

trade associations: Groups with a common interest who join together to promote their trade, craft, and standards of construction.

trade union: An alliance of workmen organized for the purpose of securing standardized privileges for all its members.

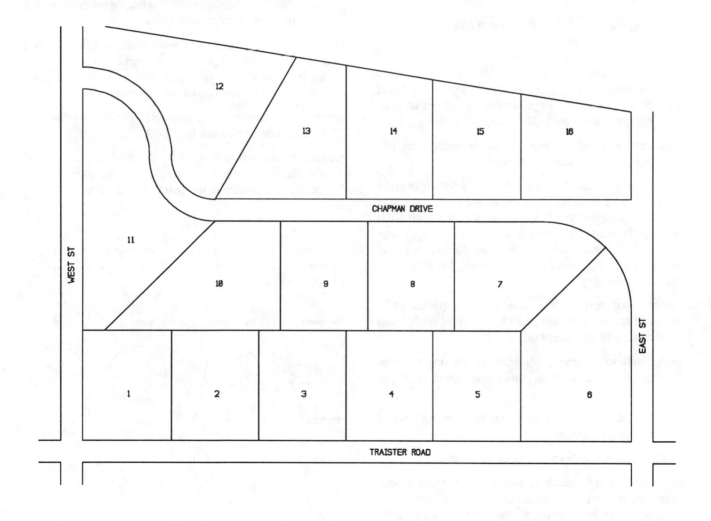

Figure T-27: Plot plan of a tract development.

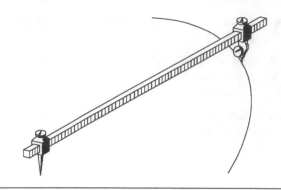

Figure T-28: Trammel or beam compass.

trammel: A beam compass, in which the head slides along a straight bar. It is tightened by setscrews, and is used to strike radii too large for the capacity of an ordinary compass. See Fig. T-28.

transducer: A device by means of which energy can flow from one or more media to another.

transformer: A static device consisting of windings and a magnetic core designed for transferring electrical power from the primary circuit to the secondary circuit to 1) step-up the secondary voltage at less current or 2) stepdown the secondary voltage at more current; with the voltage-current product being constant for either primary or secondary.

transept: That portion of a church which passes transversely between the nave and choir at right angles, and so forms a cross on the plan.

transfer switch: A device for transferring one or more load conductor connections from one power source to another.

transfer molding: Another name for injection molding of thermosetting materials.

transfer: To remove from one place to another.

transformer, safety isolation: Inserted to provide a non-grounded power supply such that a grounded person accidentally coming in contact with the secondary circuit will not be electrocuted.

transformer, vault-type: Suitable for occasional submerged operation in water.

transformer-rectifier: Combination transformer and rectifier in which input in ac may be varied and then rectified into dc.

transformer, potential: Designed for use in measuring high voltage: normally the secondary voltage is 120V.

transformer, power: Designed to transfer electrical power from the primary circuit to the secondary circuit(s) to 1) step-up the secondary voltage at less current or 2) stepdown the secondary voltage at more current; with the voltage-current product being constant for either primary or secondary.

transient: 1) Lasting only a short time; existing briefly; temporary. 2) A temporary component of current existing in a circuit during adjustment to a changed load, different source voltage, or line impulse.

transistor: An active semiconductor device usually with three or more terminals. Fig. T-29.

transit: transmission line: A long electrical circuit, usually starting at the generating plant, travelling cross-country, and ending at the point of utilization.

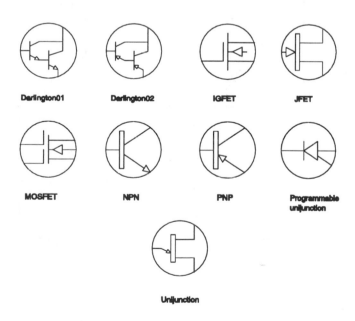

Figure T-29: Symbols for various transistors.

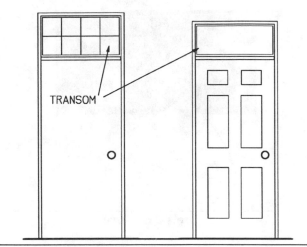

Figure T-30: Transom.

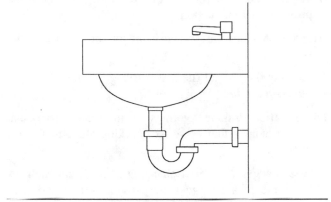

Figure T-31: Drain trap.

transmission: Transfer of electric energy from one location to another through conductors or by radiation or induction fields.

transom bar: The horizontal member which divides an opening into two parts.

transom: A small window over a door, or other window. See Fig. T-30.

transposition: Interchanging position of conductors to neutralize interference.

trap: A water-filled, U-shaped pipe that will allow water and wastes to pass through, but prevents gases and vermin inside the DWV system from slipping backwards into the house. See Fig. 31.

Trap door: A door or cover used to close off an opening in floor, ceiling, or roof.

traveler: 1)A pulley complete with suspension armor frame to be attached to overhead line structures during stringing. 2) Conductors between two three-way switches. See *three-way switch*.

tread: The flat portion of a step, as that portion on which the foot is placed when mounting the stairs.

treillage: A trellis; latticework for supporting vines.

trellis: An ornamental lattice made up of wooden strips to support vines; a summerhouse or the like made of latticework. See Fig. 32.

trench: A long narrow excavation in the ground, as a trench dug for the laying of pipes.

T rest: A T-shaped tool support used in hand turning.

trestle: 1) Usually a horizontal beam with four braced legs, used in pairs to support a horizontal board. 2) A braced frame forming the support of a table top. See *sawhorse*.

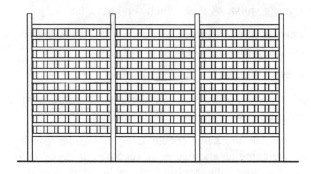

Figure T-32: Trellis.

triangular truss: A very popular truss for short spans especially for roof support.

triangle: A plane figure consisting of three closed sides and three angles.

triangular file: Metal file that is triangular in section. See *three-square file*.

triangulation: Taking sightings during a land survey that form triangles that serve as checking stations along a traverse.

trickle charge: A rectifier changing ac to dc, and delivering same to a storage battery for a certain period of time, usually at a very minute rate.

triforium: The space between the sloping roof over the aisle and the aisle vaulting in a church.

trim: 1) The front cover assembly for a panel, covering all live terminals and the wires in the gutters, but providing openings for the fuse cutouts or circuit breakers mounted in the panel; may include the door for the panel and also a lock. 2) Ornamental parts of wood or metal; used to cover the joints between jambs and plaster wall around a door or window. Also moldings and other finishings about a door, window, etc., either internal or external.

trimmer arch: The rather flat arch such as is used to support a hearth.

trimmers: (See trimming joists.)

trimming joist: A joist which supports a header joist.

triode: A three-electrode electron tube containing an anode, a cathode, and a control electrode.

triplex: Three cables twisted together.

trolley wire: Solid conductor designed to resist wear due to rolling or sliding current pickup trolleys.

trough: Another name for an "auxiliary gutter", which is a sheet metal enclosure of rectangular cross-section, used to supplement wiring spaces at meter centers, distribution centers, switchboards and similar points in wiring systems where splices or taps are made to circuit conductors. The single word "gutter" is also used to refer to this type of enclosure.

Figure T-33: Trowel.

trowel: 1) A tool used by masons to finish cement, lay mortar, and the like. See Fig. T-33. 2) To smooth cement or mortar.

trunk feeder: A feeder connecting two generating stations or a generating station and an important substation.

truss: A built-up framework of triangular units for supporting loads over long spans. See Fig. T-34.

truss clip: Metal connector used to fasten parts of a roof truss together.

trussed beam: A beam stiffened by a truss rod.

trussed: Framed structural pieces consisting of triangles in a single plane for supporting loads over spans.

truss plate: A neavy-gauge, pronged metal plate that is pressed into the sides of a wood truss at the point where two more members are to be joined together.

trust deed: A conveyance of real estate to a third person to be held for the benefit of a cestuique trust (beneficiary).

try square: A small square used in testing squareness of work; also used to lay off right angles.

T square: A draftsman's tool, consisting of a blade attached at right angles to a head, which is at least twice as thick as the blade; used for ruling parallel, horizontal lines. See Fig. T-35 on page 310.

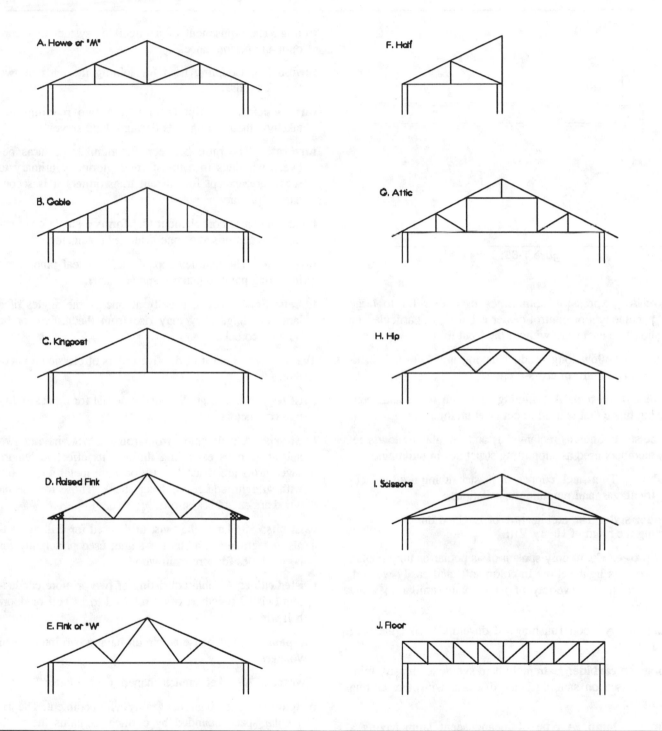

A. Howe or "M"

F. Half

B. Gable

G. Attic

C. Kingpost

H. Hip

D. Raised Fink

I. Scissors

E. Fink or "W"

J. Floor

Figure T-34: Various types of roof trusses used in building construction.

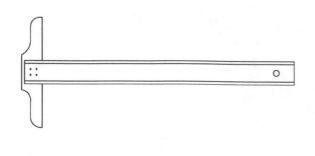

Figure T-65: T-square.

tub: An expression sometimes used to refer to large panelboards or control center cabinets, in particular the box-like enclosure without the front trim.

tube: A hollow long product having uniform wall thickness and uniform cross-section.

tube-cutting tool: A clamping tool with roller and rotating blade that is used to cut steel tubing.

tube steel: Square, rectangular, and circular hollow steel members used as supporting structures in buildings.

tubing: Thin steel, copper, or plaster tubing used in electrical, gas, and plumbing installations.

Tudor style: The architecture of England during the reign of Henry VIII.

tulip tree: Commonly spoken of as poplar or tulip poplar. Wood is light yellow in color, soft, and easily worked; it has a wide variety of uses, many similar to white pine.

tung oil: A wood finishing oil developed from the tung nut.

tungsten carbide: Extremely hard synthetic abrasive mineral used on sanding belts, discs, and in other cutting tools.

tungsten lamp: A type of incandescent lamp having a filament of fine tungsten wire.

tuning: The adjustment of a circuit or system to secure optimum performance.

turnbuckle: Coupling used for joining the ends of two threaded rods.

turning point: A point from which two readings are taken on the leveling rods during a land survey.

turn ratio: The ratio between the number of turns between windings in a transformer; normally primary to secondary, except for current transformers it is secondary to primary.

turn: The basic coil element that forms a single conducting loop comprised of one insulated conductor.

turpentine: The distilled sap of the long-leaf pine. Used in mixing paint to make it spread easier.

turrets: Small towers, usually at one of the angles of a large building. They may rise from the ground or be built on corbels.

Tuscan: The plainest of the five orders of classic architecture.

twist test: A test to grade round material for processibility into conductors.

twist drill: A drill made from round stock, having two helical grooves extending through its effective length. Twist drills are made for use on both metal and wood, with straight, square, or taper shank. Those for use on wood are called "twist bits."

twist bits: Similar to the twist drills used for drilling metals, but ground at a sharper angle; used principally for boring holes for screws in wood.

twisted cable: A cable consisting of two or more conductors twisted together; often referred to as bell or doorbell wire.

two-phase: A polyphase ac circuit having two interrelated voltages.

T wrench: A socket wrench shaped like the letter T.

tympanum: Used interchangeably with pediment. The triangular space bounded by cornice, contains the pediment.

U

U bolt: A bolt shaped like the letter U threaded at both ends; used mainly to secure conduit in place. See Fig. U-1.

Figure U-1: U bolt.

U clamp: A clamp shaped like the letter U; used for clamping down work on power-tool beds.

UBC: United Brotherhood of Carpenters and Joiners of America.

udometer: a rain gauge.

UHF (ultra high frequency): 300 MHz to 3GHz.

UL (Underwriters Laboratories): A nationally known laboratory for testing a product's performance with safety to the user being prime consideration; UL is an independent organization, not controlled by any manufacturer; the best known Lab for testing electrical products. See Fig. U-2.

Figure U-2: UL label.

ultimate set: The difference between the length of a specimen plate or bar, before testing and at the moment of fracture, and given i percentage of the length.

ultimate strength: The highest unit stress that can be sustained, this occurring just at or just before rupture.

ultrasonic: Sounds having frequencies higher than 20 KHz, which is at the upper limit of human hearing.

ultrasonic cleaning: Immersion cleaning aided by ultrasonic waves which cause microagitation.

ultrasonic detector: A device that detects ultrasonic noise such as that produced by corona or leaking gas. Also used in burglar-alarm systems to detect motion.

ultraviolet: Invisible radiation waves with frequencies shorter than wavelengths of visible light and longer than X-rays. Within the wave length range 10 to 380 nanometers.

umber: A chestnut-brown hydrated ferric oxide containing manganese oxide and clay. burnt umber produces a reddish-brown color. Both are used as pigments.

unbuttoning: Demolition of steel-frame buildings by breaking off the rivet heads.

unearned increment: An increase in value of real estate due to no effort on the part of the owner; often due to increase in population.

undercoat: Base coat (primer) to accept finish coat of paint.

undercuring: Insufficient hardening of a glue due to low temperature or too short a hardening period.

undercurrent: Less than normal operating current.

undercut: To cut back underneath, such as the kickspace in kitchen cabinets. See Fig. U-3.

undercut tenon: A tenon with its shoulder cut slightly out of square to ensure that it bears on the mortised piece.

underground cable: A single or multiple conductor cable sheathed in lead or other waterproof materials for installation below grade. Underground cable used to feed a building electric service is shown in Fig. U-4.

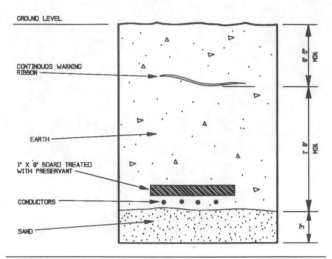

Figure U-4: Cross section of underground cable.

underlayment: A support surface installed beneath a finish surface to provide a smooth or level base. See Fig. U-5.

underpinning: Excavating under walls or erections, and filling the space with concrete or timbering without disturbing the structure.

under-ridge tiles: Plain tiles laid in the top course of a roof below the ridge tiles. Under-ridge tiles are slightly shorter than the ridge tiles.

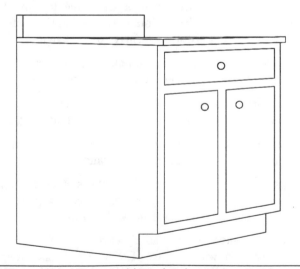

Figure U-3: Kitchen cabinet showing undercut at base.

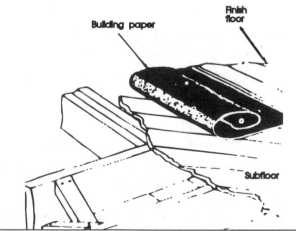

Figure U-5: The building paper is the underlayment in this application.

undertone: The color obtained when a colored pigment is reduced with a large proportion of white pigment.

undervoltage: Less than normal operating voltage.

underwriter: Name given to representative of an inspection organization who examines electrical installations for life and fire hazards.

Underwriters' Laboratories: See *UL*.

undressed lumber: Lumber that is sawn but not planed.

ungrounded: Not connected to the earth intentionally.

Unified thread: A relatively new screw thread that was designed to replace the Whitworth thread in English-speaking countries.

uniform load: Includes the weight of the structure itself and any load evenly spread over it.

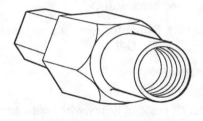

Figure U-6: Typical pipe union.

unilateral contract: One in which one party makes an express undertaking, without receiving in return any promise of performance from the other.

union: A coupling or connection for threaded conduit. See Fig. U-6. Also see *labor union*.

unit heater: Electric, oil- or gas-fired heater, usually suspended from ceiling, to heat relatively large spaces. See Fig. U-7.

unit magnetic pole: One that will repel an equal and like pole with a force of one dyne at a distance of 1 cm. A dyne is the force which, acting upon a mass of 1 gram during 1 second, gives the mass a velocity of 1 cm per second.

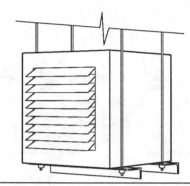

Figure U-7: Electric unit heater.

unit of magnetic flux: The total number of lines of force set up in a magnetic substance. Treated as a magnetic current flowing in a magnetic circuit.

unit of bond: The smallest length of a brick course that repeats itself.

unit of magnetic intensity: Magnetomotive force. Magnetic pressure that drives lines of force through a magnetic circuit.

unit price: Cost per item of material or work, as installing an electrical system at a given price per outlet.

unit stress: The stress on a unit of section area, usually expressed in pounds per square inch.

universal: General; all-reaching; total; entire.

universal joint: A type of coupling that permits the free rotation of two shafts whose axes are not in a straight line.

universal motor: Electric motor that will operate on both ac and dc. See Fig. U-8 on page 314.

universal plane: A hand-operated metal plane with many replaceable cutting edges, each of a different shape; used to cut different moldings, tongue-and-groove joints, and the like.

universal saw table: A saw table that may be tilted to permit sawing at a bevel.

unsound knot: A knot in wood that is not as hard as the wood surrounding it.

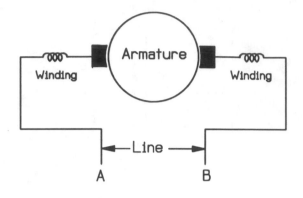

Figure U-8: Schematic wiring diagram of a universal motor.

upflow furnace: Electric, oil- or gas-fired furnace where the supply air is discharged at the top.

upright: Something standing upright, as a piece of timber in a building.

upset: To shorten or thicken metal by hammering or by pressure.

uranium: A mineral that provides one of the chief sources of radium.

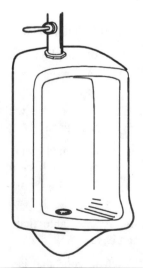

Figure U-9: Typical urinal.

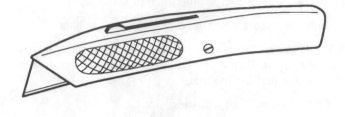

Figure U-10: Utility knife.

urban: Area within a city or the immediate surrounding area. See *rural*.

URD (Underground Residential Distribution): A single phase cable usually consisting of an insulated conductor having a bare concentric neutral.

urethane foam: Type of insulation that is foamed in between inner and outer walls.

urinal: Plumbing fixture for receiving and draining away male urine. See Fig. U-9.

usury: Charging more than the real rate of interest for the use of money.

U-tie: A wall tie made of heavy wire bent into a U shape.

utility: A public service such as a telephone or electric company. A power company furnishing electric service to users is a utility.

utility knife: Cutting knife with a retractable and replaceable blade, used for scoring drywall, floor tile, and the like. See Fig. U-10.

utilization equipment: Equipment that uses electric energy for mechanical, chemical, heating, lighting, or other useful purposes.

U value: Factor used to determine heat loss and heat gain in heating, ventilating, and air conditioning calculations. It is the rate of heat flow through building parts; that is, walls, roof, floor, windows, doors, and the like. The lower the U value, the greater the insulation value.

UURWAW: United Union of Roofers, Waterproofers and Allied Workers.

V

V.A.: Veterans' Administration.

va: Volt-amperes; Volts times amperes, which equals watts.

vacuum: A space from which the air or any matter has been exhausted. Reduction in pressure below atmospheric pressure.

vacuum cleaner: A motor-driven fan machine used for sucking up dirt and dust from rugs, shops, and job sites.

vacuum heating: A steam-heating system for buildings in which a vacuum pump is connected to the return main. It removes condensate and air from the radiators and returns the water to the boiler feed tank.

vacuum pump: A pump used to remove condensation and air from the return main of a heating system in order to (a) create a vacuum and (b) return the condensate to the boiler or to a receiving tank.

vacuum switch: A switch with contacts in an evacuated enclosure.

valance: Ornamental frame at the top of a widow to conceal drapery hardware and lighting. See Fig. V-1.

valance lighting: An arrangement of lighting fixtures behind a valance. Up-light reflects off ceiling for general room illumination, while down-light is provided for drapery accent. See Fig. V-2.

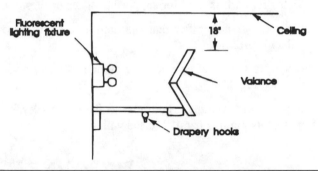

Figure 1: Valances are used at the top of a window to conceal drapery hardware and lighting.

Figure 2: Valance lighting is frequently used in formal living areas.

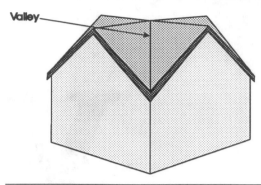

Figure 3: A roof valley is the gutter formed when two roofs meet.

valley: The gutter or angle formed by the meeting of two roof slopes. See Fig. V-3.

valley board: A wide board fixed on, and parallel to, the valley rafter. It is used for supporting slates or tiles.

valley jack: A jack rafter that fits onto a valley rafter or valley board.

valley rafter: The rafter extending along and under a valley. See Fig. V-4.

valley shingle: A shingle laid next to the roof valley, cut so that its grain is parallel to the valley.

V.A. loan: Loan guaranteed by the Veteran's Administration under Service-men's Readjustment Act of 1944, as amended; only honorably discharged veterans and their widows are eligible.

value: Marketable price of equipment, property, materials, and other objects.

value-cost contract: A cost-reimbursement contract in which the contractor receives a larger fee when his final costs are low than when his final costs are high.

valve: A device for regulating the flow of a liquid or gas through pipes.

valve, expansion: Type of refrigerant control that is used to maintain a pressure difference between high side and low side pressure in a refrigeration mechanism. The valve operates by pressure in the low or suction side.

valve, solenoid: Valve actuated by magnetic action by means of an electrically energized coil.

Vandyke brown: A native earth, being a clay stained with a bituminous compound. It is used in mixed paints on account of its deep shade.

vane: A weathercock for indicating the direction of the wind. Vanes are used in ductwork to help guide the flow of air.

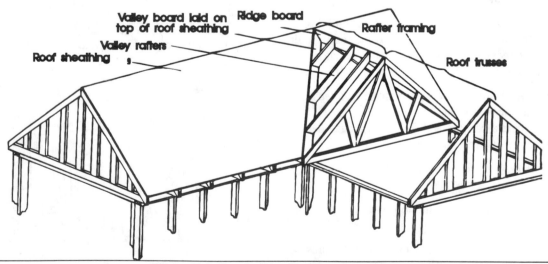

Figure V-4: Framing details of a valley roof.

vanishing point: A term used in perspective drawings to designate the point toward which all parallel receding lines converge or meet. See Fig. V-5.

vanity: A lavatory (wash basin) or make-up table, usually with an enclosed cabinet below.

vanity lighting: Incandescent lights installed around the perimeter of a vanity mirror. See Fig. V-6:

vapor: Word usually used to denote vaporized refrigerant rather than gas.

vapor barrier: Thin plastic or metal foil sheet used to prevent water vapor from penetrating the interior of building or insulating material.

vapor heating: A type of heating system consisting of a two-pipe gravity return system of steam circulation in which provision is made to retard or prevent the passage of steam from the radiator into the return main, and in which the air from the system, as well as the condensed water, is carried back to a point near the boiler. At this point the air is expelled from the mains and the water is returned to the boiler.

vaporize: To convert into a state of gas or vapor.

vapor lock: Condition where liquid is trapped in line because of a bend or improper installation that prevents the vapor from flowing.

vapor retarder: Material used to retard the movement of water vapor into walls. Vapor retarders are applied over the warm side of exposed walls or as a part of batt or blanket insulation. They usually have perm value of less than 1.0.

vapor safe: Constructed so that a device or apparatus may be operated without hazard to its surroundings in hazardous areas.

vapor, saturated: A vapor condition that will result in condensation into liquid droplets as vapor temperature is reduced.

vapor-tight: So enclosed that vapor will not enter.

vara: Spanish term of measurement, being 33⅓ inches.

variable: A mathematical quantity which may change in value while that of the others remain constant.

variable motion: When a body moves over equal spaces in unequal times, the motion is said to be variable.

variable resistance: A resistance in an electrical circuit that can be adjusted or changed to different values.

variation order: A written order from the building owner/architect/engineer authorizing an increase of the work above the amount shown in the original contract.

varnish: A solution of certain gums, in alcohol or oil, used for producing a hard, shiny finish on a surface. Varnishes may be either clear or colored.

variable speed drive: A motor having an integral coupling device that permits the output speed of the unit to be easily varied through a continuous range.

varying speed motor: A motor in which the speed varies with the load, ordinarily decreasing as the load increases. This type of motor, however, is not the same as an adjustable speed motor.

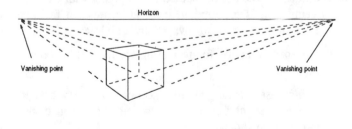

Figure V-6: Vanishing point laid out on drawing prior to drawing a perspective view of a building.

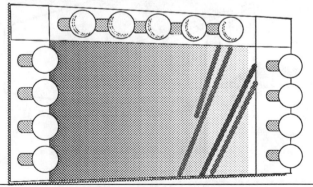

Figure V-6: One type of vanity lighting.

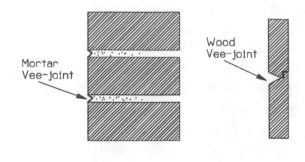

Figure V-7: Typical V-joints.

Vault: An arched ceiling or roof; an enclosed space covered by a vault, as in transformer vault.

V block: V-shaped groove in a metal block used to hold a shaft.

vee gutter: A valley gutter.

vee joint: 1) A small chamfer on the face edge of boards made so that when two boards, when joined, from a V at their junction to reduce shrinkage. See Fig. V-7. 2) A slight concave horizontal V formed in a mortar joint by tooling.

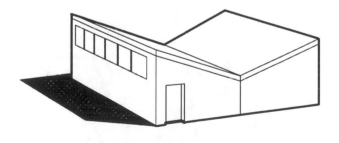

Figure V-8: Vee roof.

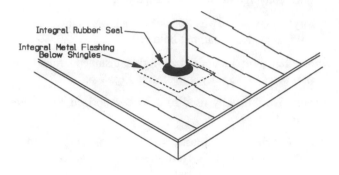

Figure V-9: Vent flashing used for weatherproofing a vent-through-roof.

vee roof: The shape formed when two lean-to or shed roofs meet at a valley. See Fig. V-8.

vee tool: A parting tool.

Venetian red: A very brilliant, red pigment for paint made by heating ferrous sulphate in the presence of lime.

Vent: A pipe or duct, or a screened or louvered opening, which provides an inlet or outlet for the flow of air. Common types of roof vents include ridge vent, soffit vents, and gable-end vents.

vent flashing: Collar made of sheet metal that fits over exposed vent pipe and is attached to the roof (under the roof covering) to prevent water from entering the roof opening. See Fig. V-9

ventilation: The act or method of supplying with fresh air. Ventilation is required in most attic areas to facilitate the removal of moisture vapor and condesate.

ventilator: A device for providing fresh air to a room or other space.

vent increaser: Enlarged pipe section for vents-through-roof to prevent freezing and for more efficient dispersion of vent fumes.

vent pipe: Any small ventilating pipe running from various plumbing fixtures to the vent stack.

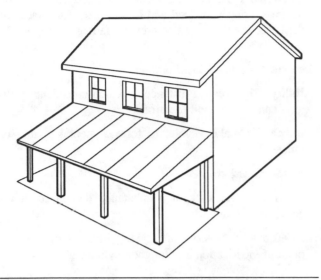

Figure V-10: Veranda

vent stack: The vertical pipe connecting with the vent pipes and extending through the roof. See Fig V-9.

veranda: An open portico extending along the outside of a building. See Fig. V-10.

verdigris: Green basic acetate of copper formed as a protective patina over copper exposed to the air. It may be of any color from brown to black in congested areas.

verge: The edge of the tiling, slate, or shingles projecting over the gable of a roof.

verge boards: Boards suspended from the verge of a gable.

verge fillet: A batten fixed on a gable wall to the ends of the roof battens to offer a finished look to the roof beyond the shingles.

verify: To prove to be accurate.

vermiculated: Stones, etc., worked so as to have the appearance of having been eaten into by worms.

vermiculite: Lightweight inert material resulting from expansion of mica granules at high temperatures that is used as an aggregate in plaster.

vermillion: A brilliant red, slightly orange-colored pigment composed of mercuric sulphide.

vernier: A small movable auxiliary scale for obtaining fractional parts of the subdivisions of a fixed scale. See *Vernier caliper*.

vernier calipers: Precision measuring instrument for taking inside and outside lienal measurements.

vertical: Plumb, perpendicular, upright.

vertical grain: The edge grain of quarter-sawn wood.

vertical sash: A sash window.

vertical weld: Welding joint made with perpendicular (up and down) motions.

vestibule: A small entrance room, either to a building or to a room within the building. See Fig. V-11.

viaduct: A large masonry, bridgelike structure for carrying a roadway or railroad over a valley, gorge, or the like.

vibrating screed: A vibrating power tool used in finishing concrete. The vibrating action of the tool tends to consolidate and level concrete in forms.

vice: A screwed metal or wooden clamp fixed to a workbench and used for holding workpieces. See *vise*.

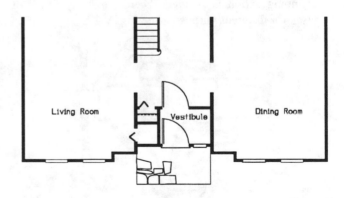

Figure V-11: Partial floor plan of a building showing vestibule.

view: On architectural and other types of working drawings, the identification of the way the building is shown; that is, floor plan, front elevation, rear elevation, side elevation, sectional, detail, etc. Also orthographic view, isometric view, perspective view, etc.

vignette: A vine-like ornament used in architectural design.

vinyl: Strong thermoplastic material used for a variety of building materials including floor tile, pipe for plumbing installations and conduit for electrical installations.

vinyl-coated nail: Nail with a vinyl coating to prevent rusting.

vinyl-tile cutter: Tool used for cutting vinyl and asphalt floor tile.

vitrified brick: Brick that has been surfaced glazed.

viscosity: Term used to describe resistance of flow of fluids. The density of fluid, gauged by the rate at which it flows through a gauge pipe of standard length and diameter.

viscous: Refers to thick or slow-flowing liquids.

vise: A mechanical contrivance for holding a piece of wood or metal while it is being worked on. It consists essentially of two jaws, one fixed and one movable, the movable jaw being operated by a screw by means of which clamping action is secured. There are many forms of vises for different purposes. See Fig. V-12.

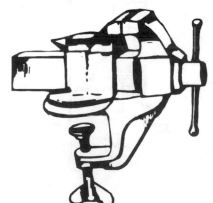

Figure V-12: Bench vise.

vise clamps: False jaws for a vise. They are usually made of brass, copper, or lead and are used over the faces of the hardened steel jaws to prevent damage to the workpiece.

vise grips: A type of locking pliers.

visibility: The greatest distance at which conspicuous objects can be seen and identified.

visualize: To make visible; to form a mental picture or image.

vitreous: Glasslike.

vitrification: The state of a substance that is fused together by burning.

vitrified brick: A very hard paving brick burned to the point of vitrification and toughen by annealing.

vitrolite: Glass wall tile used in bathrooms and for facing buildings. Made in a variety of colors.

vixen file: A flat file with curved cuts, particularly efficient on soft metals and for filing round work.

V joint: One of several types of masonry joints; this type forming a "V."

VMA: Valve Manufacturers Association of America.

voids: Vacant spaces such as occur between broken particles of a substance, as stone, coal, etc.

volatile: Capable of being rapidly vaporized.

volatile liquid: A liquid that vaporizes readily.

volatile thinner: A liquid that evaporates readily, used to thin or reduce the consistency of finishes without altering the relative volumes of pigments and nonvolatile vehicles.

volt: The unit of electromotive force; electrical pressure. One volt sends a current of 1 ampere through a resistance of 1 ohm.

voltage: The electrical property that provides the energy for current flow; the ratio of the work done to the value of the charge moved when a charge is moved between two points against electrical forces.

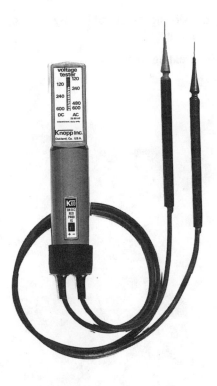

Figure V-13: One of several types of voltmeters currently used to test electrical circuits.

voltage drop: The voltage consumed in forcing a current through a certain resistance or group of resistances connected in a single circuit.

voltage regulator: A device to decrease voltage fluctuations to loads.

voltage, signal: Voltages to 50 volts.

voltage to ground: In grounded circuits the voltage between the given conductor and that point or conductor of the circuit that is grounded; in ungrounded circuits, the greatest voltage between the given conductor and any other conductor of the circuit.

voltaic cell: Primary cell. Name given to the cell first discovered by the Italian physicist, Alessandro Volta (1745-1827). Sometimes called a galvanic cell. It is a cell in which two dissimilar metals are immersed in a solution that is capable of acting chemically more upon one than on the other, to produce a difference of potential (voltage) across the metals.

volt-ampere: Volt(s) multiplied by ampere(s). Equals watt(s). Volt-amperes is currently used by the National Electrical Code to designate power in an electrical circuit. Formally, the term *watts* was used.

voltmeter: An instrument for determining or measuring voltage. See Fig. 13.

volume: The cubic contents of an object; space.

volume estimating: Determining approximate construction costs by multiplying the cubic feet of a building by a factor that is known to be reasonably accurate. See *area estimating.*

volute: The spiral ornament of Ionic and Corinthian capitals.

VOM (volt-ohm-multimeter): A commonly used electrical test instrument to test voltage, current, resistance, and continuity.

vossoir: One of the wedge-shaped blocks of stone of which an arch is composed. See Fig. 14.

VTR: Vent through roof. Used as an abbreviation for vent-through-roof on plumbing drawing and spoken *V-T-R* on the job; for example, "We just finished installing the flashing around the V-T-Rs." See Fig. V-21; also see Fig. 22 which shows how VTRs are indicated on a plumbing riser diagram.

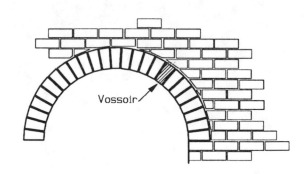

Vossoir

Figure 20: Vossoir used in an arch.

vulcanizing: Treating crude rubber with sulphur at a high temperature to increase its strength and elasticity. The joining of two pieces of rubber by the application of cement and heat.

V weld: See welding joints.

waferboard: A structural panel made with pressed wood chips (wafers) and resin. It has a strength similar to that of plywood and is used for sheathing, paneling, and crating.

wainscot: A lining of interior walls, usually paneled with a material different from the upper portion of the wall. See Fig. W-1.

wainscoting cap: The molding at the top of a wainscoting.

wainscoting: Lining or paneling interior walls with wood.

wainscot oak: Quarter-sawn oak used for wall paneling.

waist: The part of an object with the least thickness; the narrow part of an object.

waiver: The renunciation, abandonment, or surrender of some claim, right, or privilege.

waler: Horizontal bracing used to align and support concrete forms. Also called *wale, whalers, rangers.* See Fig. W-2.

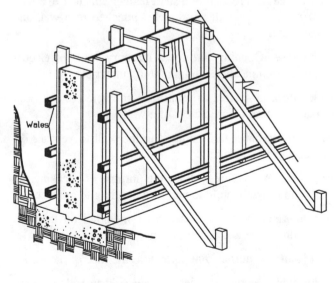

Figure W-1: Wainscot.

Figure W-2: Wales used in concrete form.

walk-through: Inspection of a building or other construction project by the owners, architect/engineer, and contractor (or their respective representatives).

walkway: A permanent platform, usually with handrails, to allow walking on a slanted roof.

wall: A vertical, load-bearing structure used to enclose a building. Non-load-bearing interior walls, used to divide areas, are called *partitions*. The thickness of insulation or jacket of cable.

wall beam: A reinforced bearing wall that spans the full width of the building from column support to column support.

wall bed: Any one of the various types of beds that swing or fold into the wall or closet when not in use. Commonly used in small apartments.

wallboard: Sheets or panels (usually 4' × 8') for surfacing ceilings and walls. Wallboards include plywood, gypsum, plasterboard, hardboard, and fiberboard.

wall box: Metal box built into masonry walls to provide a bearing for beams or joists.

wall braces: Wood or metal braces used at corners of walls to strengthen the structure.

wall bracket: Frequently a shaft hanger attached to a wall or post; in general, any bracket attached to a wall and used as a support.

wall chase: Opening through a masonry wall for electrical, communication, and plumbing materials.

wall column: A steel or reinforced concrete column built partly within the thickness of a wall.

wall fan: Exhaust fan mounted in the wall to vent air directly to the outside.

wall, fire: A dividing wall for the purpose of checking the spread of fire from one part of a building to another.

wall heater: Electric or gas heater mounted on the wall — either recessed or surface-mounted.

wall joint: A mortar joint parallel to the face of the wall.

wallpaper: Decorated printed paper sold in rolls for wall covering. There are many types of wallpaper available, and each has its own particular surface and thickness characteristics. There are pulp papers that have patterns printed directly on them. These are the least expensive wallpapers. Other wallpapers are known as ground papers. These receive a coated treatment or ground before the pattern is printed on them. They are usually of a better quality than pulp papers. Some wallpapers are oil printed; this makes them washable.

- **Embossed Papers:** Embossed papers are made from two papers that are gummed and pressed together and then passed through steel rollers. These rollers have raised and depressed areas on them. Where one roller has a raised area, the other has a corresponding depressed area. Together, these rollers produce an irregular surface on the paper. Embossed patterns are limited only by the patterns and designs of the steel rollers. Embossed papers are also known as duplex papers.

- **Flock Paper:** Flock paper is one that has very small strands of silk and wool glued to its surface. The raised paper surface is first treated with a gold size or other tacky substance before the fibers are applied.

- **Vinyl Wall Coverings:** Vinyl wall coverings can be made to provide the rustic look of cedar paneling. Such coverings come in many patterns and designs. A stucco effect that accents mediterranean and Spanish styles is achieved by the adobe vinyl wall covering. The name give to the rustic look of cedar paneling is "Planks," because its texture is like that of wood planks.

- **Fabric Wall Coverings:** Fabric wall coverings are made from silk, tapestry, burlap, and canvasses. These fabrics come in natural or dyed finishes and are washable. They are often treated with a tough glazed waterproof coating.

wall plate: 1) A flush-mounted receptacle or switch plate used to cover the outlet box and wiring device. 2) A horizontal timber on a wall for bearing the ends of joists, girders, etc., and for distributing the weight.

wall plug: An electric outlet located in or on the wall for the purpose of providing a source of current. See Fig. W-3.

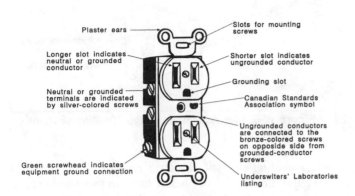

Figure W-3: Wall plug and its related connections.

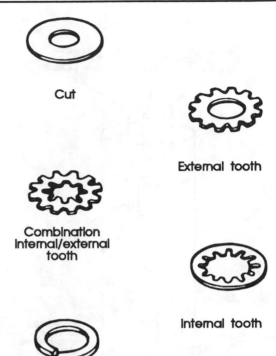

Cut

External tooth

Combination
Internal/external
tooth

Internal tooth

Split lock

Figure W-4: Several types of washers used on
building construction projects.

wall, retaining: Used to hold back a bank or solid mass of soil or water at the sides of the basement.

wall socket: See *wall plug*.

wall tile: A tile made from such materials as terra cotta, glass, concrete — glazed or unglazed — for use in obtaining a smooth or decorative face on a wall.

wane: A defect in a timber or plank.

Warerite: A type of laminated plastic.

warding file: A very thin, flat file used principally by locksmiths.

warp: To permanently distort or twist out of shape as by moisture or heat.

warranty deed: One that contains a covenant that the grantor will protect the grantee against any claimant.

washability: The ability of paint to withstand washing without damage; the ease in which dirt can be removed from a coat of paint.

washer: A small, flat, perforated disk, used to secure the tightness of a joint, screw, etc. See Fig. W-4.

waste-disposal unit: A small electrically driven garbage grinder connected to the kitchen sink drain to grind and dispose of kitchen food waste. See *garbage disposal*.

waste: Liquid-borne waste free of fecal matter.

waste pipe: A plumbing pipe that carries only waste. See Fig. W-5.

water bar: A bar or strip inserted in a joint between wood and stone sills of a window to prevent passage of water.

water closet: A toilet.

water-cooled condenser: Condensing unit that is cooled through the use of water.

water cooling tower: A wood, metal, or concrete structure used for cooling condenser water by evaporation.

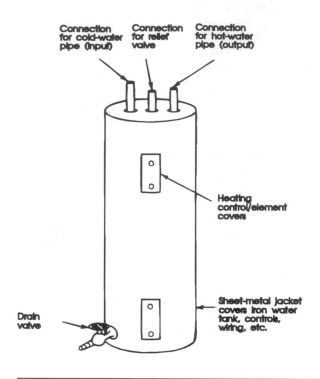

Figure W-5: Typical domestic hot-water heater.

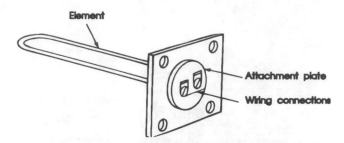

Figure W-6: Electric immersion-type heating element.

water hammer: Sounds like the blows of a hammer occurring when the flow of water in a pipe is suddenly checked. Overpressures causing water hammer also endanger the soundness of the water supply system. Air chambers installed in the cold- and hot-water pipes do away with water hammer and the overpressures caused by it.

water heater: Apparatus consisting of a tank, heating elements, thermostats to control the heating elements, and a safety valve. In operation, cold water enters the tank from the domestic cold-water line and is then heated by the heating elements. During this process, the heated water rises to the top of the tank where it is discharged through the hot-water outlet pipe. Thermostats are utilized to sense the temperature of the water and to control the heating elements to maintain the temperature of the water leaving the tank for use in the home. See Fig. W-5.

water-heating element: Electric resistor heater used in conjunction with water-heater tanks to provide domestic hot water. Two types of water-heating elements are in general use: the immersion heater and the external element. Of these two types, the immersion type is by far the most common. With this type of heater, the heating element is installed through an opening in the tank wall so that it is in direct contact with the water in the tank. The external element straps around the tank like a belt and heats the water through the wall of the tank. With this latter method, the tank itself must first be heated and then the heat from the metal tank is transferred to the water. See Fig. W-6.

water main: Water supply pipe for public use.

water meter: Measuring device used on public water-supply lines to determine the amount of water used by customers. See Fig. W-7.

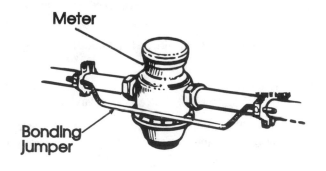

Figure W-7: Water meter.

water pipe: Pipe used to transmit water from one place to another. See *water main*.

waterproof cement: Cement that is impervious to water.

waterproofing: Means of protection to resists moisture.

water putty: A powder which, when mixed with water, makes an excellent filler for cracks, nail holes, etc. Not suitable for glazing.

water seasoning: Soaking timber for approximately two weeks and then air-drying it.

water softener: A device for eliminating from domestic water supplies the calcium and magnesium sulfates or bicarbonates which render soap valueless for cleansing purposes; also any chemical added to water to accomplish this purpose.

water spots: Slight discolorations that appear in the film of lacquer enamels and oftentimes appear to go down very deep into the film. Usually caused by sealing in moisture.

water supply system: The water service pipe, water distributing pipes, and the necessary connecting pipes and fittings, control valves and appurtenances in or adjacent to the building or premises.

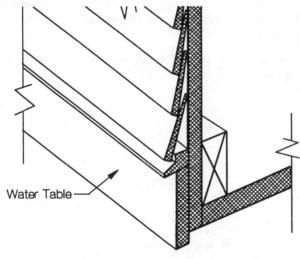

Water Table

Figure W-8: Water table used to throw rain water away from wall.

water table: 1) A projecting sloping member around a building to throw rain water away from the wall. 2) Distance from surface of ground to a depth at which natural ground water is found. See Fig. W-8.

water-cooled condenser: Condensing unit that is cooled through the use of water.

waterblocked cable: A multiconductor cable having interstices filled to prevent water flow or wicking.

waterproof: Moisture will not interfere with successful operation.

waterproofing walls: Making them impervious to water or dampness, by mixing a compound with the concrete, or by applying the compound to the surface.

watertight: So constructed that moisture will not enter the enclosing case or housing.

watt (W): The practical unit of electrical power; currently referred to as *volt-amperes*.

watt-hour meter: A meter which measures and registers the integral, with respect to time, of the active power in a circuit. A typical watt-hour meter consists of combination of coils, conductors, and gears — all encased in a housing. The coils are constructed on the same principle as a split-phase induction motor, in that the stationary current coil and the voltage coil are placed so that they produce a rotating magnetic field. The disc near the center of the meter is exposed to the rotating magnetic field. The torque applied to the disc is proportional to the power in the circuit, and the braking action of the eddy currents in the disc makes the speed of rotation proportional to the rate at which the power is consumed. The disc, through a train of gears, moves the pointers on the register dials to record the amount of power used directly in kilowatt hours (kWh). See Fig. W-9.

Figure W-9: Watt-hour meter used to measure the active power in an electrical circuit.

watt hour: A unit of measure of electrical work, equal to one watt expended for one hour.

wattmeter: An instrument for measuring electric power in watts; the unit of electrical energy, volts times amperes, combining therefore the functions of a voltmeter and an ammeter.

wave: A disturbance that is a function of time or space or both.

waveform: The geometrical shape as obtained by displaying a characteristic of the wave as a function of some variable when plotted over one primitive period.

wave length: The length in meters of one complete sine wave of an alternating current.

wave-tooth saw: Saw blade for use in jigsaws, hacksaw frames, and similar types of saws. Available in 32 pitch (teeth per inch) and is used for cutting tubing and thin sheet metal.

wavy grain: Term sometimes used to signify curly attractive grain in lumber.

wax: An organic salt (ester) of a high monatomic alcohol and a high fatty acid; one example is beeswax.

wax finish: A very smooth finish secured by polishing a wood surface with wax prepared for the purpose.

ways: Longitudinal guides upon which the work or a table bearing the work may slide, such as the ways of a lathe.

wear and tear: Depreciation in value due to use.

W beam: Wide-flanged steel beam designation. The nominal height varies between 4″ and 36″.

weather: To season, dry, injure, or alter in any way the condition of wood, stone, or other material through exposure to the weather.

weatherboard: Term sometimes used in place of *siding* or *sheathing*.

weatherboards: Boards used as an outside covering of buildings, nailed on so as to overlap and shed the rain.

weatherhead: The conduit fitting at a conduit used to allow electrical conductor entry, but prevent weather entry. See Fig. W-10.

Figure W-10: Weatherhead being used to allow electrical conductor entry, but prevents any weather entry.

weathering: 1)The slope given to offsets, buttresses, and the upper surface of cornices and moldings, to throw off rain. 2) The wearing away of the surface of timbers caused by exposure to the elements.

weatherproof: So constructed or protected that exposure to the weather will not interfere with successful operation.

weather strip: A strip of metal, wood, or other material used around doors and windows to prevent draughts.

weathervane: A weathercock.

web: An open braid; central portion of an I beam. See Fig. W-11.

webbing: Made from jute fiber and used mainly as a spring support on furniture frames.

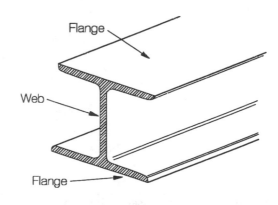

Figure W-11: Areas of an I-beam.

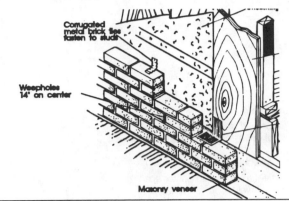

Figure W-12: Location of weepholes.

wedge: A piece of wood or metal, V shaped in longitudinal section, used for producing strong pressure, or for splitting a substance apart.

weephole: Small gaps or holes at the foot of a cavity wall, or in a wood sill, to allow water to escape outwards. See Fig. W-12.

weight correction factor: The correction factor necessitated by uneven geometric forces placed on cables in conduits; used in computing pulling tensions and side-wall loads.

weir: A dam or barrier by means of which the water of a stream is held back in order to provide a sufficient head of water for power purposes.

weld: The joining of materials by fusion or recrystallization across the interface of those materials using heat or pressure or both, with or without filler material.

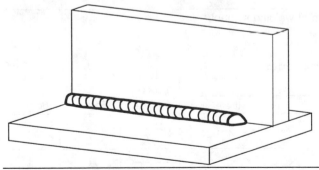

Figure W-13: Weld bead used to join two pieces of metal.

weld bead: A single welding pass made on the surface of the metal to be joined. See Fig. W-13.

welded steel fabric: Wire mesh welded together in square or rectangular configurations; for use as reinforcement in concrete slabs.

welded wire fabric: See *welded steel fabric*.

welding: Uniting of pieces of iron or steel by fusion accomplished by the oxyacetylene, electric (Fig. W-14), or hammering (forging) process.

welding cables: Flexible cable used for leads to the rod holders of arc-welders, usually consisting of size 4/0 flexible copper conductors.

welding rod: A metal rod in various diameters from ¼ to ½ inch in diameter; used for flowing into the joint to be welded. Welding rods are of different composition according to the class of work on which they are to used.

welding symbols: Graphic symbols used on working drawings to indicate the type and size of weld. See Fig. W-15 for an explanation of welding symbols; Fig. W-16 for a complete listing of welding symbols.

welding transformer: A step-down transformer used to produce sufficient instantaneous current to fuse the metals, in contact, through which it is flowing.

Figure W-14: Process of welding.

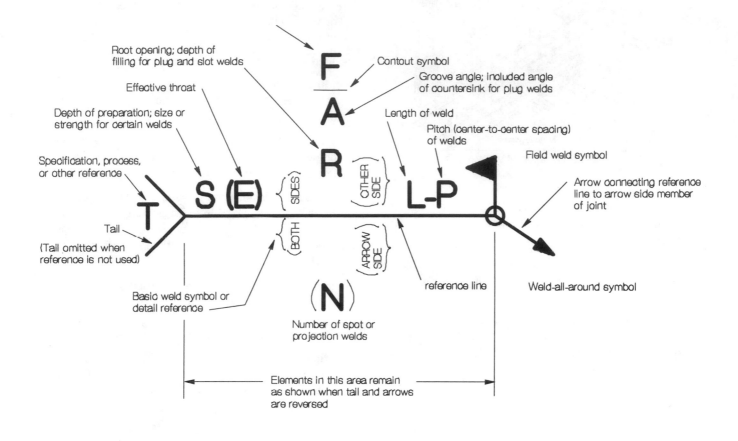

Root opening; depth of
filling for plug and slot welds

Contout symbol

Groove angle; included angle
of countersink for plug welds

Effective throat

Length of weld

Depth of preparation; size or
strength for certain welds

Pitch (center-to-center spacing)
of welds

Specification, process,
or other reference

Field weld symbol

Arrow connecting reference
line to arrow side member
of joint

Tail

(Tail omitted when
reference is not used)

Basic weld symbol or
detail reference

reference line

Weld-all-around symbol

(N)

Number of spot or
projection welds

Elements in this area remain
as shown when tail and arrows
are reversed

T S (E) {SIDES} {BOTH} R {OTHER SIDE} {ARROW SIDE} L-P

F
A
R

Figure W-15: Interpretation of welding symbol.

Basic Weld Symbols and Their Location Significance											
Location Significance	Fillet	Groove						Back or Backing	Flange	
		V	Bevel	U	J	Flare V	Flare Bevel		Edge	Corner
Arrow Side										
Other Side										
Both Sides								Not Used	Not Used	Not Used

Figure W-16: Welding symbols developed and standardized by the American Welding Society.

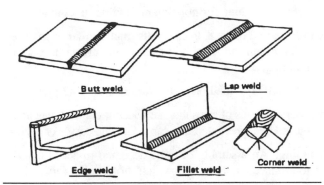

Butt weld   Lap weld

Edge weld   Fillet weld   Corner weld

Figure W-17: Typical weld joints.

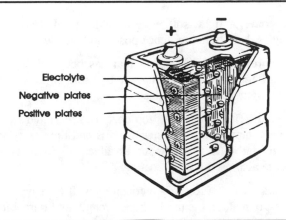

Electolyte
Negative plates
Positive plates

Figure W-19: Parts of a wet-cell battery.

weld joints: Various types of welds used to join pieces of metal together. See Fig. W-17.

well: 1) Driven, bored or dug hole in the ground from which water is pumped. 2) The space between the flights of a stair.

wellhole: The small space enclosed on three sides by three flights of a stair.

welt: A seam in metal roofing.

western framing: Platform framing.

Western Union splice: The electrical connection made by paralleling the bared ends of two conductors and then twisting these bared ends, each around the other. See Fig. W-18.

wet bulb: Device used in measurement of relative humidity.

wet cell battery: A battery having a liquid electrolyte. See Fig. W-19.

wet locations: Exposed to weather or water spray or buried.

wet rot: Decay of lumber, due to moisture and warmth.

wet vent: A vent that also receives the discharge of wastes other than from water closets.

Wheatstone bridge: A method of measuring resistances by the pro-portion existing between the resistance of three known adjustable resistances and the one to be found, all forming the arms of the bridge.

wheelbarrow: A container made from wood, steel, or similar material with a single wheel in front and two handles behind, by which it is lifted and pushed forwards. The operator lifts approximately ¼ of the total load in the container.

whetstone: A stone used for sharpening cutting tools.

whetting: The rubbing of a tool on an oilstone for the purpose of improving its cutting edge. See Fig. W-20.

Figure W-18: Western Union splice.

Figure W-20: Whetting a carpenter's chisel on a whetstone.

white cedar: A light, soft, very durable wood. Used for shingles, boat building, fence posts, and wooden ware.

white cement: Portland cement that has been selectively ground to avoid contamination by iron or to which white pigment has been added.

white coat: The hard, white top coat of a plastered wall. It is a composition of plaster of Paris and lime putty, to which marble dust is sometimes added. Gypsum plasters are also used for top coating.

white oak: The hardest of American oaks. It is heavy and close-grained, and is used where strength and durability are required.

white pine: A straight-grained softwood, light in color; used extensively in cabinetry and for interior trim.

white spruce: An inexpensive wood largely used for framing, flooring, etc.

white wash: Lime slaked in water and applied with a brush or as a spray. Salt is sometimes added to make it adhere better, and bluing may be added to give a whiter tone.

whitting: Pulverized chalk; when mixed with oil into a paste form, it becomes putty.

Whitworth thread: The standard English thread, having rounded tops and bottoms and an included angle of 55 degrees.

wicket: A small door set within a larger door.

wiggler: A device use on accurate work for exactly locating a centerpunch mark on work to be drilled, directly in line with the centerline of the drill spindle. Used also for accurately truing work in a lath chuck.

wild cherry: A reddish brown, moderately heavy, hard, strong, wood that does not warp or split in seasoning; much used in furniture, fine panels, and the like. Sometimes referred to as *wild black cherry*.

William and Mary: This furniture style, named after the English monarchs, succeeded the late Jacobean; it is characterized by lighter and more graceful designs with outlines combining the straight line and the curve; turned uprights and underpinnings, woven cane surfaces, and club-shaped foot.

winch: A machine for hoisting or hauling; a windlass.

windbreak: Fencing, trees, walls, and similar items used to break up the wind and shelter a court yard, road, or building.

wind load: The load on a structure due to wind pressure.

winders: Treads of steps that are wider at one end than at the other. Used where steps are carried around curves or angles.

winding: An assembly of coils designed to act in consort to produce a magnetic flux field or to link a flux field.

winding stair: A stair which changes directions by means of winders or a landing and winders. The wellhole is very wide and the balustrade follows the curve with only a newel at the bottom.

windlass: A machine for hoisting or hauling.

window: An opening in the wall of a building closed with transparent material inserted in a frame, placed so as to admit light, and made to open for air. See Fig. W-21.

window, bay: A window projecting outward from the face of the wall and built up from the ground.

window jack: A small portable platform which fits over a window sill and projects outward beyond it.

window, oriel: A large bay window projected from a wall and supported by a bracket or corbel.

window seat: A seat built below, or in the recess of a window. See Fig. 22.

window unit: Commonly used when referring to air conditioners that are placed in a window.

windowhead: The upper portion of a window frame.

windowsill: The lower or base framing of a window opening.

wing: A section of a building extending out from the main part.

wing nut: A form of nut which is tightened or loosened by two thin flat wings extending from opposite sides.

winged dividers: Dividers having a flat metal wing attached to one leg and projecting through the other. A setscrew on the slotted leg permits locking the dividers at a desired dimension.

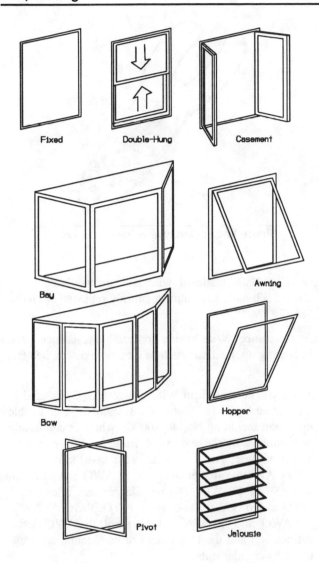

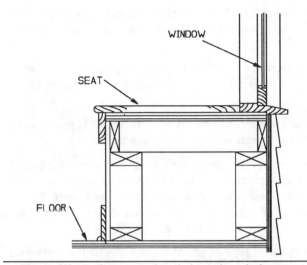

Figure W-22: Cross-sectional view of a window seat.

wire: A slender rod or filament of drawn metal: the term may also refer to insulated wire.

wire bar: A cast shape which has a square cross section with tapered ends.

wire brush: A hand brush fitted with wire or thin strips of steel instead of bristles. Used for removing rust, dirt, or foreign matter from a surface. See Fig. W-23.

wire, building: That class of wire and cable, usually rated at 600V, which is normally used for interior wire of buildings.

Figure W-23: Wire brush.

Figure W-21: Common types of windows.

wiped joint: A lead joint in which molten solder is poured upon two pieces to be jointed until they are of the right temperature. The joint is then wiped up by hand with a moleskin or cloth pad while the solder is in a plastic condition.

wire connectors: Metal or plastic device used to join two or more electrical conductors. Many types are available.

wire, covered: A wire having a covering not in conformance with NE Code standards for material or thickness.

wire drawing: The process by which wire is made; as by drawing metal through a hole in a steel plate.

wire gauge: A notched plate having a series of gauged slots, numbered according to the sizes of the wire and sheet metal manufactured; used for measuring the diameter of wire. Most widely used in the United States is the American Wire Gauge (AWG).

wire glass: A type of window glass in which is imbedded wire of coarse mesh, to prevent scattering of fragments should the glass be broken.

wire, hookup: Insulated wire for low voltage and current in electronic circuits.

wire mesh: Welded or brazed wire joined in a grid for use as reinforcement in concrete slabs.

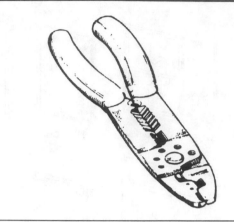

Figure W-25: One type of wire strippers.

wire nails: Nails made of wire and are available in all sizes. Such nails have almost entirely replaced cut nails. See Fig. W-24.

wire, resistance: Wire having appreciable resistance; used in heating applications such as electric toasters, heaters, etc.

wire sizes: The American Wire Gauge (AWG) is used in the United States to identify the sizes of wire and cable up to and including No. 4/0 (0000), which is commonly pronounced in the electrical trade as "four-aught." These numbers run in reverse order as to size; that is, No 14 AWG is smaller than No. 12 AWG and so on up to size No. 1 AWG. The next larger size after No. 1 AWG is No. 1/0 ("one-ought") AWG, then 2/0 AWG, 3/0 AWG, and 4/0 AWG. At this point, the AWG designations end and the larger sizes of conductors are identified by circular mils.

wire stripper: A special tool for removing the insulation from electrical conductors. See Fig. W-25.

wireway: A sheet metal trough used for housing electrical conductors. It is sometimes referred to as "auxiliary gutter."

wiring diagram: Drawing that uses lines, symbols, and notes to convey information — usually in diagrammatic form. See Fig.26.

withe: See wythe.

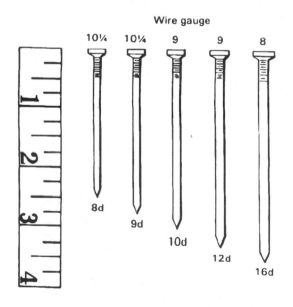

Figure W-24: Typical wire nails in common use.

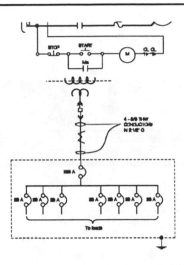

Figure W-26: Typical wiring diagram.

WM: Wattmeter.

wood alcohol: See *methanol*.

wood chisel: Tool used to remove sections of wood, such as for installing door hinges. Rough cutting is done with the bevel side down, while finish cuts are taken with the bevel side up. See Fig. W-27.

wood filler: 1) Sealer used to fill pores in wood. 2)Substance used to cover defects (holes and cracks) in wood.

wood finishing: Preparing the wood's surface to receive a finish, and applying paint, stain, or varnish; also polishing when certain kinds of finish are desired.

wood flour: Fine sawdust that is sometimes used as an extender for glues, wood fillers, and in explosives.

wood joint: Point and method of fastening pieces of wood together. See Fig. W-28 on page 336.

wood lathe: Power tool used for turning (rounding) wood pieces such as chair and table legs, dowels, and the like.

woodruff key: A semicircular or semielliptical key flattened on the sides, for use in a keyway cut by bringing a milling-machine rotary cutter against the material. See Fig. W-29 on page 336.

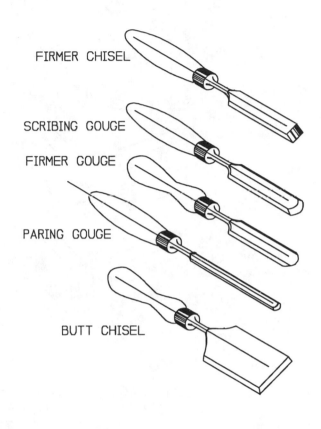

Figure W-27: Several types of wood chisels in common use.

wood screws: Made in oval-, round-, and flathead types. Length is measured from largest bearing diameter of head to the point of the screw. Gimlet points are standard, and screws are made bright galvanized and blued. The thread normally extends $7/10$ the length and the included angle of the head of flathead wood screws is 82 degrees.

wood shingle: Roof covering that is sawn rather than split as in wooden shakes.

wood turning: The art of shaping pieces of wood on a lathe.

woodwork: Things made of wood.

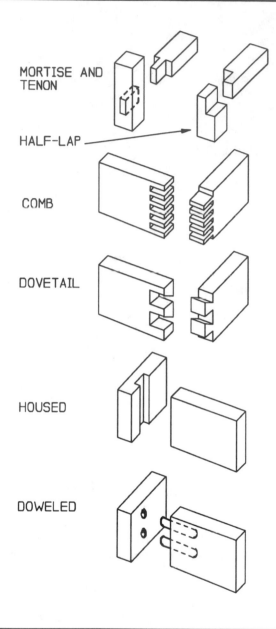

MORTISE AND TENON

HALF-LAP

COMB

DOVETAIL

HOUSED

DOWELED

Figure W-28: Types of wood joints in common use.

woodworking: The trade in which things are made out of wood.

work: Force times distance; pound-feet.

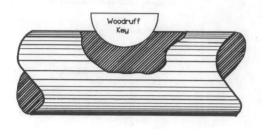

Figure W-29: Woodruff key.

workability: The ease at which lumber may be worked. For example, spruce and poplar are easier to work than most oaks; redwood is easier to work than ash or beech.

work function: The minimum energy required to remove an electron from the Fermi level of a material into field-free space; Units—electron volts.

work hardening: Hardening and embrittlement of metal due to cold working.

working drawing: A drawing containing all dimensions and instructions necessary for successfully carrying a job to completion.

working edge: See *working face*.

working face: The face side of timber.

working load: The ordinary load to which a structure is subjected; not necessarily the maximum load, but the average or mean load.

working-unit stress: The ultimate stress divided by the factor of safety.

Workmen's Compensation Insurance: Insurance for employees covering injury, illness, or death that occurs while working on the job site, or as a direct result from job performance.

worm gearing: Gearing composed of worms and worm wheels.

worm hole: Any hole bored by insects such as beetles in timber.

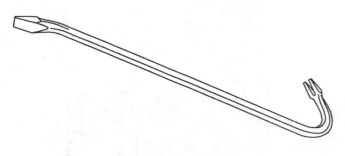

Figure 30: Wrecking bar.

worm threads: Acme type threads, having an included angle of 29 degrees, but deeper than the standard acme thread.

wreath: Section of a handrail curved in both vertical and horizontal planes ad used to connect the side of a newel post with the ascending run of the handrail.

wreath piece: The curved section of the handrail string of a curved or winding stair. Also simple wreath.

wrecking bar: A steel bar usually from 1 to 2 feet in length, with one end drawn to a thin edge, the other curved to a claw. See Fig. 30.

wrench: A tool for exerting a twisting strain, as in tightening a nut or bolt. Common types are adjustable-end wrenches, monkey wrenches, box wrenches, T wrenches, and socket wrenches. See Fig. W-31.

wrinkling: A gathered or wrinkled film in paint or lacquer that is caused by applying heavy coats, abnormal heat or humidity, or the application of an elastic film over a surface.

writ of execution: A writ that authorizes and directs the proper officer of the court (usually the sheriff) to carry into effect the judgment or decree of the court.

wrought iron: Iron that has had the major portion of its carbon, as well as the foreign elements which would effect its working value, removed. Used extensively in ornamental iron work.

W truss: Common roof truss used mainly in residential construction. See Fig. W-32.

wye: A fitting, either cast or wrought iron, (PVC) plastic, copper, or other suitable material, that has one side outlet at any other angle than 90 degrees. The angle is usually 45 degrees unless otherwise specified.

wye level: See *Y level.*

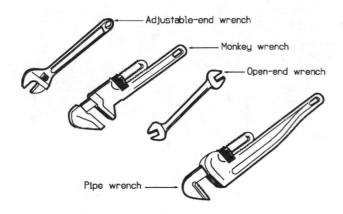

Figure W-31: Various types of wrenches used on building construction projects.

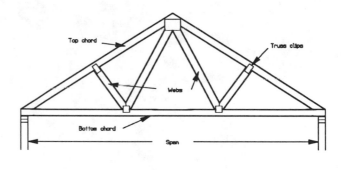

Figure W-32: W trusses are frequently used on residential construction.

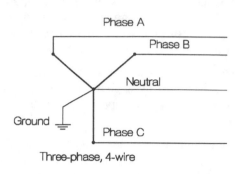

Figure W-33: Wye-connection diagram.

Wye connection: Three-phase transformer connection whereas all three phases are connected at a central point, forming a "Y" configuration. See Fig. W-33.

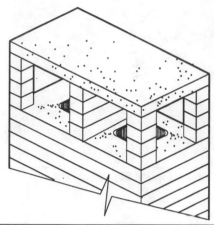

Figure W-34: A partition is sometimes used between chimney flues to prevent downdrafts.

wythe: Partition between flues of a chimney. See Fig. W-34.

xalostockite: A mineral with a pale variety of grossularite, occurring in white marble.

xanthic: A term derived from xanthine or xanthic acid pertaining to a yellow or yellowish color.

xanthic acid: An unstable organic acid with the type formula ROCSSH where R is the group, the methyl and ethyl esters of which are colorless, oily liquids with a penetrating odor. Its copper salts are bright yellow.

X-brace: Double diagonal cross-bracing forming the letter X. See Fig. X-1.

X-braced chair: A chair with X-shaped underbracing or stretchers. See Fig. X-1.

X-joint: Two wooden braces joined so as to form an X shape. See Fig. X-1.

xenolith: A rock fragment different from the igneous rock in which it is embedded. Found in some stones used for building construction.

xenon: A chemical element, Xe, atomic number 54. It is a member of the family of noble gases, group O in the periodic table. Its main use is to fill a type of flashbulb employed in photography and called electronic speed light.

xerography: A photo process for making copies of printed, written, or pictorial material.

xerox: See *xerography*.

XL: Extra large.

X-mark: A pencil or chalk mark on the face edge of a board to show that other edges are to be trued from this face.

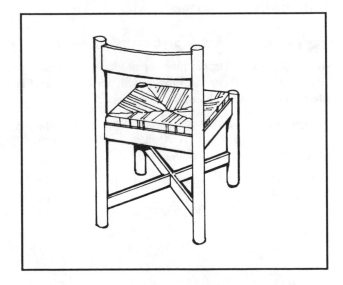

Figure X-1: Chair frame with X-shaped stretchers.

Xmfr: Abbreviation for transformer. Used mainly on electrical drawings.

x-ray: 1) Penetrating short wavelength electromagnetic radiation created by electron bombardment in high voltage apparatus; produce ionization when the rays strike certain materials. The rays pass through most objects as though they were transparent. 2) A popular name for Roentgen rays. A form of radiant energy sent out when the cathode rays of a Crookes tube strike upon the opposite walls of a tube or upon any object in the tube. Used in building construction to determine the quality of a welded joint. See Fig. X-2.

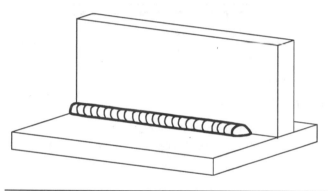

Figure X-2: The quality of welded joints may be determined by x-ray examination.

x-ray photograph: A radiograph made with x-rays.

x-ray tube: An electronic tube for producing x-rays. It is essentially a cathode-ray tube in which a metal target is bombarded with high-energy electrons.

X-shaped chair: A chair of ancient origin, the supporting structure of which is X shaped, often elaborately decorated, frequently folding.

x-stretcher: A stretcher having the form of an X. See *X-braced chair*.

x-unit: A unit of length, approximately 10^{-11} cm, formerly used for measuring x-ray wavelengths and crystal dimensions.

xylem: The botanical name for wood.

xylene: One of a group of three isometric aromatic hydrocarbons. The three isometric sylenes and ethylbenzene all have in common a molecular weight of 106.2 and the simplified formula, C_8H_{10}. Originally designated "coal tar" hydrocarbons, these compounds have, since World War II, been almost exclusively produced from petroleum by hydroforming or catalytic reforming of appropriate naphtha fractions.

xylograph: A wood carving.

xylol: An aromatic hydrocarbon distilled from coal tar at about 140° C; used as a solvent for synthetic resins and gums.

xylotomy: The art of cutting sections of wood for microscopic examination.

Xylonite: A highly inflammable mixture of camphor and pyroxylin; used as an ivory substitute in wood inlays.

xyst: Shaded or covered walk, as a promenade. See Fig. X-3.

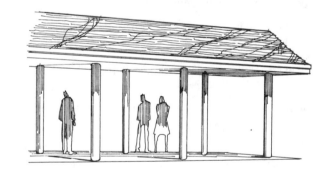

Figure X-3: Xyst or promenade.

Y

Y : See *wye*.

Yale lock: A common cylinder lock.

yankee screwdriver: A ratchet type screwdriver in which the blade is rotated by pushing the handle inward. A ratchet button provides a means to reverse the rotation of the blade. Good for light work, but this type screwdriver should never should be used when removing tight screws or for other heavy work. See Fig. Y-1.

yard: 1) A common unit of linear measure in English speaking countries, equal to 3 feet or 36 inches, and equivalent to 0.9144 meter. 2) The ground that immediately adjoins or surrounds a house, public building or other structure. 3) An enclosed area outdoors, often paved and surrounded by or adjacent to a building.

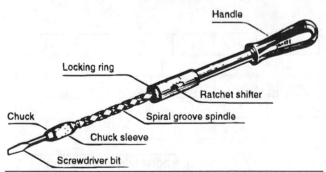

Figure Y-1: Spiral ratchet screwdriver, often called "yankee" screwdriver.

yardage: Relates to cubic yards of earth excavated. Also used in specifying cubic yards of concrete.

yard lumber: Lumber graded according to its size, length, and intended use, and stocked in a lumber yard.

Y connection: A branching connection applied to a three-phase electrical circuit or transformer; sometimes referred to as *star* connection. See Fig. Y-2 on page 342.

year ring: Also called "growth ring," or "annual ring." Rings of tubes or cells by means of which the sap is conveyed throughout a tree. They are clearly visible on a cross section of a log. Each ring represents a year of growth. See *annual ring*.

yellow brass: An alloy of 70 parts copper and 30 parts zinc. It is an inferior alloy used where strength is not essential.

yellow ocher: A color derived from the mineral earth ocher, and used as a pigment in paint, enamel, and similar substances.

yellow pine: An evergreen of two principal varieties, long leaf and short leaf. The wood of the long-leaf pine is dense, heavy, and very strong; mostly used in the form of heavy timbers. The short-leaf pine is brittle, not nearly so strong, and is less expensive; used for studding, joists, cheap flooring, etc.

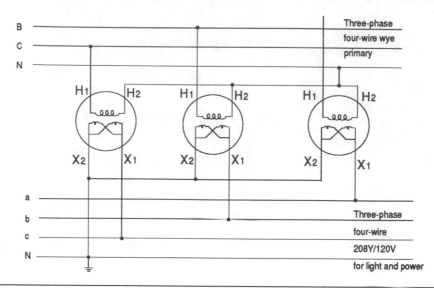

Figure Y-2: Y-Y three-phase transformer connection.

yelm: A double handful of reeds or straw laid on a roof, such as on thatch roofs.

yew: A slow-growing, medium-size evergreen tree. The wood is close grained, hard, and flexible. Its color is orange red to brown.

yield: The volume of lime putty of a given consistency obtained from a predetermined weight of quicklime.

yield point: That unit stress at which the specimen begins to stretch without increase in the load.

yield strength: The load required to produce a permanent stretch or elongation of a material.

yield value: The lowest pressure at which a plastic will flow. Below this pressure the plastic behaves as an elastic solid; above this pressure as a viscous liquid.

yoke: 1) The horizontal top member of a window frame. See Fig. Y-3. 2) A Y-shaped pipe fitting used for connecting branch pipes with a main soil pipe. 3) A measure or area of land equal to over 50, but less than 60 acres.

Yorkshire bond: A brickwork bond having random, widely spaced headers. See Fig. Y-4.

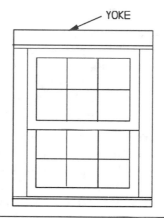

Figure Y-3: Window-frame yoke.

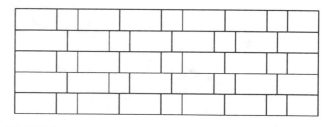

Figure Y-4: Yorkshire bond brickwork pattern.

Z

zaffer: An artificial mixture containing cobalt oxide and usually silica. Used to produce a blue color in glass and in ceramic glazes.

zax: A hatchetlike tool for cutting and punching nail holes in slate.

Z-bar: A metal bar with a z-shaped section used as a wall tie.

zebrawood: A very large tree, native to the west coast of Africa. Wood is hard, heavy, and tough; its light-colored background and parallel dark-brown stripes makes it valuable for unusual effects on fine furniture and paneling. Sometimes referred to as *zebrano*.

zein: A protein from maize. When dissolved in alcohol it gives a tough film that is used as a substitute for shellac.

zeolites: Minerals that are used in the base-exchange process of water softening.

zero: The numeral 0; a cipher; the lowest point. For example, "the voltmeter shows a reading of zero," meaning no voltage is present.

Z-flashing: Metal flashing used between the horizontal gaps formed at the intersection of two types of siding material. For example, a stucco-finish gable end and a wood-siding lower wall should be flashed at their junction as shown in Fig. Z-1.

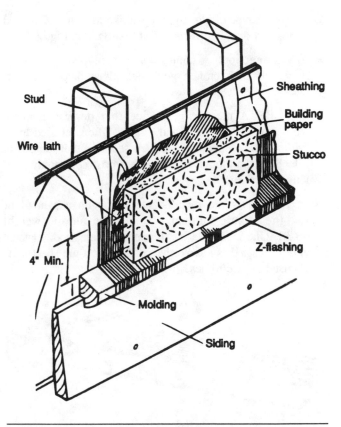

Figure Z-1: Z-flashing used at the junction of stucco and wood siding. *Courtesy U.S. Dept. of Agriculture.*

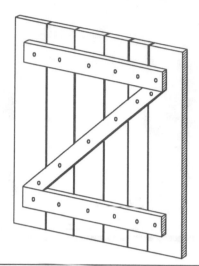

Figure Z-2: Door with z-frame bracing.

Z-frame: A type of bracing, shaped like the letter "Z," used to support doors constructed of boards. See Fig. Z-2.

zigzag connection: A connection of polyphase windings in which each branch generates phase-displaced voltages.

zigzag rule: A measuring device that derives it name from the way it opens and closes. Often referred to as *folding rule*. Made up in 6-inch sections that total from 2 to 8 feet, although 6 feet is the most commonly used length. See Fig. Z-3.

zinc: A bluish-white metallic element with a symbol of Zn, the atomic number 30, and the atomic weight 65.38. Zinc is familiar as the negative electrode material in dry cells and as a protective coating for some sheet metals used in electronics.

Figure Z-3: Six-foot folding (zigzag) rule.

zinc-alkaline battery: A relatively new type of storage battery introduced some years ago. Due to its high cost, it has gained very little commercial appeal. However, space projects are utilizing this type of battery due to its greater power output compared to its size and weight. Compared with the conventional lead-acid cell, the zinc cell has a watt-hour output per pound of cell over five times the output of the lead cell, while its watt-hour output per cubic inch of cell is three to four times as much.

zinc-carbon battery: A primary cell in which the negative electrode is zinc and the positive electrode is carbon, and which may be either wet or dry. See Fig. Z-4.

zinc chloride (ZnCl$_2$): A white deliquescent salt, obtained by the solution of zinc or zinc oxide, in hydrochloric acid, or by burning zinc in chlorine. Used as a soldering flux.

zinc dust: powdered zinc that is used in priming paints for use on galvanized iron.

zinc oxide: A nonpoisonous permanent white pigment that prevents chalking in other pigments.

zinc white: Zinc oxide used as a pigment in paint.

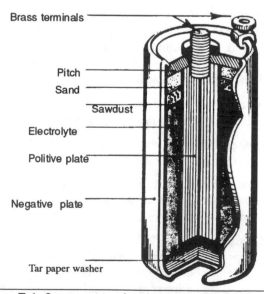

Figure Z-4: Components of a zinc-carbon battery.

zinox: A hydrated oxide of zinc used in the manufacture of enamels.

zip strip: A T-shaped vinyl strip used to form expansion joints in concrete slabs. Zip strips are inserted into the slab immediately after pouring the concrete and finishing. Once the concrete hardens, the flat top portion of the strip is peeled away to expose the concealed joint.

zonal cavity method: A calculation method used to determine the amount of illumination produced by a given installation of lighting fixtures in a particular room or other indoor area.

The zonal cavity method of calculating average illumination levels assumes each room or area to consist of the following three separate cavities: ceiling cavity, room cavity, and floor cavity.

Figure Z-5 shows that the *ceiling cavity* extends from the lighting fixture plane upward to the ceiling. The *floor cavity* extends from the work plane downward to the floor, while the *room cavity* is the space between the lighting fixture plane and the working plane.

If the lighting fixtures are recessed or surface-mounted on the ceiling, there will be no ceiling cavity and the ceiling cavity reflectance will be equal to the actual ceiling reflectance. Similarly, if the work plane is at floor level, there will be no floor cavity and the floor cavity reflectance will be equal to the actual floor reflectance. The geometric proportions of these spaces become the "cavity ratios."

Cavity ratio: Rooms are classified according to shape by *ten* cavity ratio numbers. The basic formula for obtaining cavity ratios in rectangular-shaped rooms is:

$$Cavity\,Ratio = \frac{5 \times Height\,(Length + Width)}{Length \times Width}$$

where height is the height of the cavity under consideration — that is, ceiling, floor, or room cavity.

There are five key steps in using the zonal-cavity method:

- Determine the required level of illumination.

- Determine the coefficient of utilization.

- Determine the maintenance factor.

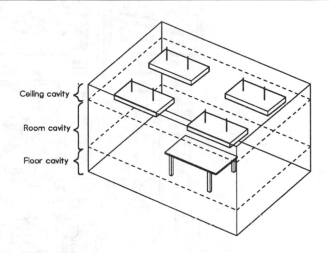

Figure Z-5: The three room cavities used with the zonal-cavity method.

- Calculate the number of lamps and lighting fixtures required.

- Determine the location of the lighting fixtures.

The general applications of the zonal cavity method are to determine the number of lighting fixtures that are required to produce a given lighting level in footcandles and to determine what lighting level will be produced by a given number of lighting fixtures (illuminaires).

zone: 1) Portion of a heating or air conditioning system separated for a certain temperature or other condition. 2) Portion of a town, city, county, state, or governmental subdivision set aside for a specific use.

zone control: Heating, ventilating, air conditioning (HVAC) control system that has more than one temperature control. Used to regulate air temperature, humidity, and the like in different areas of a building. Motorized dampers are used to control air distribution (in ductwork), while motorized valves are used to regulate the flow of liquids in a hot-water heating system. In some cases, entire separate systems may be used for each area of the building as the means of control. See Fig. Z-6 on page 346.

zoning: Division of land or subdivisions of land into areas that are regulated for the different types of usage; that is, residential, commercial, industrial, and the like.

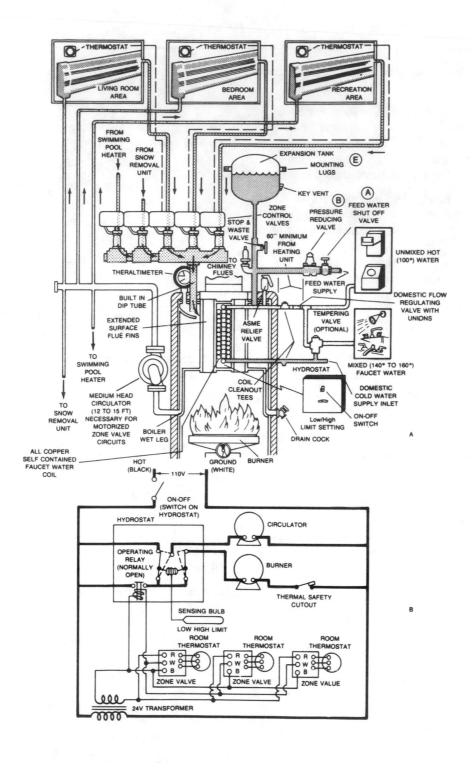

Figure Z-6: Typical hot-water system operating diagram. Note that this system is divided into five zones: living room area, bedroom area, recreation area, swimming pool, and snow removal unit.

Appendices

Contents

Appendix A

Conversion Factors

Multiply	by	to obtain
acres	43,560	square feet
acres	4047	square meters
acres	1.562×10^{-3}	square miles
acres	5645.38	square varas
acres	4840	square yards
amperes	0.11	abamperes
atmospheres	76.0	cm of mercury
atmospheres	29.92	inches of mercury
atmospheres	33.90	feet of water
atmospheres	10.333	kg per sq. meter
atmospheres	14.70	pounds per sq. inch
atmospheres	1.058	tons per sq. foot
British thermal units	0.2520	kilogram-calories

Multiply	by	to obtain
British thermal units	777.5	foot-pounds
British thermal units	3.927×10^{-4}	horsepower-hours
British thermal units	1054	joules
British thermal units	107.5	kilogram-meters
British thermal units	2.928×10^{-4}	kilowatt-hours
Btu per min.	12.96	foot-pounds per sec.
Btu per min.	0.02356	horsepower
Btu per min.	0.01757	kilowatts
Btu per min.	17.57	watts
Btu per sq. ft./Min.	0.1220	watts per sq. inch
bushels	1.244	cubic feet
bushels	2150	cubic inches
bushels	0.03524	cubic meters
bushels	4	pecks
bushels	64	pints (dry)
bushels	32	quarts (dry)
centimeters	0.3397	inches
centimeters	0.01	meters
centimeters	393.7	mils
centimeters	10	millimeters
centimeter-grams	980.7	centimeter-dynes
centimeter-grams	10^{-5}	meter-kilograms
centimeter-grams	7.233×10^{-5}	pound-feet
centimeters of mercury	0.01316	atmospheres
centimeters of mercury	0.4461	feet of water
centimeters of mercury	136.0	kg per sq. meter
centimeters of mercury	27.85	pounds per sq. meter
centimeters of mercury	0.1934	pounds per sq. inch

Multiply	by	to obtain
centimeters per second	1.969	feet per minute
centimeters per second	0.03281	feet per second
centimeters per second	0.036	kilometers per hour
centimeters per second	0.6	meters per minute
centimeters per second	0.02237	miles per hour
centimeters per second	3.728×10^{-4}	miles per minute
cubic centimeters	3.531×10^{-5}	cubic feet
cubic centimeters	6.102×10^{-2}	cubic inches
cubic centimeters	10	cubic meters
cubic centimeters	1.308×10^{-6}	cubic yards
cubic centimeters	2.642×10^{-4}	gallons
cubic centimeters	10	liters
cubic centimeters	2.113×10^{-3}	pints (liq)
cubic centimeters	1.057×10^{-3}	quarts (liq)
cubic feet	62.43	pounds of water
cubic feet	2.832×10^{4}	cubic cm
cubic feet	1728	cubic inches
cubic feet	0.02832	cubic meters
cubic feet	0.03704	cubic yards
cubic feet	7.481	gallons
cubic feet	28.32	liters
cubic feet	59.84	pints (liq)
cubic feet	29.92	quarts (liq)
cubic feet per minute	472.0	cubic cm per sec.
cubic feet per minute	0.1247	gallons per sec.
cubic feet per minute	0.4720	liters per sec.
cubic feet per minute	62.4	lb of water per min.
cubic inches	16.39	cubic centimerters

Multiply	by	to obtain
cubic inches	5.787×10^{-4}	cubic feet
cubic inches	1.639×10^{-5}	cubic meters
cubic inches	2.143×10^{-5}	cubic yards
cubic inches	4.329×10^{-3}	gallons
cubic inches	1.639×10^{-2}	liters
cubic inches	0.03463	pints (liq)
cubic inches	0.01732	quarts (liq)
cubic yards	7.646×10^{5}	cubic centimerters
cubic yards	27	cubic feet
cubic yards	46,656	cubic inches
cubic yards	0.7646	cubic meters
cubic yards	202.0	gallons
cubic yards	764.6	liters
cubic yards	1616	pints (liq)
cubic yards	807.9	quarts (liq)
cubic yards per minute	0.45	cubic feet per sec.
cubic yards per minute	3.367	gallons per second
cubic yards per minute	12.74	liters per second
degrees (angle)	60	minutes
degrees (angle)	0.01745	radians
degrees (angle)	3600	seconds
dynes	1.020×10^{-3}	grams
dynes	7.233×10^{-5}	poundals
dynes	2.248×10^{-6}	pounds
ergs	9.486×10^{-11}	kilograms
ergs	1	dyne-centimeters
ergs	7.376×10^{-8}	foot-pounds
ergs	1.020×10^{-3}	gram-centimeters

Multiply	by	to obtain
ergs	10^{-7}	joules
ergs	2.390×10^{-11}	kilogram-calories
ergs	1.020×10^{-8}	kilogram-meters
feet	12	inches
feet	0.3048	meters
feet	0.36	varas
feet	3	yards
feet of water	0.02950	atmospheres
feet of water	0.8826	inches of mercury
feet of water	304.8	kg per sq. meter
feet of water	62.43	pounds per sq. ft.
feet of water	0.4335	pounds per sq. inch
foot-pounds	1.286×10^{-3}	British thermal units
foot-pounds	1.356×10^{7}	ergs
foot-pounds	5.050×10^{-7}	horsepower hours
foot-pounds	1.356	joules
foot-pounds	3.241×10^{-4}	kilogram-calories
foot-pounds	0.1383	kilogram-meters
foot-pounds	3.766×10^{-7}	kilowatt-hours
foot-pounds per minute	1.286×10^{-3}	Btu per minute
foot-pounds per minute	0.01667	foot pounds per sec.
foot-pounds per minute	3.030×10^{-5}	horsepower
foot-pounds per minute	3.241×10^{-4}	kg-calories per min.
foot-pounds per minute	2.260×10^{-5}	kilowatts
foot-pounds per sec.	7.717×10^{-2}	Btu per minute
foot-pounds per sec.	1.818×10^{-3}	horsepower
foot-pounds per sec.	1.945×10^{-2}	kg-calories per min.
foot-pounds per sec.	1.356×10^{-3}	kilowatts

Multiply	by	to obtain
gallons	8.345	pounds of water
gallons	3785	cubic centimeters
gallons	0.1337	cubic feet
gallons	231	cubic inches
gallons	3.785×10^{-3}	cubic meters
gallons	4.951×10^{-3}	cubic yards
gallons	3.785	liters
gallons	8	pints (liq)
gallons	4	quarts (liq)
gallons per minute	2.228×10^{-3}	cubic ft per sec.
gallons per minute	0.06308	liters per second
grains (troy)	1	grains (av)
grains (troy)	0.06480	grams
grains (troy)	0.04167	pennyweights (troy)
grams	980.7	dynes
grams	15.43	grains (troy)
grams	10^{-3}	kilograms
grams	10^{3}	milligrams
grams	0.03527	ounces
grams	0.03215	ounces (troy)
grams	0.07093	poundals
grams	2.205×10^{-3}	pounds
horsepower	42.44	Btu per min
horsepower	33,000	foot-pounds per min.
horsepower	550	foot-pounds per sec.
horsepower	1.014	horsepower (metric)
horsepower	10.70	kg-calories per min.
horsepower	0.7457	kilowatts

Multiply	by	to obtain
horsepower	7.457	watts
horsepower (boiler)	33,520	Btu per hour
horsepower (boiler)	9,804	kilowatts
horsepower-hours	2547	British thermal units
horsepower-hours	1.98×10^6	foot-pounds
horsepower-hours	2.684×10^6	joules
horsepower-hours	641.7	kilogram-calories
horsepower-hours	2.737×10^5	kilogram-meters
horsepower-hours	0.7457	kilowatt-hours
inches	2.540	centimeters
inches	10^3	mils
inches	0.03	varas
inches of mercury	0.03342	atmospheres
inches of mercury	1.133	feet of water
inches of mercury	345.3	kg per sq meter
inches of mercury	70.73	pounds per sq ft.
inches of mercury	0.4912	pounds per sq in.
inches of water	0.07355	inches of mercury
inches of water	25.40	kg per sq. meter
inches of water	0.5781	ounces per sq in.
inches of water	5.204	pounds per sq ft.
inches of water	0.03613	pounds per sq in.
kilograms	980,665	dynes
kilograms	10^3	grams
kilograms	70.93	poundals
kilograms	2.2046	pounds
kilograms	1.102×10^{-3}	tons (short)
kilogram-calories	3.968	British thermal units

Multiply	by	to obtain
kilogram-calories	3086	foot-pounds
kilogram-calories	1.558×10^{-3}	horsepower-hours
kilogram-calories	4183	joules
kilogram-calories	426.6	kilogram-meters
kilogram-calories	1.162×10^{-3}	kilowatt-hours
kg-calories per min.	51.43	foot pounds per sec.
kg-calories per min.	0.09351	horsepower
kg-calories per min.	0.06972	kilowatts
kilometers	10^5	centimeters
kilometers	3281	feet
kilometers	10^3	meters
kilometers	0.6214	miles
kilometers	1093.6	yards
kilowatt-hours	3415	British thermal units
kilowatt-hours	2.655×10^6	joules
kilowatt-hours	1.341	horsepower-hours
kilowatt-hours	3.6×10^6	joules
kilowatt-hours	860.5	kilogram-calories
kilowatt-hours	3.671×10^5	kilogram-meters
kilowatts	56.92	Btu per min.
kilowatts	4.425×10^4	foot-pounds per min.
kilowatts	737.6	foot-pounds per sec.
kilowatts	1.341	horsepower
kilowatts	14.34	kg-calories per min.
kilowatts	10^3	watts
log10N	2.303	logEN or ln N
logEN or lnN	0.4343	log10N
meters	100	centimeters

Multiply	by	to obtain
meters	3.2808	feet
meters	39.37	inches
meters	10^{-3}	kilometers
meters	10^3	millimeters
meters	1.0936	yards
miles	1.609×10^5	centimeters
miles	5280	feet
miles	1.6093	kilometers
miles	1760	yards
miles	1900.8	varas
miles per hour	44.70	centimeters per sec.
miles per hour	88	feet per minute
miles per hour	1.467	feet per second
miles per hour	1.6093	kilometers per hour
miles per hour	0.8684	knots per hour
miles per hour	0.4470	M per sec.
months	30.42	days
months	730	hours
months	43,800	minutes
months	2.628×10^6	seconds
ounces	8	drams
ounces	437.5	grains
ounces	28.35	grams
ounces	0.0625	pounds
ounces per sq inch	0.0625	pounds per sq inch
pints (dry)	33.60	cubic inches
pints (liq)	28.87	cubic inches
pounds	444,823	dynes

Multiply	by	to obtain
pounds	7000	grains
pounds	453.6	grams
pounds	16	ounces
pounds	32.17	poundals
pounds of water	0.01602	cubic feet
pounds of water	27.68	cubic inches
pounds of water	0.1198	gallons
pounds of water per min.	2.669×10^{-4}	cubic feet per sec.
pounds per cubic foot	0.01602	grams per cubic cm.
pounds per cubic foot	16.02	kg per cubic meter
pounds per cubic foot	5.786×10^{-4}	pounds per cubic inch
pounds per cubic foot	5.456×10^{-9}	pounds per mil foot
pounds per square foot	0.01602	feet of water
pounds per square foot	4.882	kg per sq. meter
pounds per square foot	6.944×10^{-3}	pounds per sq. inch
pounds per square inch	0.06804	atmospheres
pounds per square inch	2.307	feet of water
pounds per square inch	2.036	inches of mercury
pounds per square inch	703.1	kg per sq. meter
pounds per square inch	144	pounds per sq. foot
quarts	32	fluid ounces
quarts (dry)	67.20	cubic inches
quarts (liq)	57.75	cubic inches
rods	16.5	feet
square centimeters	1.973×10^{5}	circular mils
square centimeters	1.076×10^{-3}	square feet
square centimeters	0.1550	square inches
square centimeters	10^{-6}	square meters

Multiply	by	to obtain
square centimeters	100	square millimeters
square feet	2.296×10^{-5}	acres
square feet	929.0	square centimeters
square feet	144	square inches
square feet	0.09290	square meters
square feet	3.587×10^{-8}	square miles
square feet	0.1296	square varas
square feet	1/9	square yards
square inches	1.273×10^{6}	circular mils
square inches	6.452	square centimeters
square inches	6.944×10^{-3}	square feet
square inches	10^{6}	square mils
square inches	645.2	square millimeters
square miles	640	acres
square miles	27.88×10^{6}	square feet
square miles	2.590	square kilometers
square miles	3,613,040.45	square varas
square miles	3.098×10^{6}	square yards
square yards	2.066×10^{-4}	acres
square yards	9	square feet
square yards	0.8361	square meters
square yards	3.228×10^{-7}	square miles
square yards	1.1664	square varas
temp. (degs.C)+17.8	1.8	temp.(degs.F)
temp. (degs.F)-32	5/9	temp.(degs.C)
tons (long)	2240	pounds
tons (short)	2000	pounds
yards	0.9144	meters

Appendix B

Architectural Graphic Symbols

In all large construction projects and in most of the smaller ones, an architect is commissioned to prepare complete working drawings and specifications for the project. These drawings usually include:

- A plot plan indicating the location of the building on the property.

- Floor plan showing the walls and partitions for each floor level.

- Elevations of all exterior faces of the building.

- Several vertical cross sections to indicate clearly the various floor levels and details of footings, foundation, walls, floors, ceilings, and roof construction.

- Large-scale drawings showing such construction details as may be required.

Many symbols are used in preparing architectural drawings. These symbols represent types of material, structural elements, or construction details. Most of these symbols are standardized and are familiar to all architects and workers. If any variations are encountered, a legend or symbol list is provided to clarify the situation. Whenever new material or equipment is employed for which there are no standard symbols, the items are identified by numbered or lettered circles and described completely by note, detail, or in the written specification form. Special symbols are also used to designate various structural, electrical, plumbing, and heating, ventilating, and air conditioning systems.

The following are the most common symbols used on architectural working drawings. Anyone involved in building construction, in any capacity, should have a good knowledge of these symbols.

Architectural Material Symbols

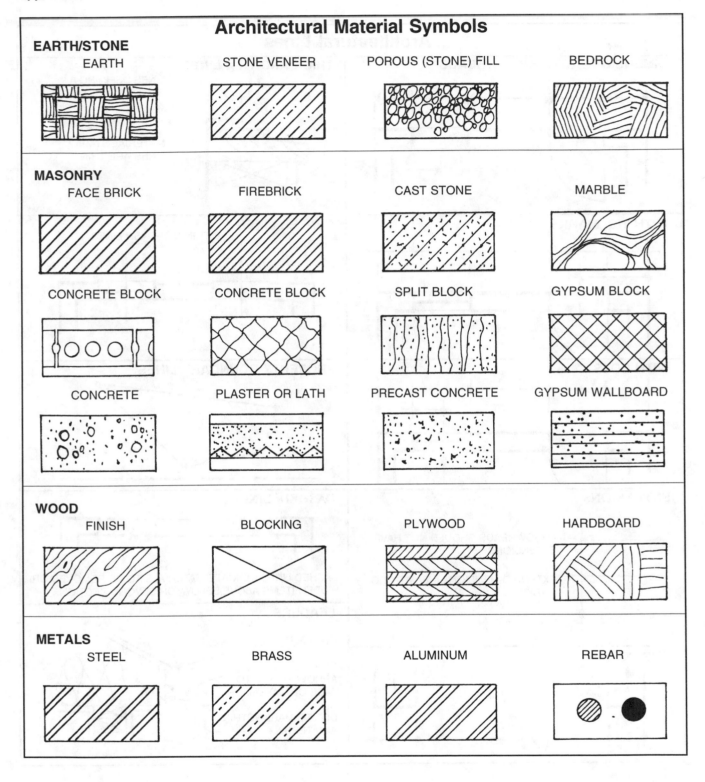

EARTH/STONE

| EARTH | STONE VENEER | POROUS (STONE) FILL | BEDROCK |

MASONRY

| FACE BRICK | FIREBRICK | CAST STONE | MARBLE |

| CONCRETE BLOCK | CONCRETE BLOCK | SPLIT BLOCK | GYPSUM BLOCK |

| CONCRETE | PLASTER OR LATH | PRECAST CONCRETE | GYPSUM WALLBOARD |

WOOD

| FINISH | BLOCKING | PLYWOOD | HARDBOARD |

METALS

| STEEL | BRASS | ALUMINUM | REBAR |

Architectural Lines

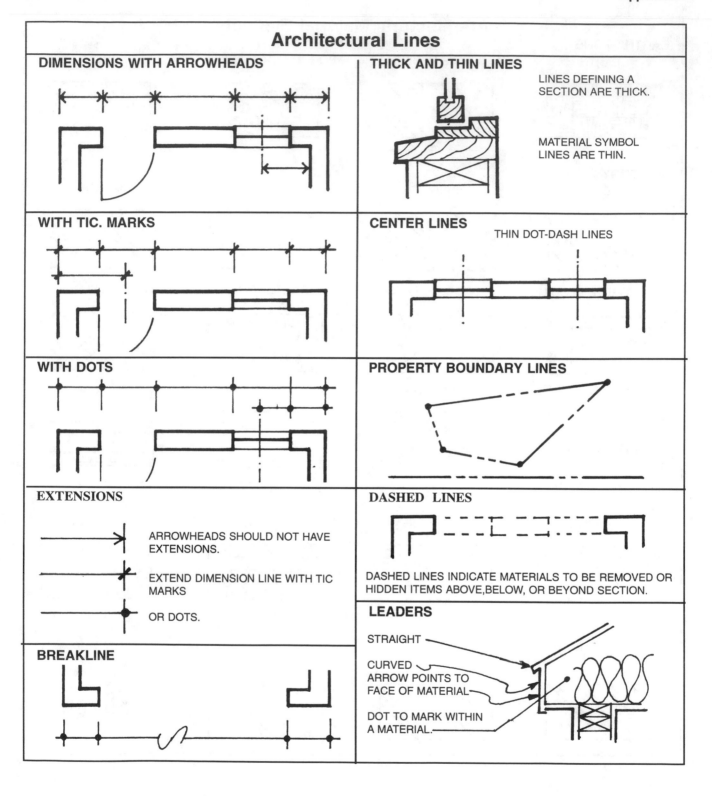

DIMENSIONS WITH ARROWHEADS

THICK AND THIN LINES

LINES DEFINING A SECTION ARE THICK.

MATERIAL SYMBOL LINES ARE THIN.

WITH TIC. MARKS

CENTER LINES

THIN DOT-DASH LINES

WITH DOTS

PROPERTY BOUNDARY LINES

EXTENSIONS

ARROWHEADS SHOULD NOT HAVE EXTENSIONS.

EXTEND DIMENSION LINE WITH TIC MARKS

OR DOTS.

DASHED LINES

DASHED LINES INDICATE MATERIALS TO BE REMOVED OR HIDDEN ITEMS ABOVE, BELOW, OR BEYOND SECTION.

LEADERS

STRAIGHT

CURVED
ARROW POINTS TO FACE OF MATERIAL

DOT TO MARK WITHIN A MATERIAL.

BREAKLINE

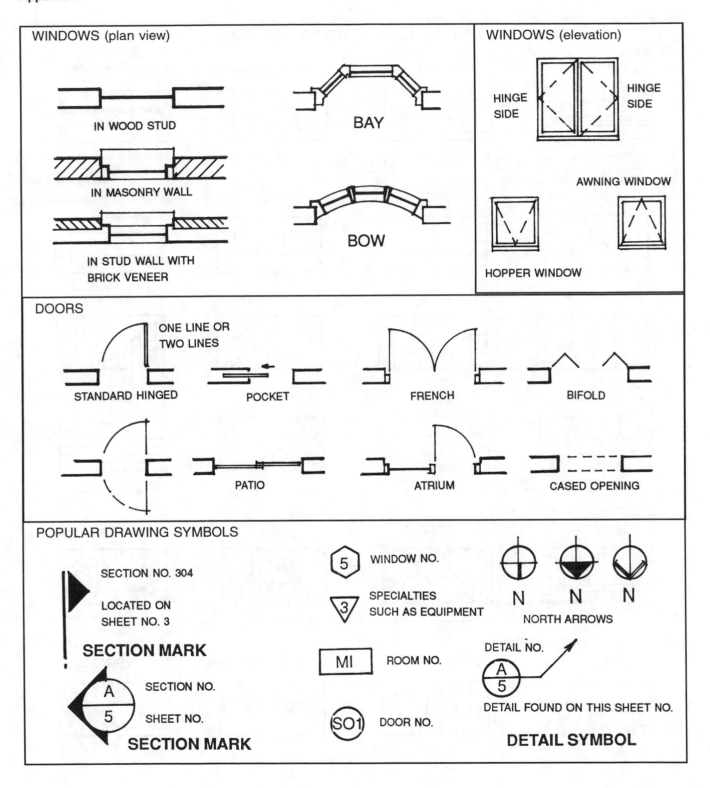

WINDOWS (plan view)

IN WOOD STUD

IN MASONRY WALL

IN STUD WALL WITH
BRICK VENEER

BAY

BOW

WINDOWS (elevation)

HINGE
SIDE

HINGE
SIDE

AWNING WINDOW

HOPPER WINDOW

DOORS

ONE LINE OR
TWO LINES

STANDARD HINGED

POCKET

FRENCH

BIFOLD

PATIO

ATRIUM

CASED OPENING

POPULAR DRAWING SYMBOLS

SECTION NO. 304

LOCATED ON
SHEET NO. 3

SECTION MARK

A
5

SECTION NO.

SHEET NO.

SECTION MARK

5   WINDOW NO.

3   SPECIALTIES
     SUCH AS EQUIPMENT

MI   ROOM NO.

SO1   DOOR NO.

N   N   N

NORTH ARROWS

DETAIL NO.

A
5

DETAIL FOUND ON THIS SHEET NO.

DETAIL SYMBOL

Architectural Material Symbols

GLASS

SECTION PLATE GLASS BLOCK PLASTIC VENEER

COMMON ELEVATIONS

GLASS BRICK BLOCK BLOCK (STACK BOND)

METAL ROOF CERAMIC TILE CONCRETE METALS

FIELD STONE WOOD INSECT SCREEN SHINGLE ROOF

INSULATION

RIGID BATT BLOWN FOAM

MISCELLANEOUS FINISHES

CARPET ACOUSTIAL TILE WATERPROOF MEMBRANE TERRAZZO

Appendix C

Electrical Graphic Symbols

The purpose of an electrical drawing is to show the complete design and layout of the electrical systems for lighting, power, signal and communication systems, special raceways, and related electrical equipment. In preparing such drawings, the electrical layout is shown through the use of lines, symbols, and notation which should indicate, beyond any question or any doubt, exactly what is required.

Many engineers, designers, and draftsmen use symbols adapted by the United States of America Standards Institute (USASI). However, no definite standard schedule of symbols is always used in its entirety. Consulting engineering firms quite frequently modify these standard symbols to meet their own needs. Therefore, in order to identify the symbols properly, the engineer provides, on one of the drawings or in the written specifications, a list of symbols with a descriptive note for each — clearly indicating the part of the wiring system which it represents.

The following list of electrical symbols are currently recommended by USASI. It is evident from these symbols that many have the same basic form, but, because of some slight difference, their meaning changes. For example, the outlet symbols each have the same basic form — a circle — but the addition of a line or an abbreviation gives each an individual meaning. A good procedure to follow in learning symbols is to first learn the basic form and then apply the variations for obtaining different meanings.

ELECTRICAL SYMBOLS.

SWITCH OUTLETS

Single-Pole Switch	S
Double-Pole Switch	S_2
Three-Way Switch	S_3
Four-Way Switch	S_4
Key-Operated Switch	S_K
Switch and Fusestat Holder	S_FH
Switch and Pilot Lamp	S_P
Fan Switch	S_F
Switch for Low-Voltage Switching System	S_L
Master Switch for Low-Voltage Switching System	S_{LM}
Switch and Single Receptacle	$\ominus$S
Switch and Duplex Receptacle	$\ominus$S
Door Switch	S_D
Time Switch	S_T
Momentary Contact Switch	S_{MC}
Ceiling Pull Switch	Ⓢ
"Hand-Off-Auto" Control Switch	HOA
Multi-Speed Control Switch	M
Push Button	⊡

RECEPTACLE OUTLETS

Where weather proof, explosion proof, or other specific types of devices are to be required, use the upper-case subscript letters. For example, weather proof single or duplex receptacles would have the uppercase WP subscript letters noted alongside of the symbol. All outlets should be grounded.

Single Receptacle Outlet	⊖
Duplex Receptacle Outlet	⊜
Triplex Receptacle Outlet	
Quadruplex Receptacle Outlet	
Duplex Receptacle Outlet - Split Wired	
Triplex Receptacle Outlet - Split Wired	
250 Volt Receptable Single Phase Use Subscript Letter to Indicate Function (DW-Dishwasher; RA-Range, CD - Clothes Dryer) or numeral (with explanation in symbol schedule)	
250 Volt Receptacle Three Phase	
Clock Receptacle	Ⓒ
Fan Receptacle	Ⓕ
Floor Single Receptacle Outlet	
Floor Duplex Receptacle Outlet	
Floor Special-Purpose Outlet	*
Floor Telephone Outlet - Public	
Floor Telephone Outlet - Private	

Example of the use of several floor outlet symbols to identify a 2, 3, or more gang floor outlet:

Underfloor Duct and Junction Box for Triple, Double or Single Duct System as indicated by the number of parallel lines.

Example of use of various symbols to identify location of different types of outlets or connections for underfloor duct or cellular floor systems:

Cellular Floor Header Duct

*Use numeral keyed to explanation in drawing list of symbols to indicate usage.

CIRCUITING

Wiring Exposed (not in conduit) ——E——

Wiring Concealed in Ceiling or Wall ————————

Wiring Concealed in Floor — — — —

Wiring Existing* – – – – – –

Wiring Turned Up ————————o

Wiring Turned Down ————————●

Branch Circuit Home Run to Panel Board. 2 1

Number of arrows indicates number of circuits. (A number at each arrow may be used to identify circuit number.)**

BUS DUCTS AND WIREWAYS

Trolley Duct*** | T | | T |

Busway (Service, Feeder, or Plug-in)*** | B | | B |

Cable Trough Ladder or Channel*** | C | | C |

Wireway*** | W | | W |

PANELBOARDS, SWITCHBOARDS AND RELATED EQUIPMENT

Flush Mounted Panelboard and Cabinet***

Surface Mounted Panelboard and Cabinet***

Switchboard, Power Control Center, Unit Substations (Should be drawn to scale)***

Flush Mounted Terminal Cabinet (In small scale drawings the TC may be indicated alongside the symbol)*** [TC]

Surface Mounted Terminal Cabinet (In small scale drawings the TC may be indicated alongside the symbol)*** [TC]

Pull Box (Identify in relation to Wiring System Section and Size)

Motor or Other Power Controller (May be a starter or contactor)***

Externally Operated Disconnection Switch***

Combination Controller and Disconnection Means***

POWER EQUIPMENT

Electric Motor (HP as indicated) (¼)

Power Transformer

Pothead (Cable Termination)

Circuit Element, e.g., Circuit Breaker |CB|

Circuit Breaker

Fusible Element

Single-Throw Knife Switch

Double-Throw Knife Switch

Ground

Battery

Contactor |C|

Photoelectric Cell |PE|

Voltage Cycles, Phase Ex: 480/60/3

Relay |R|

Equipment Connection (as noted) (▲)

*Note: Use heavy-weight line to identify service and feeders. Indicate empty conduit by notation CO (conduit only).
**Note: Any circuit without further identification indicates two-wire circuit. For a greater number of wires, indicate with cross lines, e.g.:

——|||—— 3 wires; ——||||—— 4 wires, etc.

Neutral wire may be shown longer. Unless indicated otherwise, the wire size of the circuit is the minimum size required by the specification. Identify different functions of wiring system, e.g., signalling system by notation or other means.
***Identify by Notation or Schedule

REMOTE CONTROL STATIONS FOR MOTORS OR OTHER EQUIPMENT

Pushbutton Station	PB
Float Switch - Mechanical	F
Limit Switch - Mechanical	L
Pneumatic Switch - Mechanical	P
Electric Eye - Beam Source	⧄→
Electric Eye - Relay	⧄
Temperature Control Relay Connection (3 Denotes Quantity)	R₃
Solenoid Control Valve Connection	S
Pressure Switch Connection	P
Aquastat Connection	A
Vacuum Switch Connection	V
Gas Solenoid Valve Connection	G
Flow Switch Connection	F
Timer Connection	T
Limit Switch Connection	L

LIGHTING

	Ceiling	Wall
Surface or Pendant Incandescent Fixture PC = pull chain)	TYPE / Ⓥ WATTS	SWITCH PCⓄ CIRCUIT
Surface or Pendant Exit Light	⊗	─⊗
Blanked Outlet	Ⓑ	─Ⓑ
Junction Box	Ⓙ	─Ⓙ
Recessed Incandescent Fixtures	▢	
Surface or Pendant Individual Fluorescent Fixture	▭	

Surface or Pendant Continous-Row Fluorescent Fixture (Letter indicating controlling switch)

⟵ Fixture No.
⟵ Wattage
Symbol not needed at each fixture

*Bare-Lamp Fluorescent Strip ├─┼─┼─┤

ELECTRIC DISTRIBUTION OR LIGHTING SYSTEM, AERIAL

Pole**	○
Street or Parking Lot Light and Bracket**	☼─
Transformer**	△
Primary Circuit**	─────
Secondary Circuit**	─────
Down Guy	──→
Head Guy	──●─
Sidewalk Guy	──○→
Service Weather Head**	─◁

ELECTRIC DISTRIBUTION OR LIGHTING SYSTEM, UNDERGROUND

Manhole**	M
Handhole**	H
Transformer Manhole or Vault**	TM
Transformer Pad**	TP
Underground Direct Burial Cable (Indicate type, size and number of conductors by notation or schedule)	─ ─ ─ ─
Underground Duct Line (Indicate type, size, and number of ducts by cross-section identification of each run by notation or schedule. Indicate type, size, and number of conductors by notation or schedule.	─ ─⊏─
Street Light Standard Fed From Underground Circuit**	⊗

*In the case of continuous-row bare-lamp fluorescent strip above an area-wide diffusing means, show each fixture run using the standard symbol; indicate area of diffusing means and type by light shading and/or drawing notation.
**Identify by Notation or Schedule

SIGNALLING SYSTEM OUTLETS

INSTITUTIONAL, COMMERCIAL, AND INDUSTRIAL OCCUPANCIES

I Nurse Call System Devices (any type)
Basic Symbol.

(Examples of Individual Item Identifiction Not a part of Standard)

Nurses' Annunciator (add a number after it as +①24 to indicate number of lamps) — ①

Call Station, single cord, pilot light — ②

Call Station, double cord, microphone speaker — ③

Corridor Dome Light, 1 lamp — ④

Transformer — ⑤

Any other item on same system - use numbers as required. — ⑥

II Paging System Devices (any type)
Basic Symbol.

(Examples of Individual Item Identification. Not a part of Standard)

Keyboard — ◇1

Flush Annunciator — ◇2

2-Face Annunciator — ◇3

Any other item on same system - use numbers as required — ◇4

III Fire Alarm System Devices (any type) including Smoke and Sprinkler Alarm Devices
Basic Symbol.

(Examples of Individual Item Identification. Not a part of Standard)

Control Panel — □1

Station — □2

10" Gong — □3

Pre-signal Chime — □4

Any other item on same system - use numbers as required. — □5

IV Staff Register System Devices (any type)
Basic Symbol.

(Examples of Individual Item Identification. Not a part of Standard)

Phone Operators' Register — ◇1

Entrance Register - flush — ◇2

Staff Room Register — ◇3

Transformer — ◇4

Any other item on same system - use numbers as required. — ◇5

V Electric Clock System Devices (any type)
Basic Symbol.

(Examples of Individual Item Identification. Not a part of Standard)

Master Clock — ⬡1

12" Secondary - flush — ⬡2

12" Double Dial - wall mounted — ⬡3

18" Skeleton Dial — ⬡4

Any other item on same system - use numbers as required. — ⬡5

VI Public Telephone System Devices
Basic Symbol.

(Examples of Individual Item Identification. Not a part of Standard)

Switchboard — ◀1

Desk Phone — ◀2

Any other item on same system - use numbers required. — ◀3

VII   Private Telephone System
       Devices (any type)

       Basic Symbol.

       (Examples of Individual Item Identi-
       fication. Not a part of Standard)

Switchboard

Wall Phone

Any other item on same system -
   use numbers as required.

VIII   Watchman System Devices
        (any type)

       Basic Symbol.

       (Examples of Individual Item Identi-
       fication. Not a part of Standard)

Central Station

Key Station

Any other item on same system -
   use numbers as required.

IX   Sound System

       Basic Symbol.

       (Examples of Individual Item Identi-
       fication. Not a part of Standard)

Amplifier

Microphone

Interior Speaker

Exterior Speaker

Any other item on same system -
   use numbers as required.

X   Other Signal System Devices

       Basic Symbol.

       (Examples of Individual Item Identi-
       fication. Not a part of Standard)

Buzzer

Bell

Pushbuttom

Annunciator

Any other item on same system
   use numbers as required.

RESIDENTIAL OCCUPANCIES

Signalling system symbols for use in identifying
standardized residential-type signal system items
on residential drawings where a descriptive symbol list is not included on the drawing. When
other signal system items are to be identified, use
the above basic symbols for such items together
with a descriptive symbol list.

Pushbutton

Buzzer

Bell

Combination Bell-Buzzer

Chime

Annunciator

Electric Door Opener

Maid's Signal Plug

Interconnection Box

Bell-Ringing Transformer

Outside Telephone

Interconnecting Telephone

Television Outlet

Appendix D

Mechanical Graphic Symbols

Graphic symbols, used on working drawings for mechanical systems, fall under the HVAC (heating, ventilating, air conditioning) and plumbing trades. Most of these drawings are highly diagrammatical and are used to locate pipes, fixtures, ductwork, equipment, etc. within the building or around the premises.

The exact method of showing mechanical layouts on drawings for building construction will vary with the engineer or consulting firm. The following description is typical for most mechanical drawings.

HVAC Drawings: The purpose of an HVAC drawing is to show the location of air-handling and other HVAC equipment; to show the sizes and routes of the various ductwork (both supply and return air); to show the air outlets, including the volume and velocity of air from each; and other necessary information required in a duct system. Circulating hot-water heating systems and cold-water chillers also fall under this building trade.

Plumbing Drawings: Plumbing drawings cover the complete design and layout of the plumbing system and show floor-play layouts, cross-sections of the building, and detailed drawings. The purpose of a plumbing draw-

ing is to show the location, size of pipe, valves, and other necessary information required for the workmen to do their job. In preparing such drawings, symbols are used to simplify the drawing — both for the draftsman preparing them and the workers on the job who must read them.

In general, pipe is used in plumbing systems for transporting liquids, gases, and solids. The following are examples of each:

- *Liquids:* Water supply to buildings; fuel oil supply to buildings; liquid waste away from buildings.

- *Gases:* Steam in heating systems; LP or natural gas to buildings; air from air compressors; air for ventilation in sewage drain pipes.

- *Solids:* Human waste.

Figure AD-1 shows a list of plumbing symbols currently being used by one consulting engineering firm, while Fig. AD-2 shows HVAC symbols used by the same firm. The symbols to follow in Fig. AD-3 are mechanical symbols (both plumbing and HVAC) recommended by Construction Specifications Institute and the American Society of Mechanical Engineers. A basic knowledge of these symbols will give anyone associated with building construction the ability to approach his or her work more intelligently, even if their main concern is with a trade other than plumbing or HVAC. Learn the basic form; then apply the variations to obtain the different meanings. Also become familar with the many abbreviations used on working drawings; this includes both mechanical and architectural abbreviations, as well as abbreviations used by other building trades.

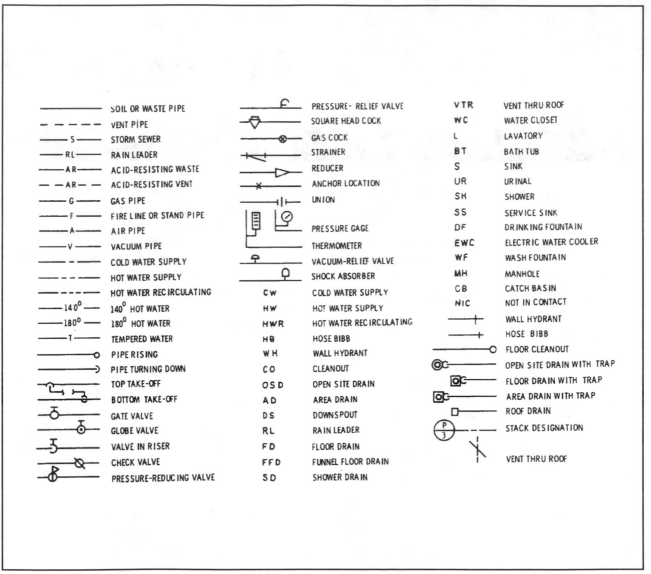

Fig. AD-1: Plumbing drawing symbols that frequently appear in a symbol list or legend on mechanical working drawings.

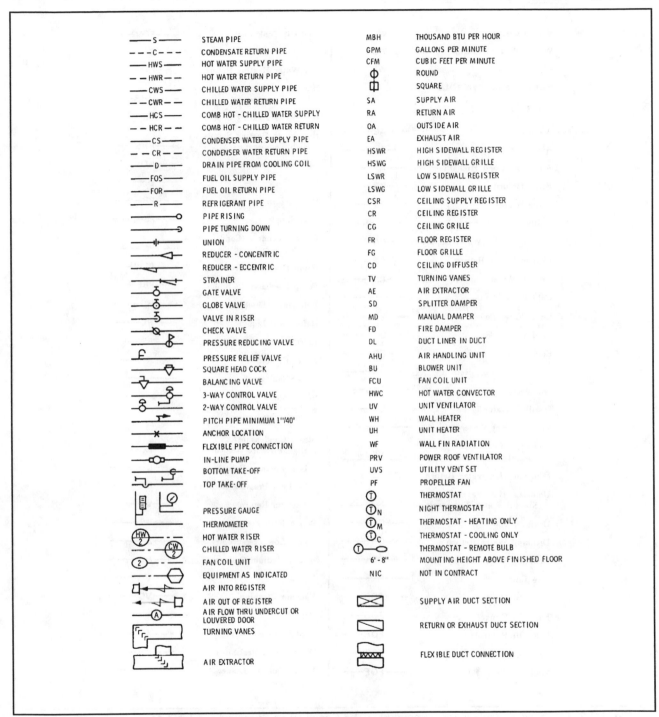

——S——	STEAM PIPE	MBH	THOUSAND BTU PER HOUR
———C———	CONDENSATE RETURN PIPE	GPM	GALLONS PER MINUTE
——HWS——	HOT WATER SUPPLY PIPE	CFM	CUBIC FEET PER MINUTE
——HWR——	HOT WATER RETURN PIPE		ROUND
——CWS——	CHILLED WATER SUPPLY PIPE		SQUARE
——CWR——	CHILLED WATER RETURN PIPE	SA	SUPPLY AIR
——HCS——	COMB HOT - CHILLED WATER SUPPLY	RA	RETURN AIR
——HCR——	COMB HOT - CHILLED WATER RETURN	OA	OUTSIDE AIR
——CS——	CONDENSER WATER SUPPLY PIPE	EA	EXHAUST AIR
——CR——	CONDENSER WATER RETURN PIPE	HSWR	HIGH SIDEWALL REGISTER
——D——	DRAIN PIPE FROM COOLING COIL	HSWG	HIGH SIDEWALL GRILLE
——FOS——	FUEL OIL SUPPLY PIPE	LSWR	LOW SIDEWALL REGISTER
——FOR——	FUEL OIL RETURN PIPE	LSWG	LOW SIDEWALL GRILLE
——R——	REFRIGERANT PIPE	CSR	CEILING SUPPLY REGISTER
	PIPE RISING	CR	CEILING REGISTER
	PIPE TURNING DOWN	CG	CEILING GRILLE
	UNION	FR	FLOOR REGISTER
	REDUCER - CONCENTRIC	FG	FLOOR GRILLE
	REDUCER - ECCENTRIC	CD	CEILING DIFFUSER
	STRAINER	TV	TURNING VANES
	GATE VALVE	AE	AIR EXTRACTOR
	GLOBE VALVE	SD	SPLITTER DAMPER
	VALVE IN RISER	MD	MANUAL DAMPER
	CHECK VALVE	FD	FIRE DAMPER
	PRESSURE REDUCING VALVE	DL	DUCT LINER IN DUCT
	PRESSURE RELIEF VALVE	AHU	AIR HANDLING UNIT
	SQUARE HEAD COCK	BU	BLOWER UNIT
	BALANCING VALVE	FCU	FAN COIL UNIT
	3-WAY CONTROL VALVE	HWC	HOT WATER CONVECTOR
	2-WAY CONTROL VALVE	UV	UNIT VENTILATOR
	PITCH PIPE MINIMUM 1"/40'	WH	WALL HEATER
	ANCHOR LOCATION	UH	UNIT HEATER
	FLEXIBLE PIPE CONNECTION	WF	WALL FIN RADIATION
	IN-LINE PUMP	PRV	POWER ROOF VENTILATOR
	BOTTOM TAKE-OFF	UVS	UTILITY VENT SET
	TOP TAKE-OFF	PF	PROPELLER FAN
			THERMOSTAT
			NIGHT THERMOSTAT
	PRESSURE GAUGE		THERMOSTAT - HEATING ONLY
	THERMOMETER		THERMOSTAT - COOLING ONLY
	HOT WATER RISER		THERMOSTAT - REMOTE BULB
	CHILLED WATER RISER	6' - 8"	MOUNTING HEIGHT ABOVE FINISHED FLOOR
	FAN COIL UNIT	NIC	NOT IN CONTRACT
	EQUIPMENT AS INDICATED		
	AIR INTO REGISTER		
	AIR OUT OF REGISTER		SUPPLY AIR DUCT SECTION
	AIR FLOW THRU UNDERCUT OR LOUVERED DOOR		RETURN OR EXHAUST DUCT SECTION
	TURNING VANES		
	AIR EXTRACTOR		FLEXIBLE DUCT CONNECTION

Fig. AD-2: Mechanical symbols used by one consulting engineering firm.

Wall Hydrant	WH	Fuel Oil Tank Vent	---FOV---
Wet Bulb Temperature	WB	Gas	——G——
Wide Flange	WF	Hot Water Heating Return	---HR---

MECHANICAL SYMBOLS

PIPES

		Hot Water Heating Return High Temperature	---HHR---
		Hot Water Heating Supply	——HS——
Acetelyne	——AC——	Hot Water Heating Supply High Temperature	——HHS——
Air Compressed	——A——	Nitrous Oxide	——NO——
Air Compressed High Pressure	——HA——	Oxygen	——O——
Brine Return	---BR---	Pool Return	---PR---
Brine Supply	——BS——	Pool Supply	——PS——
Chilled or Hot Water Return	---C/HR---	Pump Discharge (Condensate or Vacuum Pump)	——PC——
Chilled or Hot Water Supply	——C/HS——	Pump Discharge Feedwater	——PFW——
Chilled Water Return	---CR---	Pump Discharge to Boiler	——PBF——
Chilled Water Supply	——CS——	Refrigerant Discharge	——HG——
Cold Water	————————	Refrigerant Liquid	——RL——
Cold Water Lake	——┼——	Refrigerant Suction	---RS---
Cold Water Soft	——S——	Sewer Combined *	——┼——
Condensate Drain	——CD——	Sewer Sanitary *	————————
Condenser Water Return	---CWR---	Sewer Storm *	—— —— ——
Condenser Water Supply	——CWS——	*Sewers may be wider line outside building	
Distilled Water	——DW——		
Domestic Hot Water	—— —— ——	Steam Condensate High Pressure Return	---HC---
Domestic Hot Water (Special Water Temperature)	——180——	Low Pressure Return	---C---
Domestic Hot Water Return	—— —— ——	Medium Pressure Return	---MC---
Domestic Hot Water Return (Special Water Temperature)	——180——	Vacuum Return	---VC---
Drain	——D——	Steam High Pressure	——HS——
Drinking Water Return	——DR——	Steam Low Pressure	——S——
Drinking Water Supply	——DS——	Steam Medium Pressure	——MS——
Fire Line	——F——	Vacuum	——V——
Fuel Oil Discharge	——FOD——	Vacuum Cleaning	——VC——
Fuel Oil Gauge	——FOG——	Vent	—— —— ——
Fuel Oil Return	---FOR---	Vent (Acid Resistant	---AR---
Fuel Oil Suction	——FOS——	Waste (Acid Resistant)	——AR——
Fuel Oil Tank Fill	——FOF——	Waste, Soil, or Leader	————————

Fig. AD-3: Mechanical symbols recommended by the Construction Specifications Institute.

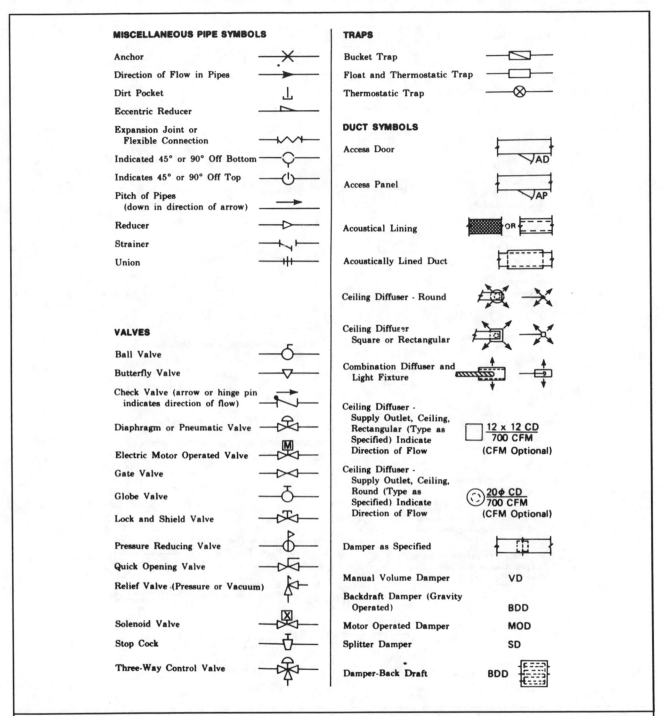

MISCELLANEOUS PIPE SYMBOLS

Anchor

Direction of Flow in Pipes

Dirt Pocket

Eccentric Reducer

Expansion Joint or
   Flexible Connection

Indicated 45° or 90° Off Bottom

Indicates 45° or 90° Off Top

Pitch of Pipes
   (down in direction of arrow)

Reducer

Strainer

Union

VALVES

Ball Valve

Butterfly Valve

Check Valve (arrow or hinge pin
   indicates direction of flow)

Diaphragm or Pneumatic Valve

Electric Motor Operated Valve

Gate Valve

Globe Valve

Lock and Shield Valve

Pressure Reducing Valve

Quick Opening Valve

Relief Valve (Pressure or Vacuum)

Solenoid Valve

Stop Cock

Three-Way Control Valve

TRAPS

Bucket Trap

Float and Thermostatic Trap

Thermostatic Trap

DUCT SYMBOLS

Access Door

Access Panel

Acoustical Lining

Acoustically Lined Duct

Ceiling Diffuser - Round

Ceiling Diffuser
   Square or Rectangular

Combination Diffuser and
   Light Fixture

Ceiling Diffuser -
   Supply Outlet, Ceiling,
   Rectangular (Type as
   Specified) Indicate
   Direction of Flow

12 x 12 CD
700 CFM
(CFM Optional)

Ceiling Diffuser -
   Supply Outlet, Ceiling,
   Round (Type as
   Specified) Indicate
   Direction of Flow

20φ CD
700 CFM
(CFM Optional)

Damper as Specified

Manual Volume Damper   VD

Backdraft Damper (Gravity
   Operated)   BDD

Motor Operated Damper   MOD

Splitter Damper   SD

Damper-Back Draft   BDD

Fig. AD-3: Mechanical symbols recommended by the Construction Specifications Institute. (*cont.*)

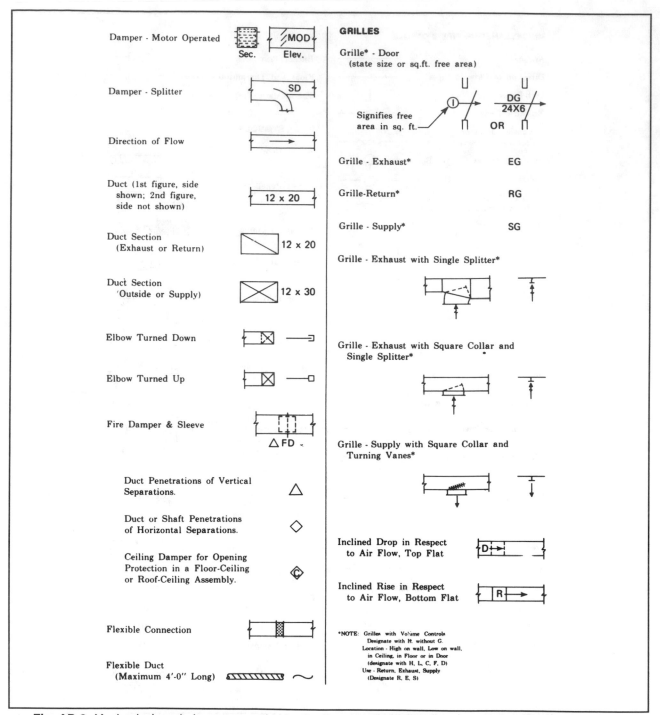

LOUVERS

Louver (1st Figure, Side shown; 2nd Figure, side not shown) — 36 x 24

Power or Gravity Exhaust

Power or Gravity Intake

Power or Gravity Roof Ventilator - Louvered

Radius Type Elbow (Splitter as specified)

Sound Trap — ST

Transformation

Turning Vanes

Unit Heater (Downblast)

Unit Heater (horizontal)

Unit Ventilator (If recessed show amount)

Utility Set

Cabinet Heater (If recessed show amount)

Air Handling Unit

EQUIPMENT

Air Vent or Relief Valve M=Manual A=Automatic — MAV AAV

Baseboard — WALL

Convector — SURFACE SEMI-RECESSED FULLY RECESSED

Drain, Area — AD

Drain, Floor — FD

Drain, Roof — RD

Existing

Fire Hydrant — FH

Humidistat — H

Outlet, Air — A

Outlet, Gas — G

Outlet, Vacuum — V

Pressure Tap & Gauge Cock — P

Pressure Tap & Gauge Cock w/0 - 30 PSI Gauge — P 0-30

Thermometer Well — T

Thermometer Well w/0 - 100°F Thermometer — T 0-100

Thermostat — (Insulated Base) I G (Guard) (Quantity of 4 T RA (Reverse Devices Controlled) Acting)

Wall Fin — WALL

Wall Hydrant — WH

Fig. AD-3: Mechanical symbols recommended by the Construction Specifications Institute. (*Cont.*)

A	Air	LRA	Locked Rotor Amps
AC UNIT	Package Air Conditioning Unit	MA	mixed air
ACU	Air Conditioning Unit	MBH	One Thousand British Thermal Units Per
A/E	Architect/Engineer	OA	Outside Air
AHU	Air Handling Unit (AC or H&V)	OV	Outlet Velocity
		P	Pump
Air BB	Air Baseboard	PD	Pressure Drop
BB	Baseboard	RA	Return Air
BT	Bath Tub	RHC	Reheat Coil
CD	Ceiling Diffuser	S	Sink
CH	Cabinet Heater	SA	Supply Air
CI	cast-iron	SH	Sensible Heat
CO	Cleanout	SH	Shower
Conc	Concrete	SP	Static Pressure
CONV	Convector	TEV	Thermal Expansion Valve
DPT	Dew Point Temperature	TH	Total Heat
EA.	Exhaust Air	TOL	Thermal Overload
EAT	Entering Air Temperature	TS	Tip Speed
ELF	Exhaust Fan	U	Urinal
EWE	Electric Water Cooler	UC	Undercut
FAD	Fire Damper	UH	Unit Heater
FL	Full Load Amps	UNEX	Unexacavated
FTR	Finned Tube Radiation	US	Utility Set
H&V	Heating and Ventilating	W	Unit Ventilator
HB	Hose Bibb	VCP	Vitreous Clay Pipe
HOA	Hand-Off-Auto	VP	Velocity Pressure
HVAC	Heating	VTR	Vent Through Roof
I	Invert	W.H.	Water Heater
L	Lavatory	WB	Wet Bulb Temperature
LAT	Leaving Air Temperature	WC	Water Closet
LH	Latent Heat	WF	Wide Flange
LHR	Latent Heat Ratio	WH	Wall Hydrant

Fig. AD-4: Mechanical abbreviations. Also see Appendix E.

Appendix E

Abbreviation	Term
A	
A	Air
A	Area
A	Astragal
AB	Anchor Bolt
AB SW	Air-break Switch
ABBR	Abbreviate
ABCB	Air-break Circuit Breaker
ABRSV	Abrasive
ABRSV RES	Abrasive-resistant
ABS	Aboslute
ABSORB	Absorption
ABSTR	Abstract
ABV	Above
AC	Alternating Current
AC	Armored Cable

Abbreviation	Term
AC UNIT	Package Air Conditioning Unit
ACCEL	Accelerate
ACCESS	Accessory
ACIRC	Air Circulating
ACLD	Air-cooled
ACR	Across
ACS	Access
ACST	Acoustic
ACT	Activity
ACTE	Actuate
ACTG	Actuating
ACTL	Actual
ACTR	Actuator
ACTVT	Activate
ACTVTR	Activator
ACU	Air Conditioning Unit
AD	Air Drawn
AD	Area Drain

Abbreviation	Term
ADD	Addition
ADDT	Additive
ADH	Adhesive
ADJ	Adjacent
ADJT	Adjustable
ADPTR	Adapter
ADSORB	Adsorbent
ADV	Advance
AE	Air Extractor
A/E	Architect/Engineer
AGT	Agent
AHR	Anchor
AHU	Air Handling Unit (AC or H&V)
AIL	Aileron
AINDTN	Air Induction
Air BB	Air Baseboard
AIR COND	Air-Condition(er)
AIRCLNR	Air Cleaner
AL	Air Lock
AL	Aluminum
ALC	Alcohol
ALIGN	Alignment
ALK	Alkaline
ALLOW	Allowance
ALTN	Alternate
ALTRN	Aleration
ALY	Alloy
ALY STL	Alloy Steel
AMB	Ambient
AMNA	Ammonia
AMP	Ampere
AMPTD	Amplitude
AMR	Advance Material Request
AMT	Amount
ANCFIL	Anchored Filament
ANDA	Anodize
ANLR	Angular
ANLR	Annular

Abbreviation	Term
ANN	Annunciator
ANS	American National Standard
ANT	Antenna
AO	Access Opening
AP	Acidproof
AP	Air Passage
APA	Axial Pressure Angle
APC	Acoustical Plastic Ceiling
APERT	Aperture
APP	Appearance
APP	Approximate
APPL	Application
APPROX	Approximate
APPV	Approve
APPVL	Approval
APPX	Appendix
APRCH	Approach
APT	Apartment
APU	Auxiliary Power Unit
APVD	Approved
APVD EQL	Approved Equal
AQL	Aceptable Quality Level
AR	As Required
ARCH	Architect
ARCW	Arc Weld
ARM	Armature
ARM	Armored
ARST	Arrester
ARTF	Artificial
AS	Automatic Sprinkler
ASB	Asbestos
ASPH	Asphalt
ASPHRS	Asphalt Roof Shingles
ASPRTR	Aspirator
ASTR	Astronomical
ASV	Angle Stop Valve
ASW	Auxiliary Switch
ASWG	American Steel Wire Gauge

Abbreviation	Term
AT	Acoustical Tile
AT	Ashphalt Tile
ATB	Asphalt-Tile Base
ATCH	Attachment
ATF	Asphalt-tile Floor
ATTEN	Attenuation
ATTEN	Attenuator
AUTO	Automatic
AUTO OVLD	Automatic Overload
AUTOSTR	Automatic Starter
AUX	Auxiliary
AUXR	Auxiliary Register
AVDP	Avoirdupois
AVE	Avenue
AVG DIA	Average Diameter
AW	Above Water
AWG	American Wire Gauge
AWN	Awning
AX FL	Axial Flow
AXP	Axial Pitch
AZ	Azimuth

B

Abbreviation	Term
B	Base
B	Bath(room)
B	Bit
B TO B	Back-to-Back
B&S	Bell and Spigot
BARR	Barrier
BAT	Batter
BATT	Batten
BB	Baseboard
BB	Brass Bolt
BC	Beginning of Curve
BC	Between Centers
BC	Bookcase
BC	Bottom Chord

Abbreviation	Term
BCL	Basic Contour Line
BCL	Broom Closet
BD	Board
BDO	Bow Door
BDRY	Boundary
BDY	Boundary
BDZR	Bulldozer
BE	Back End
BEV	Bevel
BF	Backface
BF	Board Foot
BF	Both Faces
BF	Bottom Face
BH	Brinell Harness
BHD	Bulkhead
BHN	Brinell Harness Number
BI	Black Iron
BITUM	Bituminous
BK	Book
BKGD	Background
BL	Base line
BL	Bend Line
BL	Billet
BL	Blade
BL	Bottom Layer
BL	Boundary Line
BL	Building Line
BL STL	Billet Steel
BLDG	Building
BLDR	Bleeder
BLK	Blank
BLK	Block
BLKG	Blocking
BLKT	Blanket
BLO	Blower
BLR	Boiler
BLT	Bolt
BLTIN	Built-in

Abbreviation	Term
BLVD	Boulevard
BLW	Below
BLWT	Blowout
BM	Beam
BM	Bench Mark
BMVP	Barrier, Moisture Vapor-proof
BNCH	Bench
BND	Bonded
BNSH	Burhish
BOD	Blackout Door
BOT	Bottom
BP	Back Pressure
BP	Base Plate
BP	Between Perpendiculars
BP	Bolted Plate
BPL	Bearing Plate
BR	Bedroom
BR	Bend Radius
BR	Braided
BR	Branch
BR	Brush
BRDG	Bridge
BRG	Bearing
BRK	Break
BRK	Brick
BRKR	Breaker
BRKT	Bracket
BRT	Bright
BS	Backsight
BS	Both Sides
BSC	Basic
BSHG	Bushing
BSMT	Basement
BSP	Ball Stop
BT	Bath Tub
BTU	British Thermal Unit
BTWLD	Butt Weld
BU	Blower Unit

Abbreviation	Term
BU	Bushel
BUR	Built-up Roofing
BVC	Beginning Vertical Curve
BWP	Barrier, Waterproof
BWS	Beveled Wood Siding
BYP	Bypass

C

Abbreviation	Term
°C	Degrees Celsius
C	Carbon
C	Centigrade
C	Courses
C	Cycle
C TO C	Center-to-Center
CA	Cable
CA	Cold Air
CAB	Cabinet
CAL	Caliber
CAL	Calibrate
CAL	Calorie
CAN	Canopy
CANTIL	Cantilever
CAP	Capacitor
CAP	Capacity
CAP	Capital
CARP	Carpentry
CARP	Carpet
CAT	Catalogue
CAV	Canvas
CAV	Cavity
CB	Catch Basin
CB	Circuit Breaker
CBAL	Counterbalance
CBD	Carbide
CBORE	Counterbore
CC	City Council
CC	Color Code

Abbreviation	Term
CC	County Council
CCA	Circuit Court of Appeals
CCB	Concrete Block
CCF	Concrete Floor
CD	Cable Duct
CD	Ceiling Diffuser
CD	Cold-Drawn
CD	Condesate Drain
CE	Civil Engineer
CEL	Celsius
CEM	Cement
CENT	Centigrade
CER	Ceramic
CF	Cement Floor
CF	Cooling Fan
CFLG	Counterflashing
CFM	Cubic Feet Per Minute
CFS	Cubic Feet Per Second
CH	Cabinet Heater
CH	Case Harden
CH	Chain
CH	Church
CHAM	Chamfer
CHAN	Channel
CHAP	Chapter
CHD	Chord
CHG	Charge
CHM	Chimney
CI	Cast Iron
CI	Circuit Interrupter
CIP	Case-Iron Pipe
CIR	Circle
CIRC	Circular
CIV	Civil
CJ	Construction Joint
CKT	Circuit
CL	Center line
CL	Clearance

Abbreviation	Term
CL	Close
CL	Crane Load
CL GL	Clear Glass
CLD	Cleared
CLE	Closed End
CLG	Ceiling
CLJ	Control Joint
CLKJ	Caulked Joint
CLO	Closet
CLOS	Closure
CLP	Clamp
CLPR	Caliper
CLR	Clear
CLR	Collar
CLR	Cooler
CLT	Cleat
CM	Common Meter
CMBSTR	Combustor
CMET	Coated Metal
CMPS	Compass
CNCL	Concealed
CNCV	Concaved
CND	Conduit
CNDCT	Conductor
CNDS	Condesate
CNTR	Counter
CNVR	Conveyor
CO	Cased Opening
CO	Change Order
CO	Cleanout
CO	Colbalt
CO	Company
CO	County
CO	Cutoff
COAX	Coaxial
COD	Cash (or collect) on Delivery
COEF	Coefficient
COL	Colonial

Abbreviation	Term	Abbreviation	Term
COL	Column	CR	Credit
COLL	Collector	CRCLT	Circulate
COM	Common	CRCMF	Circumference
COMB	Combination	CRE	Corrosion-Resistant
COMB	Combustion	CRES	Corrosion-Resistant Steel
COMBL	Combustible	CRG	Carriage
COML	Commercial	CRK	Crank
COMP	Compacted	CRN	Crane
COMP	Compound	CRN	Crown
COMPA	Compressed Air	CRS	Coarse
COMPL	Complete	CRS	Cold-Rolled Steel
COMPOS	Composition	CRSHD STN	Crushed Stone
CONC	Concrete	CRSN	Corrosion
COND	Condenser	CRSV	Corrosive
CONST	Construction	CRTG	Centering
CONT	Continuous	CRV	Curve
CONTR	Contact	CS	Carbon Steel
CONTR	Contractor	CS	Case
CONV	Convector	CS	Cast Steel
COOL	Coolant	CS	Control Switch
COR	Corner	CSB	Concrete Splash Block
COR	Cornice	CSG	Casing
CORR	Corrugate	CSK	Countersink
COT	Cotter	CT	Ceramic Tile
COT	Cotter Pin	CTB	Ceramic Tile Base
COV	Cutoff Valve	CTD	Coated
CP	Clay Pipe	CTF	Ceramic Tile Floor
CP	Conrete Pipe	CTG	Coating
CPA	Certified Public Accountant	CTL	Central
CPC	Cement Plaster Ceiling	CTLST	Catalyst
CPL	Cement Plaster	CTR	Center
CPLG	Coupling	CTR	Contour
CPM	Critical Path Method	CTWALK	Catwalk
CPNTR	Carpenter	CU	Copper
CPPD	Capped	CU	Cubic
CPT	Critical Path Technique	CU IN	Cubic Inches
CR	Chromium	CU YD	Cubic Yards
CR	Cold-Rolled	CULV	Culvert

Abbreviation	Term	Abbreviation	Term
CUR	Current	DENS	Density
CUTS	Cut Stone	DEPR	Depression
CV	Check Valve	DEPT	Department
CVNTL	Conventional	DET	Detail
CVX	Convex	DET	Detent
CW	Cold Water	DEV	Development
CW	Continuous Wave	DEV LG	Developed Length
CWP	Circulating Water Pump	DF	Douglas Fir
CYL	Cyindrical	DF	Drinking Fountain
CSMT	Casement	DF	Drive Fit
		DF	Drop Forge

D

Abbreviation	Term	Abbreviation	Term
°DEG	Degree	DFT	Draft
D	Deep	DFT	Drift
D	Density	DFTG	Drafting
D	Dryer	DFTSMN	Draftsman
D&M	Dressed and Matched	DHMY	Dehumidify
DAD	Double-Acting Door	DHW	Double-Hung Windows
DAT	Datum	DIA	Diameter
DB	Decibel	DIAG	Diagonal
DB	Distribution Box	DIAG	Diagram
DB	Dry Bulb	DIAL PHTH	Diallyl Phthalate
DBL	Double	DIAPH	Diaphragm
DBL ACT	Double-Actaing	DIFF	Differential
DBLF	Double Face	DIFFUS	Diffusing
DBLW	Double-Wall	DIL	Dilute
DC	Direct Current	DIM	Dimension
DCL	Door Closer	DIR	Direct
DD	Deep Drawn	DIR	Direction
DD	Disconnecting Device	DISC	Disconnect
DD	Dutch Door	DISCH	Discharge
DDR	Direct Drive	DIST	Distance
DE	Double End	DIV	Divide
DEC	Decimal	DIV	Division
DECAL	Decalcomania	DK	Deck
DECR	Decrease	DL	Dead Load
DED	Dedendum	DL	Drawing List
DEG	Degree	DLO	Daylight Opening
		DLY	Delay

Abbreviation	Term
DMGZ	Demagnetize
DML	Demolition
DMPR	Damper
DMR	Dimmer
DN	Down
DNDFT	Downdraft
DNG	Dining
DNG R or DR	Dining Room
DNTRD	Denatured
DO	Ditto
DOZ	Dozen
DP	Dampproofing
DP	Depth
DP	Dew Point
DP	Diameter Pitch
DP	Dripproof
DP	Dual-Purpose
DP SW	Double-Pole Switch
DPDT	Double-Pole Double-Throw
DPDT SW	Double-Pole Double-Throw Switch
DPNL	Distribution Panel
DPST	Double-Pole Single-Throw
DPST SW	Double-Pole Single-Throw Switch
DPT	Dew Point Temperature
DPV	Dry Pipe Valve
DR	Door
DR	Drain
DR	Drill
DR	Drill Rod
DR	Drive
DRK	Derrick
DRM	Dormer
DRS	Dressed (lumber)
DS	Disconnect Switch
DS	Downspout
DS	Draft Stop
DSG	Double Strength Glass
DSGN	Design

Abbreviation	Term
DSPEC	Design Specification
DST	Door Stop
DSW	Door Switch
DT	Drain Tile
DT	Dust-Tight
DTCH	Detached
DTY CY	Duty Cycle
DUP	Duplicate
DVL	Develop
DVTL	Dovetail
DW	Dishwasher
DW	Double Weight
DW	Drywall
DW	Dumbwaiter
DWEL	Dwelling
DWG	Drawing
DWL	Design Water Line
DWL	Dowel
DWR	Drawer
DX	Duplex
DYNMT	Dynamite

E

Abbreviation	Term
E	East
E	Enamel
E TO E	End-To-End
E&SP	Equipment and Spare Parts
EA	Each
EA	Exhaust Air
EAT	Entering Air Temperature
ECC	Eccentric
ECD	Estimated Completion Date
ECO	Electronic Checkout
ECO	Engineering Change Order
EDGW	Edgewise
EF	Each Face
EF	Exhaust Fan

Abbreviation	Term	Abbreviation	Term
EF	Extra Fine (threads)	ENGRG	Engineering
EFF	Effective	ENLG	Enlarge
EFF	Efficiency	ENLGD	Enlarged
EHP	Effective Horsepower	ENRGZ	Energize
EJCTR	Ejector	ENT	Entrance
EJN	Ejection	ENTR	Entrance
EL	Electroluminescent	ENV	Envelope
EL	Elevation	ENVIR	Environment
ELAS	Elastic	EP	Electric Panel
ELAS	Elasticity	EP	Explosion-Proof
ELB	Elbow	EPD	Electric Power Distribution
ELCTC	Electric Contact	EPT	External Pipe Thread
ELCTD	Electrode	EPWR	Emergency Power
ELCTLT	Electrolytic	EQ	Equal
ELEC	Electric	EQL	Equally Spaced
ELEK	Electronic	EQPT	Equipment
ELEM	Elementary	EQUIP	Equipment
ELEV	Elevate	EQUIV	Equivalent
ELEV	Elevator	ERCG	Erecting
ELEX	Electronics	ERCR	Erector
ELL	Elbow	ERECT	Erection
ELMCH	Electromechanical	ES	Electrostatic
ELONG	Elongation	ES	Equal Section
ELP	Elliptical	ESC	Escutcheon
ELPNEU	Electropneumatic	EST	Estimate
EM	Electromagnetic	ET	Edge Thickness
EM	Expanded Metal	EVAP	Evaporator
EMB	Emboss	EVLTN	Evaluation
EMER	Emergency	EWC	Electric Water Cooler
EMF	Electromotive Force	EX	Example
EMT	Electrical Metallic Tubing	EX	Extra
EMUL	Emulsion	EXC	Excavate
ENAM	Enamel	EXCH	Exchange
ENCL	Enclose	EXEC	Executive
ENCL	Enclosure	EXH	Exhaust
ENCSD	Encased	EXHV	Exhaust Vent
ENG	Engine	EXIST	Existing
ENGR	Engineer	EXP	Expand

Abbreviation	Term	Abbreviation	Term
EXP	Expansion	FAO	Finish All Over
EXP	Exposed	FB	Face Brick
EXP JT	Expansion Joint	FB	Flat Bar
EXPLD	Explode	FB	Fuse Block
EXPSR	Exposure	FBCK	Firebrick
EXST	Existing	FBDC	Fiberboard, Corrugated
EXT	Extension	FBDS	Fiberboard, Solid
EXT	Exterior	FBM	Feet Board Measure
EXT	Extinguish	FBR	Fiber
EXT	Extinguisher	FBRBD	Fiberboard
EXT GR	Exterior Grade	FBRS	Fibrous
EXTD	Extrude	FC	Fireclay
EXTR	Extractor	FC	Flat Cable
EXTR	Extrude	FC	Footcandle
EYLT	Eyelet	FC	Furred Ceiling
EYPC	Eyepiece	FCC	Flat Conductor Cable
EL	Elastic Limit	FCG	Facing
		FD	Fire Damper
		FD	Floor Drain
F		FD	Forced Draft
°F	Fahrenheit	FDB	Forced-Draft Blower
4P	Four-Pole	FDFL	Fluid Flow
4PDT SW	Four-Pole Double-Throw Switch	FDN	Foundation
4PST SW	Four-Pole Single-Throw Switch	FDPL	Fluid Pressure Line
4PSw	Four-Pole Switch	FDR	Feeder
4W	Four-Wire	FDR	Fire Door
4WAY	Four-Way	FDWL	Fiberboard, Double Wall
F	Fahrenheit	FED	Federal
F	Farads	FEM	Female
F	Flat	FEX	Flexible
F/D	Face or Field of Drawing	FEXT	Fire Extinguisher
F DR	Fire Door	FFILH	Flat Fillister Head
F TO F	Face-To-Face	FFL	Female Flared
F1S	Finish One Side	FG	Finish Grade
F2S	Finish Two Sides	FGD	Forged
FAB	Fabricate	FH	Fire Hose
FABL	Fire Alarm Bell	FH	Flat Head
FABX	Fire Alarm Box	FHC	Fire Hose Cabinet

Abbreviation	Term	Abbreviation	Term
FHP	Fractional Horsepower	FLTG	Floating
FHR	Fire Hose Rack	FLTR	Filter
FHY	Fire Hydrant	FLUOR	Fluorescent
FIG	Figure	FLW	Flat Washer
FIL	Filament	FM	Flow Meter
FIL	Fillet	FM	Frequency Modulation
FIL	Fillister	FMAN	Foreman
FILH	Fillister Head	FNSH	Finish
FIN	Finish	FNSH FL	Finished Floor
FIR	Fired	FORG	Forging
FIX	Fixture	FP	Faceplate
FL	Flashing	FP	Fireplace
FL	Flat	FP	Flat Point
FL	Floor	FP	Freezing Point
FL	Floor Line	FPL	Fire Plug
FL	Flow	FPM	Feet Per Minute
FL	Fluid	FPS	Feet Per Second
FL	Flush	FPRF	Firepoof
FLA	Full Load Amps	FR	Frame
FLD	Field	FR	From
FLDG	Folding	FR	Front
FLDT	Floodlight	FRAC	Fractional
FLG	Flange	FRAG	Fragmentation
FLG	Flashing	FREQ	Frequency
FLG	Flooring	FRES	Fire-Resistant
FLGSTN	Flagstone	FRWK	Framework
FLH	Flat Head	FRZR	Freezer
FLM	Flame	FS	Far Side
FLMB	Flammable	FS	Field Specification
FLMPRF	Flameproof	FSBL	Fusible
FLMT	Flush Mount	FSC	Full Scale
FLN	Fuel Line	FSN	Federal Stock Number
FLR	Filler	FST	Forged Steel
FLRD	Flared	FSTNR	Fastner
FLRG	Flaring	FT	Feet
FLRT	Flow Rate	FT	Foot
FLSW	Floor Switch	FT LB	Foot Pound Force
FLT	Filter	FTG	Fitting

Abbreviation	Term	Abbreviation	Term
FTG	Footing	GBG	Garbage
FTHRD	Female Thread	GDR	Guard Rail
FTK	Fuel Tank	GEN	Generator
FTR	Finned Tube Radiation	GEN CONT	General Contractor
FTR	Fixed Transom	GENL	General
FTR	Flat Tile Roof	GFCI	Ground-Fault Circuit Interrupter
FU	Fuse	GFU	Glazed Facing Unit
FUBX	Fuse Box	GGL	Ground Glass
FUHLR	Fuse Holder	GL	Glass
FUNL	Funnel	GL	Grade Line
FUR	Furring	GLB	Glass Block
FURN	Furnish	GLV	Globe Valve
FUS	Fuselage	GLZ	Glaze
FUT	Future	GND	Ground
FV	Front View	GRD	Ground
FV	Full Voltage	GNLTD	Granulated
FW	Fire Wall	GOVT	Government
FW	Fresh Water	GP	General-Purpose
FWD	Forward	GPC	Gypsum-Plaster Ceiling
FWP	Fresh Water Pump	GPH	Galon Per Hour
FX WDW	Fixed Window	GPH	Graphite
FXD	Fixed	GPM	Gallon Per Minute
FXTR	Fixture	GPW	Gypsum-Plaster Wall
		GR	Gear
	G	GR	Grab Rod
G	Gas	GR	Grade
G	Girder	GR	Grain
G	Grid	GR	Gross
G	Grounded (outlet)	Grad	Gradient
GA	Gauge	GRAN	Granite
GABD	Gauge Board	GRD	Grind
GAL	Gallon	GRDTN	Graduation
GAll	Gallery	GRK	Gear Rack
GALV	Galvanize	GRL	Grill
GALVI	Galvanized Iron	GROM	Grommet
GALVS	Galvanized Steel	GRTG	Grating
GAR	Garage	GRV	Groove
GARPH	Graphic	GRVD	Grooved

Abbreviation	Term	Abbreviation	Term
GRVG	Grooving	HD	Hard-Drawn
GRVR	Groover	HD	Head
GRWT	Gross Weight	HD	Heavy-Duty
GSB	Gypsum Sheathing Board	HDDRN	Hard-Drawn
GSFU	Glazed Structural Facing Unit	HDL	Handle
Gskt	Gasket	HDLS	Headless
GSV	Globe Stop Valve	HDN	Harden
GSWR	Galvanized Steel Wire Rope	HDR	Header
GT	Grease Trap	HDW	Hardware
GTV	Gate Valve	HDWD	Hardwood
GUT	Gutter	HE	Heat Exchange
GVL	Gravel	HEX	Hexagon
GVW	Gross Vehicle Weight	HEX HD	Hexagonal Head
GWB	Gypsum Wallboard	HF	High Frequency
GWT	Glazed Wall Tile	HGBN	Herringbone
GYM	Gymnasium	HGR	Hanger
GYP	Gypsum	HGT	Height
		HI HUM	High Humidity
H		HIMP	High Impact
½H	Half-Hard	HINT	High Intensity
½RD	Half-Round	HLCL	Helical
H	Hard	HM	Hollow Metal
H	Henries	HMD	Humidity
H	High	HNDRL	Hand Rail
H GALV	Hot-Galvanize	HNDWL	Handwheel
H&V	Heating and Ventilating	HNG	Hanging
HA	Hot Air	HNG	Hinge
HAZ	Hazardous	HNYCMB	Honeycomb
HB	Hose Bibb	HOA	Hand-Off-Auto
HBD	Hardboard	HOL	Hollow
HC	High Carbon	HOR	Horizontal
HC	Hollow Core	HP	High Pressure
HC	Hose Clamp	HP	Horsepower
HCL	Horizontal Center Line	HPS	High-Pressure Steam
HCS	High Carbon Steel	HPT	High Point
HCSHT	High Carbon Steel, Heat-Treated	HR	Hook Rail
HD	Hand (comb form)	HR	Hose Rack
HD	Hard	HR	Hot Rolled

Abbreviation	Term
HR	Hour
HRS	Hot Rolled Steel
HS	High-Speed
HSG	Housing
HSS	High-Speed Steel
HST	Hoist
HSTH	Hose Thread
HT	Heat
HT	High Tension
HT	Hollow Tile
HT RES	Heat-Resisting
HT SHLD	Heat Shield
HT TR	Heat Treat
HTCI	High-Tensile Cast Iron
HTD	Heated
HTG	Heating
HTNSL	High Tensile
HTR	Heater
HTS	High Tensile Strength
HTS	High-Tensile Steel
HV	High Voltage
HVAC	Heating, Ventilating, and Air Conditioning
HVR	High-Voltage Regulator (electrical distribution)
HVRNG	Hovering
HVY	Heavy
HW	Hot Water
HWL	High-Water Line
HWT	Hot Water Tank
HWY	Highway
HYD	Hydraulic
HYDR	Hydraulic
HYDRELC	Hydroelectric (electrical distribution systems)
HYDRO	Hydrostatic
HYDTD	Hydrated
Hz	Hertz

I

Abbreviation	Term
I	Invert
I	Iron
I&O	Inlet and Outlet
IACS	International Copper Standard
IC	Intergrated Circuit
ICS	International Control Station
ICTCD	Insecticide
ID	Induced Draft
ID	Inside Diameter
ID	Internal Diameter
IDCTR	Inductor
IDL	Idler
IF	Intermediate Frequency
IGN	Ignition
ILLUM	Illumination
IMP	Impedance
IMPG	Impregnate
IN	Inch(es)
IN LB	Inch-Pound
INBD	Inboard
INC	Incorporated
INCAND	Incandescent
INCL	Include; inclusive
INCOMP	Incomplete
INCR	Increase
IND	Indicate
IND	Industry
INL	Inlet
INRT	Inert
INRTG	Inert Gas
INS	Inside
INSP	Inspect
INSR	Insert
INSTL	Install; Installation
INSUL	Insulate; Insulation
INT	Integral

Abbreviation	Term
IINT	Intensifier
INTCHG	Interchangeable
INTERCOM	Intercommunication
NTL	Internal
INTMD	Intermediate
INTMT	Intermittent
INTR	Interior
INTRPT	Interrupt
IP	Iron Pipe
IPS	Inch(es) Per Second
IPS	International Pipe Standard
IPS	Iron Pipe Size
IPT	Internal Pipe Thread
IPT	Iron Pipe Thread
IR	Infrared
IR	Inside Radius
IRREG	Irregular
ISO	Isometric
ISS	Issue
IW	Indirect Waste

J

Abbreviation	Term
J	Joiner or Joist
J&P	Joists and Planks
JAP	Japanned
JB	Jamb
JC	Janitor's Closet
JCT	Junction
JCTBX	Junction Box
JFET	Junction Field-effect Transistor
JK	Jack
JKSCR	Jackscrew
JMB	Jamb
JNL	Journal
JO	Job Order
JOUR	Journeyman
JR	Junior
JT	Joint

Abbreviation	Term

K

Abbreviation	Term
K	Keel
K	Key
K	Kilohm
K	Kip (1000 lb)
KAL	Kalamein
KALD	Kalamein Door
KB	Knee Brace
KC	Kilocycle
KD	Kiln-Dried
KD	Knocked Down
KINE	Kinescope
KIT	Kitchen
KN	Knot
KN SW	Knife Switch
KO	Knockout
KPL	Kickplate
KST	Keyseat
KV	Kilovolt
KVAH	Kilovolt-Ampere Hour
KVAHM	Kilovolt-Ampere Hour Meter
KVAM	Kilovolt-Ampere Meter
KWH	Kilowatt Hour
KWY	Keyway

L

Abbreviation	Term
L	Lavatory
L	Left
L	Left
L	Line
L	Long
LA	Lightning Arrester
LAB	Laboratory
LAD	Ladder
LAG	Lagging
LAM	Laminate

Abbreviation	Term	Abbreviation	Term
LLAQ	Lacquer	LNTL	Lintel
LAT	Lateral	LOA	Length Over All
LAT	Leaving Air Temperature	LOC	Locate
AU	Laundry	LOS	Line-of-Sight
LAV	Lavatory	LP	Low Pressure
LBL	Label	LPG	Lapping
LBR	Lumber	LPW	Lumen Per Watt
LBRY	Library	LR	Ladder Rung
LC	Laundry Chute	LR	Living Room
LC	Lead Covered	LRA	Locked Rotor Amps
LC	Low Carbon	LS	Left Side
LCH	Latch	LS	Limestone
LCL	Linen Closet	LS	Loudspeaker
LCL	Local	LS	Low-Speed
LDG	Landing	LT	Laundry Tray
LDR	Leader	LT	Light
LE	Leading Edge	LT	Low Tension
LF	Linoleum Floor	LTC	Lattice
LF	Low Frequency	LTEMP	Low Temperature
LG	Length	LTH	Lath
LG	Long	LTQ	Low Torque
LGE	Large	LTR	Letter
LH	Latet Heat	LTSW	Light Switch
LH	Left Hand	LUB	Lubricate; Lubricator
LHR	Latent Heat Ratio	LV	Low Voltage
LIM	Limit	LVL	Level
LIM SW	Limit Switch	LWC	Lightweight Concrete
LIN	Linear	LWIC	Lightweight Insulating Concrete
LINOL	Linoleum	LWR	Lower
LIQ	Liquid	LWST	Lowest
LITHO	Lithograph	LYR	Layer
LK	Link		
LK WASH	Lock Washer		
LKG	Locking		

M

Abbreviation	Term	Abbreviation	Term
LKNT	Locknut	M	Magnaflux
LL	Live Load	M	Meter (Measure of Length)
LM	List of Materials	M	Meter (Instrument)
LNG	Lining	M&F	Male & Female

Abbreviation	Term	Abbreviation	Term
MA	Master	MECH	Mechanical
MA	Metal Anchor	MED	Median
MA	Mixed Air	MED	Medical
MACH	Machine	MEG	Megohm
MAG	Magnesium	MEL	Melamine
MAG	Magnetic	MEMB	Membrane
MAH	Mahogany	MEMO	Memorandum
MAINT	Maintenance	MER	Meridian
MAL	Malleable	MET	Metal
MALL	Malleable	METB	Metal Base
MAN	Manual	METD	Metal Door
MAN OVLD	Manual Overload	METF	Metal Flashing
MANF	Manifold	METG	Metal Grill
MAR	Marine	METJ	Metal Jalousie
MAS	Metal Anchor Slots	METP	Metal Partition
MASU	Machined Surface	METR	Metal Roof
MAT	Matrix	METS	Metal Strip
MATL	Material	MEZZ	Mezzanine
MATW	Metal Awning-Type Window	MF	Mastic Floor
MAX	Maximum	MF	Metered Flow
MB	Mailbox	MFD	Manufactured
MB	Model Block	MFG	Manufacturing
MBH	Thousand British Thermal Units	MFR	Manufacture
MBL	Mobile	MFSFU	Matt-Finish Structural Facing Units
MBR	Member	MG	Magnetic Armature
MC	Manhole Cover	MGL	Mogul
MC	Medicine Cabinet	MH	Manhole
MC	Megacycle	MH	Millihenries
MC	Metal-clad Cable	MHD	Masthead
MCD	Metal Covered Door	MI	Malleable Iron
MCHRY	Machinery	MI	Mile
MDC	Motor Direct-Connected	MI	Mineral-insulated Cable
MDL	Middle	MIC	Micrometer
MDM	Maximum Design Meter	MIL	Military
MDM	Medium	MIN (')	Minute
MDN	Median	MIN	Minimum
MDRL	Mandrel	MIN	Minor
MEAS	Measure	MIR	Mirror

Abbreviation	Term	Abbreviation	Term
MISC	Miscellaneous	MS	Machine Steel
MIT	Miter	MSCR	Machine Screw
MIX	Mixture	MSL	Mean Sea Level
MJ	Mastic Joint	MSNRY	Masonry
MK	Mark	MSR	Mineral-Surface Roof
MKR	Marker	MST	Machine Steel
MKUP	Makeup	MSTC	Mastic
ML	Material List	MSW	Master Switch
ML	Mold Line	MT	Maximum Torque
ML	Monolithic	MT	Metal Threshold
MLD	Molded	MT	Mount
MLDG	Molding	MTCHD	Matched
MLP	Metal Lath and Plaster	MTD	Mounted
MM	Millimeter	MTG	Mounting
MN	Main	MTHRD	Male Threaded
MNL OPR	Manually Operated	MTLC	Metallic
MNRL	Mineral	MTRDN	Motor Driven
MO	Masonry Opening	MTRS	Mattress
MO	Month	MTWF	Metal Through-Wall Flashing
MO	Motor-operated	MTZ	Motorized
MOD	Model	MULT	Multiple
MOD	Modification	MVBL	Movable
MOD	Modify	MWG	Music Wire Gage
MON	Monument	MWO	Modification Work Order
MOR	Mortar	MWP	Maximum Working Pressure
MOR T	Morse Taper	MWP	Membrane Waterproofing
MOS	Mosaic	MWV	Maximum Working Voltage
MOT	Motor	MXT	Mixture
MP	Melting Point		
MPE	Maximum Permissible Exposure		
MPG	Miles per Gallon	**N**	
MPH	Miles per Hour	N	Noon
MPT	Male Pipe Thread	N	North
MR	Marble	NA	Naval Architect
MRD	Metal Rolling Door	NA	Not Applicable
MRF	Marble Floor	NAR	Narrow
MRT	Marble Threshold	NAS	National Aircraft Standards
MRT	Mildew-Resistant Thread	NAT	Natural

Abbreviation	Term
NATL	National
NBS	National Bureau of Standards
NBS	New British Standard
NC	National Coarse (thread)
NC	Normally Closed
NCH	Notched
NCM	Noncorrosive Metal
NCOMBL	Noncombustible
NEC	National Electrical Code
NEF	National Extra Fine (thread)
NEG	Negative
NESC	National Electrical Safety Code
NEUT	Neutral
NF	National Fine (thread)
NF	Near Face
NIC	Not In Contract
NIP	Nipple
NM	Nonmetallic
NMAG	Nonmagnetic
NO	Normally Open
NO	Number
NOM	Nominal
NONFLAMB	Nonflammable
NORM	Normal
NOS	Nosing
NOZ	Nozzle
NP	Name Plate
NP	National Pipe Thread
NPL	Nameplate
NPRN	Neoprene
NPT	National Taper Pipe (thread)
NRCP	Nonreinforced-Concrete Pipe
NS	National Special (thread)
NS	Near Side
NST	Nonslip Tread
NTS	Not to Scale
NTWT	Net Weight
NYL	Nylon

Abbreviation	Term
	O
OA	Outside Air
OA	Overall
OB	Obscure
OBJV	Objective
OBS	Obsolete
OBW	Observation Window
OC	On Center
OC	Outside Circumference
OC	Overcurrent
OCB	Oil Circuit Breaker
OCC	Occupy
OCLD	Oil-Cooled
OCR	Overcurrent Relay
OCT	Octagon
OD	Outside Diameter
OE	Open End
OF	Outside Face
OFCE	Office
OFF	Office
OGL	Obscure Glass
OHM	Ohmmeter
OI	Oil-Insulated
OP	Operator
OPA	Opaque
OPN	Operation
OPNG	Opening
OPNL	Operational
OPP	Opposite
OPR	Operate
OPRT	Operator Table
OPT	Optical
OPT	Optimum
OR	Outside Radius
ORD	Ordinance
ORIG	Origin

Abbreviation	Term	Abbreviation	Term
ORIG	Original	PANB	Panic Bolt
ORLY	Overload Relay	PAR	Parallel
OSC	Oscillator	PARA	Paragraph
OSP	Operating Steam Pressure	PARAT	Partial
OUT	Outlet	PASS	Passage
OUT	Output	PAT	Patent
OUT	Outside	PATT	Pattern
OUTBD	Outboard	PB	Painted Base
OV	Outlet Velocity	PB	Pull Box
OV	Over	PB	Push Button
OVFL	Overflow	PB SW	Pull-Button Switch
OVH	Oval Head	PBD	Paperboard
OVHD	Overhead	PBD	Pressboard
OVHG	Overhanging	PC	Personal Carrier
OVHL	Overhaul	PC	Piece
OVLD	Overload	PC	Pitch Circle
OVRD	Override	PC	Pressure Controlled
OWJ	Open Web Joist	PC MK	Piece Mark
OWU	Open Window Unit	PCC	Point Of Compound Curve
OXD	Oxidized	PCH	Punch
OXY	Oxygen	PD	Pitch Diameter
OZ	Ounce	PD	Pivoted Door
		PD	Potential Difference
		PD	Pressure Drop
P		PED	Pedestal
		PEIM	Perimeter
P	Page	PEMB	Permeability
P	Pilaster	PEN	Penetration
P	Pitch	PERF	Perfect
P	Plate (electron tube)	PERF	Perforate
P	Pole	PERM	Permanent
P	Porch	PERP	Perpendicular
P	Port	PF	Power Factor
P	Pump	PF	Profile
P-P	Peak-To-Peak	PF	Pump, Fixed Displacement
P&O	Paints and Oils	PFD	Preferred
P&T	Posts and Timbers	PFN	Prefinished
PA	Pressure Angle	PG	Pressure Gauge
PAN	Pantry		

Abbreviation	Term	Abbreviation	Term
PGMT	Pigment	PM	Pulse Modulation
PH	Phase	PMP	Pump
PH BRZ	Phosphor Bronze	PMU	Plaster Mockup
PHEN	Phenolic	NEU	Pneumatic
PHH	Phillips Head	PNH	Pan Head
PHR	Preheater	PNL	Panel
PI	Point Of Intersection	PNT	Paint
PIN	Pinion	PNTGN	Pentagon
PK	Pack	PO	Production Order
PK	Peck	POB	Point Of Beginning
PKG	Packing	POL	Polish
PKT	Pocket	POLYEST	Polyester
PKWY	Parkway	POP	Popping
PL	Padlock	PORC	Porcelain
PL	Place	PORT	Portable
PL	Plain	POS	Position
PL	Plate	POSN SW	Position Switch
PL	Plug	POT	Potential
PL	Property Line	POT	Potentiometer
PLAS	Plaster	POTW	Potable Water
PLATF	Platform	PP	Panel Point
PLB	Pull Button	PP	Piping
PLBLK	Pillow Block	PP	Push-Pull
PLD	Plated	PPLN	Pipeline
PLG	Piling	PR	Pair
PLGL	Plate Glass	PR	Pipe Rail
PLGR	Plunger	PRCST	Precast
PLK	Plank	PREFAB	Prefabricated
PLMB	Plumbing	PREP	Prepare
PLR	Pillar	PRESS	Pressure
PLR	Pliers	PRI	Primary
PLRT	Polarity	PRIM	Primary
PLSTC	Plastic	PRL	Parallel
PLT	Pilot	PRM	Priming
PLTG	Plating	PRMR	Primer
PLVRZD	Pulverized	PROC	Process
PLYWD	Plywood	PROD	Production
PM	Phase Modulation	PROJ	Project

Abbreviation	Term
PROP	Propelling
PROP	Property
PROP	Proposed
PROT	Protective
PRS	Press
PRSD MET	Pressed Metal
PRV	Pressure-Reducing Valve
PS	Polystyrene
PS	Pressure Switch
PS	Pull Switch
PSI	Pounds per Square Inch
PSL	Pipe Sleeve
PSU	Power Supply Unit
PSVTV	Preservative
PSW	Potential Switch
PT	Part
PT	Pint
PT	Pipe Tap
PT	Point
PTD	Painted
PTN	Partition
PU	Pick Up
PUB	Publication
PUBN	Publication
PUL	Pulley
PUR	Purchase
PV	Plan View
PV	Pump, Variable Displacement
PVC	Polyvinyl Chloride
PVT	Pivot
PW	Plain Washer
PW	Projected Window
PWM	Pulse-Width Modulation
PWR	Power
PWR SPLY	Power Supply

Q

Abbreviation	Term
QA	Quick-Acting
QC	Quality Control
QDRNT	Quadrant
QRY	Quarry
QT	Quarry Tile
QT	Quart
QTB	Quarry-Tile Base
QTF	Quarry-Tile Floor
QTR	Quarry Tile Roof
QTR	Quarter
QTY	Quantity
QUAD	Quadrant
QUAL	Quality
QUOT	Quotation

R

Abbreviation	Term
R	Radius
R	Resistance
R	Right
R	Riser
RA	Return Air
RAACT	Radioactive
RAB	Rabbet
RACT	Reverse-Acting
RAD	Radial
RAD	Radius
RB	Roller Bearing
RB	Rubber Base
RBN	Ribbon
RBR	Rubber
RC	Rate of Change
RC	Reinforced Concrete
RCD	Reverse-Current Device
RCHT	Rachet
RCP	Reinforced-Concrete Culvert Pipe
RCP	Reinforced-Concrete Pipe
RCPT	Receptacle

Abbreviation	Term
RCPTN	Reception
RCS	Remote Control System
RCVG	Receiving
RD	Road
RD	Roof Drain
RD	Root Diameter
RD	Round
RDC	Reduce
RDCR	Reducer
RDG	Ridge
RDH	Round Head
RDL	Radial
RDTR	Radiator
RE	Reel
REAC	Reactive
REASM	Reassemble
REC	Recess
REC	Record
RECD	Received
RECHRG	Recharger
RECIRC	Recirculate
RECP	Receptacle
RECT	Rectangle
RECT	Rectifier
RED	Reduce
RED	Reducer
REDWN	Redrawn
REF	Reference
REF L	Reference Line
REFL	Reference Line
REFLD	Reflected
REFR	Refrigerator
REG	Register
REG	Regulator
REINF	Reinforce
REL	Relay
REL	Relief
REM	Remove

Abbreviation	Term
REM COV	Removable Cover
REN	Renewable
REPL	Replace
REPRO	Reproduce
REQ	Require
REQ	Requisition
REQD	Required
REQT	Requirement
RES	Resistor
RESIL	Resilient
RET	Retard
RET	Return
REV	Reverse
REV	Revise
REV	Revolution
RF	Radio Frequency
RF	Raised Face
RF	Roof
RFG	Roofing
RFGT	Refrigerant
RFRC	Refractory
RGD	Rigid
RGH	Rough
RGLTR	Regulator
RGNG	Rigging
RGTR	Register
RH	Relative Humidity
RH	Right Hand
RH	Rockwell Hardness
RHC	Reheat Coil
RHEO	Rheostat
RI	Reflective Insulation
RIB	Ribbed
RINSUL	Rubber Insulation
RIV	Rivet
RL	Roof Leader
RLD	Rolled
RLF	Relief

Abbreviation	Term
RLG	Railing
RLR	Roller
RM	Ream
RM	Room
RMR	Reamer
RMS	Root Mean Square
RND	Round
RNDN	Random
RNG	Range
RNWBL	Renewable
RO	Rough Opening
RPM	Revolution Per Minute
RPQ	Request for Price Quotation
RPS	Revolution Per Second
RPVNTV	Rust Preventative
RR	Railroad
RR	Roll Roofing
RS	Rough Sawn
RS	Rubble Stone
RSD	Raised
RST	Reinforcing Steel
RSTPF	Rustproof
RT	Raintight
RTANG	Right Angle
RTD	Retard
RTF	Rubber-Tile Floor
RTL	Reinforced Tile Lintel
RTN	Return
RTNG	Retaining
RTR	Rotor
RTTL	Rattail
RV	Rear View
RV	Relief Valve
RVA	Reactive Volt-Ampere Meter
RVLG	Revolving
RVLV	Revolve
RVM	Reactive Voltmeter
RVS	Reverse

Abbreviation	Term
RVSBL	Reversible
RVT	Rivet
RW	Right-Of-Way
RWC	Rainwater Conductor
RWD	Redwood
RY	Railway

S

Abbreviation	Term
S	Scuttle
S	Side
S	Sink
S	Soft
S	South
S CHG	Supercharge
S1S	Surfaced or dressed one side
S1S1E	Surfaced or dressed one side and one edge
S2S	Surfaced or dressed two sides
S4S	Surfacead or dressed four sides
SA	Stress Anneal
SA	Supply Air
SAC	Sprayed Acoustical Ceiling
SAF	Safety
SAN	Sanitary
SAPC	Suspended Acoustical-Plaster Ceiling
SAT	Saturate
SATC	Suspended Acoustical-Tile Ceiling
SB	Sleeve Bearing
SB	Splash Block
SB	Stove Bolt
SBW	Steel Basement Window
SC	Scale
SC	Sill Cock
SC	Smooth Contour
SC	Solid Core
SCC	Single Conductor Cable
SCD	Screen Door

Abbreviation	Term	Abbreviation	Term
SCDR	Screwdriver	SFXD	Semi-Fixed
SCH	Schedule	SG	Structural Glass
SCH	Socket Head	SGL	Single
SCHED	Schedule	SGSFU	Salt-Glazed Structural Facing Units
SCHEM	Schematic	SGW	Security Guard Window
SCR	Screw	SH	Shackle
SCR GT	Screen Gate	SH	Sheet
SCRN	Screen	SH	Sheet
SCSH	Structural Carbon Steel, Hard	SH	Sheeting
SCSM	Structural Carbon Steel, Medium	SH	Sensible Heat
SCSS	Structural Carbon Steel, Soft	SH	Shower
SCT	Structural Clay Tile	SH & T	Shower and Toilet
SCUP	Scupper	SH&RD	Shelf and Rod
SD	Shower Drain	SHELV	Shelving
SDG	Siding	SHGL	Shingle
SDL	Saddle	SHK	Shake
SE	Service Entrance Cable	SHK	Shank
SE	Single-End	SHL	Shell
SE	Special Equipment	SHL	Shellac
SEC	Second	SHLD	Shielding
SEC	Secondary	SHLD	Shoulder
SECT	Section	SHLDR	Shoulder
SEJ	Sliding Expansion Joint	SHORT	Short Circuit
SEL	Select	SHRD	Shroud
SELF CL	Self-Closing	SHTHG	Sheathing
SEP	Separate	SHTR	Shutter
SEP	Separator	SHV	Sheave
SER	Serial	SI LT	Side Light
SERR	Serrate	SIF	Single-Face
SERV	Servo	SIG	Signal
SEW	Sewage	SILS	Silver Solder
SEW	Sewer	SIM	Similar
SEW	Sewer	SJ	Slip Joint
SF	Semi-Finished	SK	Sink
SF	Soffit	SK	Sketch
SF	Spot Faced	SKT	Skirt
SFB	Solid Fiberboard	SKT	Socket
SFT	Shaft	SL	Sea Level

Abbreviation	Term
SL	Sliding
SLD	Sealed
SLD	Sliding Door
SLDR	Solder
SLDR	Soldering
SLFCLN	Self-Cleaning
SLFLKG	Self-Locking
SLFSE	Self-Sealing
SLFTPG	Self-Tapping
SLOT	Slotted
SLP	Slope
SLT	Skylight
SLT	Spotlight
SLT or S	Slate
SLV	Sleeve
SLVT	Solvent
SM	Sheet Metal
SM	Small
SMLS	Seamless
SNDPRF	Soundproof
SNM	Shielded Nonmetallic Cable
SO	Shop Order
SOC	Socket
SOL	Solenoid
SOL PLT	Solenoid Controlled, Pilot Operated
SOV	Shut Off Valve
SOV	Shutoff Valve
SP	Static Pressure
SP	PH Split Phase
SP	Shear Plate
SP	Single Pole
SP	Soil Pipe
SP	Space
SP	Spare
SP	Special Purpose
SP	Specific
SP	Speed
SP	Splashproof

Abbreviation	Term
SP	Standpipe
SP GR	Specific Gravity
SP SW	Single-Pole Switch
SPC	Suspended Plaster Ceiling
SPCL	Special
SPCR	Spacer
SPDT	Single-Pole Double-Throw
SPDT SW	Single-Pole Double-Throw Switch
SPEC	Specification
SPG	Spring
SPH	Space Heater
SPHER	Spherical
SPHN	Siphon
SPK	Spike
SPL	Special
SPL	Spiral
SPLC	Splice
SPLN	Spline
SPLY	Supply
SPNSN	Suspension
SPR	Sprinkler
SPRDR	Spreader
SPST	Single-pole Single-Throw
SPST SW	Single-Pole Single-Throw Switch
SQ	Square
SQ CG	Squirrel Cage
SQ FT	Square Foot
SQ IN	Square Inch
SQ YD	Square Yard
SQH	Square Head
SR	Slip Ring
SR	Split Ring
SRPR	Scraper
SS	Semi-Steel
SS	Service Sink
SS	Set Screw
SS	Slop Sink
SSAC	Suspended Sprayed Acoustical Ceiling

Abbreviation	Term
SSBR	Smooth-Surface Built-Up Roof
SSCR	Set Screw
SSD	Subsoil Drain
SSG	Single Strength Glass
SSK	Soil Stack
SST	Stainless Steel
SSTU	Seamless Steel Tubing
ST	Single-Throw
ST	Stairs
ST	Street
ST PR	Static Pressure
STA	Station
START	Starter
STBD	Starboard
STD	Standard
STDF	Standoff
STGR	Stringer
STIF	Stiffener
STIR	Stirrup
STK	Stack
STK	Stock
STL	Steel
STN	Stained
STN	Stainless
STN	Stone
STNLS	Stainless
STOR	Storage
STP	Strip
STR	Straight
STR	Strainer
STR	Strainer
STR	Strength
STR	Strip
STR	Structural
STRBK	Strongback
STRL	Structural
STRUCT	Structure
STS	Special Treatment Steel

Abbreviation	Term
STW	Storm Water
STWY	Stairway
SUB	Substitute
SUBMG	Submergence
SUBST	Substitute
SUBSTR	Substructure
SUM	Summary
SUP	Supply
SUPERSTR	Superstructure
SUPP	Supplement
SUPRSTR	Superstructure
SUPSD	Supersede
SUPT	Superintendent
SUPT	Superintendent
SUPV	Supervise
SURF	Surface
SURV	Survey
SUSP	Suspend
SV	Safety Valve
SVCE	Service
SW	Short Wave
SW	Spot-Weld
SW	Switch
SW	Switch
SWBD	Switchboard
SWG	Swage
SWGD	Swinging Door
SWGR	Switchgear
SWP	Safe Working Pressure
SWSG	Security Window Screen and Guard
SWT	Sweat
SYM	Symbol
SYM	Symmetrical
SYN	Synchronous
SYN	Synthetic
SYS	System

T

Abbreviation	Term	Abbreviation	Term
T	T-Bar	TER	Terazzo
T	Tee	TERB	Terazzo Base
T	Teeth	TERM	Terminal
T	Time	TEV	Thermal Expansion Valve
T	Toilet	TF	Tile Floor
T	Tooth	TFR	Top of Frame
T	Top	TGL	Toggle
T	Truss	TH	Threshold
T&B	Top and Bottom	TH	Toilet-Paper Holder
T&BB	Top and Bottom Bolt	TH	Total Heat
T&G	Tongue and Groove	THD	Thread
TAB	Tabulate	THERMO	Thermostat
TACH	Tachometer	THK	Thick
TAN	Tangent	THKNS	Thickness
TARP	Tarpaulin	THRM	Thermal
TB	Tile Base	THRT	Throat
TBE	Thread Both Ends	THRU	Through
TBG	Tubing	TIR	Total Indicator Reading
TBLR	Tumbler	TKL	Tackle
TC	Terra Cotta	TLLD	Total Load
TC	Thermocouple	TLPC	Tailpiece
TC	Thread Cutting	TM	Technical Manual
TC	Top Chord	TMBR	Timber
TC	Tray Cable	TMPD	Tempered
TCH	Temporary Construction Hole	TNG	Tongue
TD	Tile Drain	TNG	Training
TD	Time Delay	TNL	Tunnel
TDM	Tandem	TNSL	Tensile
TE	Trailing Edge	TNSN	Tension
TECH	Technical	TOB BRZ	Tobin Bronze
TEL	Telephone	TOL	Thermal Overload
TELB	Telephone Booth	TOL	Tolerance
TEM	Temper	TOPG	Topping
TEMP	Temperature	TOT	Total
TEMP	Template	TPG	Tapping
TEMP	Temporary	TPI	Teeth per Inch
TEMPL	Template	TPI	Threads per Inch
TENS	Tension	TPL	Triple

Abbreviation	Term
TPLW	Triple Wall
TPR	Taper
TR	Technical Report
TR	Towel Rack or Rod
TR	Transom
TR	Truss
TRANS	Transfer
TRANS	Transformer
TRANS	Transparent
TRANS	Transporation
TRANSV	Transverse
TRAV	Traversing
TRH	Truss Head
TRK	Track
TRLR	Trailer
TRMR	Trimmer
TRNBKL	Turnbuckle
TRND	Turned
TRNGL	Triangle
TRQ	Torque
TRSN	Torsion
TRTD	Treated
TRX	Triplex
TS	Tensile Strength
TS	Tip Speed
TS	Tool Steel
TSR	Tile-Shingle Roof
TT	Tile Threshold
TUB	Tubing
TV	Television
TW	Tile Wainscot
TW	Twisted
TYP	Typical

U

Abbreviation	Term
U	Urinal
U&L	Upper and Lower

Abbreviation	Term
UC	Undercut
UCMT	Unglazed Ceramic Mosaic Tile
UGND	Underground
UH	Unit Heater
UHF	Ultra-High Frequency
UHPFB	Untreated Hard-Pressed Fiberboard
ULT	Ultimate
UN	Union
UNC	Unified Coarse Thread
UND	Under
UNEF	Unified Extra Fine Thread
UNEX	Unexacavated
UNEXC	Unexcavated
UNF	Unified Fine Thread
UNFIN	Unfinished
UNIV	Universal
UNS	Unified Special Thread
UOS	Unless Otherwise Specified
UP	Underground Feeder Cable
UPDFT	Updraft
UPR	Upper
UR	Urinal
UR	Utility Room
URWC	Urinal Water Closet
US	Undersize
US	Utility Set
USASI	United States of America Standards Institute
USE	Underground Service-entrance Cable
USFU	Unglazed Structural Facing Units
USG	United States Gage
USS	United States Standard
USUB	Unglazed Structural Unit Base
UTIL	Utility
UTRTD	Untreated
UV	Unit Ventilator
UV	Ultraviolet
UWTR	Underwater

Abbreviation	Term
	V
V	Valve
V	Velocity
V	Vent
V	Volt
V	Voltage
VAC	Vacuum
VAL	Valley
VAP	PRF Vapor Proof
VAR	Variance
VAR	Variation
VARN	Varnish
VBR	Vibration
VC	Vitrified Clay
VCP	Vitreous Clay Pipe
VCT	Vitrified Clay Tile
VD	Vandyke
VD	Void
VD	Voltage Drop
VE	Ventilating Equipment
VENT	Ventilate
VENT	Ventilator
VERN	Vernier
VERT	Vertical
VEST	Vestibule
VF	Video Frequency
VH	Vent Hole
VHF	Very-High Frequency
VIB	Vibrate
VIT	Vitreous
VIT	Vitreous
VLT	Volute
VM	Volt per Meter
VM	Voltmeter
VOL	Volume
VP	Velocity Pressure

Abbreviation	Term
VP	Vent Pipe
VP	Vent Pipe
VR	Voltage Regulator
VRLY	Voltage Relay
VS	Vent Stack
VS	Voltmeter Switch
VSBL	Visible
VSTM	Valve Stem
VT	Vacuum Tube
VT	Vaportight
VT or VTILE	Vinyl Tile
VTR	Vent Through Roof
VTVM	Vacuum Tube Voltmeter
VULC	Vulcanize
	W
W	Wall
W	Waste
W	Watt
W	West
W	Wide
W	Width
W	Wire
W/	With
WA	Wainscot
WA	Warm Air
WASH	Washer
WATW	Wood Awning-Type Window
WB	Wet Bulb Temperature
WB	Wheel Base
WB	Wood Base
WBF	Wood Black Floor
WBL	Wood Blocking
WC	Water Closet
WCW	Wood Casement Window
WD	Width
WD	Wood

Abbreviation	Term
WD	Wood Door
WDF	Wood Door and Frame
WDF	Woodruff
WDO	Window
WDP	Wood Panel
WF	Wide Flange
WFS	Wood Furring Strips
WG	Window Guard
WG	Wire Gauge
WGL	Wire Glass
WH	Wall Hydrant
WH	Water Heater
WH	Watthour
WH	Weep Hole
WHSE	Warehouse
WI	Wrought Iron
WIC	Walk In Closet
WJ	Wood Jalousie
WK	Week
WL	Water Line
WL	Wide Load
WLB	Wallboard
WLD	Welded
WLDR	Welder
WLDS	Weldless
WM	Washing Machine
WM	Water Meter
WM	Wire Mesh
WMGR	Warm Gear
WN	Winch
WNDR	Winder
WO	Without
WP	Waste Pipe
WP	Weatherproof (insul)
WP	White Pine
WPFC	Waterproof Fan-Cooled
WPG	Waterproofing
WPR	Working Pressure

Abbreviation	Term
WR	Wall Receptacle
WR	Washroom
WR	Water Resistant
WR	Weather-Resistant
WR	Wrench
WRT	Wrought
WS	Waste Stack
WS	Weather Stripping
WSHR	Washer
WSL	Weather Seal
WSP	Working Steam Pressure
WSR	Wood-Shingle Roof
WT	Weight
WT	Wood Threshold
WTHPRF	Weatherproof
WTR	Water
WTRPRF	Waterproof
WTRTT	Watertight
WU	Window Unit
WV	Wall Vent
WV	Working Voltage
WW	Wireway
WWF	Welded Wire Fabric

X

Abbreviation	Term
X HVY	Extra Heavy
X STR	Extra Strong
XDCR	Transducer
XFMR	Transformer
XSECT	Cross Section
XSTR	Transistor
XX STR	Double Extra Strong

Y

Abbreviation	Term
YD	Yard
YR	Year

Appendix F

Trade-Related Math

Decimal/Millimeter Equivalents of Common Fractions

Fraction	Decimal	MM
$\frac{1}{64}$	.015625	.397
$\frac{1}{32}$	.03125	.794
$\frac{3}{64}$	.046875	1.191
$\frac{1}{16}$	.0625	1.588
$\frac{5}{64}$	.078125	1.984
$\frac{3}{32}$	.09375	2.381
$\frac{7}{64}$	.109375	2.778

Fraction	Decimal	MM
$\frac{1}{8}$	.125	3.175
$\frac{9}{64}$	.140625	3.572
$\frac{5}{32}$	.15625	3.969
$\frac{11}{64}$	.171875	4.366
$\frac{3}{16}$	.1875	4.763
$\frac{13}{64}$	.203125	5.159
$\frac{7}{32}$	.21875	5.556

Decimal/Millimeter Equivalent of Common Fractions

Fraction	Decimal	MM	Fraction	Decimal	MM
$\frac{15}{64}$	.234375	5.953	$\frac{1}{2}$	.5	12.700
$\frac{1}{4}$	.25	6.350	$\frac{33}{64}$	.515625	13.097
$\frac{17}{64}$	.265625	6.747	$\frac{17}{32}$	.53125	13.494
$\frac{9}{32}$	.28125	7.144	$\frac{35}{64}$	.546875	13.891
$\frac{19}{64}$	.296875	7.540	$\frac{9}{16}$	.5625	14.288
$\frac{5}{16}$	.3125	7.938	$\frac{37}{64}$	.578125	14.684
$\frac{21}{64}$	.328125	8.334	$\frac{19}{32}$	.59375	15.081
$\frac{11}{32}$	.34375	8.731	$\frac{39}{64}$	.609375	14.478
$\frac{23}{64}$	.359375	9.128	$\frac{5}{8}$	.625	15.875
$\frac{3}{8}$	.375	9.525	$\frac{41}{64}$	.640625	16.272
$\frac{25}{64}$	.390625	9.922	$\frac{21}{32}$	.65625	16.669
$\frac{13}{32}$	.40625	10.319	$\frac{43}{64}$	.671875	17.066
$\frac{27}{64}$	.421875	10.716	$\frac{11}{16}$	.6875	17.463
$\frac{7}{16}$	.4375	11.113	$\frac{45}{64}$	.703125	17.859
$\frac{29}{64}$	.453125	11.509	$\frac{23}{32}$	.71875	18.256
$\frac{15}{32}$	.46875	11.906	$\frac{47}{64}$	.734375	18.653
$\frac{31}{64}$	.484375	12.303	$\frac{3}{4}$	.75	19.050

Decimal/Millimeter Equivalent of Common Fractions

Fraction	Decimal	MM		Fration	Decimal	MM
$\frac{49}{64}$	.765625	19.447		$\frac{57}{64}$	.890625	22.622
$\frac{25}{32}$	.78125	19.844		$\frac{29}{32}$	.90625	23.019
$\frac{51}{64}$	.796875	20.241		$\frac{59}{64}$	.921875	23.416
$\frac{13}{16}$	.8125	20.638		$\frac{15}{16}$	.9375	23.813
$\frac{53}{64}$	.828125	21.034		$\frac{61}{64}$	.953125	24.209
$\frac{27}{32}$	.84375	21.431		$\frac{31}{32}$	.96875	24.606
$\frac{55}{64}$	.859375	21.828		$\frac{63}{64}$	.984375	25.003
$\frac{7}{8}$	.875	22.225		1	1.	25.400

Electronic Calculator Use

Building construction workers of the past have rated the slide rule as one of the most important instruments for making mathematical calculations, especially when dealing with square roots and trigonometric functions. The slide rule did help reduce long minutes of paper-and-pencil calculations to a few simple manipulations of the "slipstick" and "runner." While the slide rule has proved indispensable since its invention in the 1850s, the electronic pocket calculator is faster, more versatile, more compact, more accurate, and better able to solve today's building construction problems involving mathematical calculations.

The fundamentals of electronic calculator operation may readily be obtained from the handbook accompanying the device. Practically anyone can master the basic operations of an electronic calculator in a single evening. Therefore, the following sections explain how to make specific, basic calculations that are required by workers in the building trades. Actual directions are given for pressing the required keys in most examples.

A selected number of basic examples have been chosen. Many other problems are solvable by the same process.

TYPES OF CALCULATORS

Slide-rule Calculator: Slide-rule calculators were developed to replace the conventional slide rule. These compact tools perform calculations of roots, powers, reciprocals, common and natural logarithms, and trigonometry, in addition to basic arithmetic. Versatile memory functions may include STORE, RECALL, SUM TO MEMORY, and MEMORY/DISPLAY EXCHANGE. Besides these, answers are displayed to as many as 10 places to the right of the decimal sign, in standard format or in scientific notation, correlation, linear regression, trend-line analysis, and more are available in these units.

Programmable Calculators: Programmable calculators are really miniature computers with powerful capabilities. Multiuse memories provide addressable memory locations in which data can be stored for recall. Several hundred program steps or 100 or more memories are available.

Master Library Modules: The master library module available from Texas Instruments, Inc., plugs into their electronic calculator and includes many different programs for many building construction applications.

Solid-state Software Libraries: Solid-state software is an advance in micromemory technology and provides new programming versatility. It only takes seconds to drop in a module and access a program with a few key strokes. Optional solid-state software library modules are currently available for building construction problems requiring mathematical calculations.

Printing Calculators: Several calculators are available with printing attachments that imprint on a tape all actions of the calculator. Although designed mainly for office use, they are powerful when the operator needs to

Fig. AF-1: Texas Instruments, Inc. programmable calculator.

refer to notes while performing a design, estimate, or other operation.

The type of electronic calculator most often used by building trade workers are the inexpensive hand-held type. These calculators perform the four basic math functions, and most of them have square, square root, and reciprocal keys also. Some of the better ones will have all the functions mentioned for the slide-rule calculator described above.

Each may operate slightly differently from another, so always review the owner's manual to become acquainted with the type being used.

Math Symbols	
+	Plus: add
−	Minus; substract
×	Times; multiply
*	Times; multiply in computer languages
÷	Division; divide
/	Division in computer languages
Σ	Sum of
=	Equals; equal to
≠	Not equal to

Math Symbols		
±	Plus or minus; approximately equal to	
≅	Apprximately equals	
>	Is greater than	
<	Is less than	
$\sqrt{0}$	Square root	
E ^{x}	Power of	
( )	Parentheses*	Note:* Mathematical operations indicated within parentheses, brackets, or braces are to be performed first.
[ ]	Brackets*	
{ }	Braces*	

Practical Construction Applications

EXCAVATION

Problem: A rectangular building's outside dimensions are 24 feet and 32 feet. What is the volume of excavation if the depth is uniformly nine feet and the lines of excavation are one foot outside the building lines?

Solution: Since one foot is to be added to each of the building's sides, the dimensions of the hole are 9 ft. x 26 feet x 34 feet. Therefore, the volume is:

$$9 \times 26 \times 34 = 7956 \text{ cubic feet.}$$

To convert cubic feet to cubic yards, find appropriate multiplication factor from the tables in Appendix A; this conversion factor is .03704. To find cubic yards of excavation required:

$$7956 \times .03704 = 294.37 \text{ cubic yards}$$

To solve the above problem on the electronic calculator, perform the following operations:

A. Key in 9
B. Press the multiplication key (X)
C. Key in 26
D. Press the multiplication key (X)
E. Key in 34
F. Press equals key (=)
G. Read the answer displayed (7956.)
H. Press the multiplication key
I. Key in 03704 (conversion factor)
J. Press equals key (=)
K. Read the answer displayed (294.3720)

Therefore, to find the volume of any square or rectangular excavation, perform the following calculator operations:

A. Key in depth (in feet)
B. Press the multiplication key (X)
C. Key in width dimension (in feet)
D. Press the multiplication key (X)
E. Key in length dimension (in feet)
F. Press equals key (=)
G. Read the answer displayed (volume in cubic feet)
H. Press the multiplication key
I. Key in .03704 (conversion factor)
J. Press equals key (=)
K. Read the answer displayed (294.3720)

Some workers find it easier to remember the number "27" when converting cubic feet to cubic yards. Although not quite as accurate as the multiplication factor (.03704), it is accurate enough for all practical purposes. When the conversion factor "27" is used, it is divided into the cubic feet rather than multiplying. Therefore, to solve the same problem as before,

A. Key in 9
B. Press the multiplication key (X)
C. Key in 26
D. Press the multiplication key (X)
E. Key in 34
F. Press equals key (=)
G. Read the answer displayed (7956.)
H. Press the division key
I. Key in 27 (conversion factor)
J. Press equals key (=)
K. Read the answer displayed (294.6666)

FOUNDATIONS

Foundations that must carry heavy loads require careful attention to the bearing capacity of the soil. The unit soil pressure exerted by a foundation is computed by dividing the total weight by the bearing area. Safe bearing capacities in tons per square foot are:

Material	Allowable Bearing Tons per Sq. Ft.
Quicksand and alluvial soil	½
Soft clay	1
Moderately dry clay, fine sand	2
Firm and dry loam or clay	3
Compact coarse sand or stiff gravel	4
Coarse gravel	6
Gravel and sand, well-cemented	8
Good hardpan or hard shale	10
Very hard native bedrock	20
Rock under caissons	25

Problem: A building weighing 100 tons rests on a footing 18 inches wide and with a total length of 90 feet. What is the unit soil pressure?

Solution: First convert inches to feet; then find bearing area of footing.

$$\frac{18 \; inches}{12 \; (inches \; per \; foot)} = 1.5 \; feet$$

Bearing area = 1.5 x 90 = 135 sq. ft.

Unit soil pressure = $\frac{100}{135}$ = .7407 ton per sq. ft.

To perform these calculations on an electronic calculator:

A. Key in 18 (inches)
B. Press division key
C. Key in 12 (number of inches in a foot)
D. Press equals key (=)
E. Read answer displayed: 1.5000
F. Press multiplication key (X)
G. Key in 90 (length of footing)
H. Press equals key (=)
I. Read answer displayed: 135.0000 (square feet)
J. Press cancel key (C)
K. Key in 100 (weight of building)
L. Press division key
M. Key in 135 (area of footing)
N. Press equals key (=)
O. Read answer displayed: 0.7407 (ton per sq. ft.)

Problem: What area of bearing is required to support a load of 72 tons on coarse gravel?

Solution: The safe bearing on coarse gravel is 6 tons per square foot. Consequently, the required bearing area is:

$$\frac{72}{6} = 12 \; sq. \; ft.$$

To solve on the electronic calculator,

A. Key in 72 (weight of load in tons)
B. Press division key
C. Key in 6 (bearing factor for coarse gravel)
D. Press equals key (=)
E. Read answer displayed: 12.0000 (square feet)

MASONRY

Common brick: The number of brick required for an application is necessary for estimating costs, and to know the amount to order from suppliers. The traditional way to determine the number of brick required is to multiply the net cubic feet of wall by 20 brick per cubic foot, allowing for about 5% waste. This means multiplying the length by the height by the thickness of the wall. All openings, such as windows, doors, or other openings are deducted from the first figure to obtain the net cubic feet of wall.

Problem: A retaining wall 8 inches thick, 4 feet high and 12 feet long is to be built from common brick. If there are no openings, how many brick should be ordered for the job?

Solution: To obtain the volume of the wall in cubic feet, use the following equation:

Length x height x thickness = cubic feet

Therefore,

12 x 4 x 8 = 384 cubic feet

384 (cubic feet) x 20 (brick per cubic foot) = 7680

So, if the bricklayer ordered 7680 brick from a brick supplier, he or she should have a sufficient amount with perhaps a few left over.

To perform this same calculation on the electronic calculator, perform the following operations:

A. Key in 12
B. Press multiplication key
C. Key in 4
D. Press multiplication key
E. Key in 8
F. Press equals key (=)
G. Read answer displayed: 384 (cubic feet)
H. Press multiplication key
I. Key in 20
J. Press equals key
K. Read answer displayed: 7680

Problem: A bricklayer is required to lay a perimeter of common brick around a circular flagstone patio, 12 feet in diameter. If the brick are to be laid with their ends facing outward, how many brick would be required to complete the job?

Solution: Find perimeter (circumference) dimensions of the circular patio using the following equation:

$$C = pi(\pi)(3.1416) \times d$$

$$C = 3.1416 \times 12 = 37.6992 \text{ feet}$$

If a common brick (laid end ways) takes up 4 inches (including the mortar joint) first convert 37.6993 feet to inches:

37.6992 x 12 = 452.39 inches

Divide 4 inches into 452.39 inches to obtain the number of brick required to reach around the perimeter of the circular patio.

452.39/4 = 1809.56 or 1810 brick

To perform the operations for this problem on the electronic calculator:

A. Key in 3.1416 or press π key
B. Press multiplication key
C. Key in 12
D. Press equals (=) key
E. Read answer displayed: 37.6993
F. Press multiplication key
G. Key in 12 (to convert feet to inches)
H. Press equals (=) key
I. Read answer displayed: 452.39
J. Press division key
K. Key in 4
L. Read answer displayed: 1809.5616

Problem: How many square feet of flagstone is required to cover the circular patio as described in the last problem?

Solution: Since the circular patio is 12 feet in diameter, the following equation may be used to find the area:

$$Area = \frac{1}{4}\pi D^2$$

$$Area = .25 \times 3.1416 \times 12^2 = 113.10 \text{ square feet}$$

To perform these operations on an electronic calculator, perform the square operation first:

A. Key in 12
B. Press multiplication key
C. Key in 12
D. Press equals (=) key
E. Read answer displayed: 144 (12 squared)

Note: If calculator has x^2 key, enter the numeral 12 only once and press the x^2 key to obtain the answer; that is, 144.

F. Press multiplication key
G. Key in .25
H. Press multiplication key
I. Key in 3.1416 or press π key
J. Read answer displayed: 113.0976

CARPENTRY

Laying out stairs: Carpenters working on construction projects — especially residential projects — will often have the ocassion to lay out stairs. To do so:

• Determine the height or rise from the top of the floor below where the stairs will start to the top of the floor above where they will end.

• Determine the run or distance measured horizontally.

• Mark the total rise on a rod or a piece of 1 x 2 furring strip to make what is called a "story pole." Divide the height or rise by the number of risers desired; it is common to obtain a result in terms of inches and a fraction of an inch. A simple method of laying out the rises o the furring strip is to space off the total rise with a pair of compasses.

• Lay out or space off the number of treads wanted in the horizontal distance or run. There is always one less tread than there are risers. Therefore, the thickness of the tread should be deducted from the first riser to have this first step of uniform height with all the others.

Problem: The distance between two floors is 12 feet 4 inches. How many steps will be required if the rise is to be about 7¼ inches?

$$12 \text{ ft., } 4 \text{ in. } = 148 \text{ inches}$$
$$148 \text{ div } 7.25 = 20.4$$

Since the quotient is not a whole number, divide the rise of the stair by 20.

$$148/20 = 7.4$$

Looking at the table on page 464 of this book, 0.4 inches is approximately 13/32 of an inch. Consequently, 20 steps of 7 13/32 inches each are required.

To solve the above problem on the electronic calculator,

A. Key in 12 (feet)
B. Press multiplication key
C. Key in 12 (convert inches to feet)
D. Press equals (=) key
E. Read answer on panel as: 144.00
F. Press addition key (+)
G. Key in 4 (inches)
H. Press equals (=) key
I. Read answer on panel as: 148.00 (inches)
J. Press division key
K. Key in 7.25 (approx. amount of rise)
L. Press equals (=) key
M. Read answer on panel as: 20.4
N. Press C key to cancel
O. Key in 148
P. Press division key
Q. Key in 20
R. Press equals (=) key
S. Read answer as 7.4 (inches)

Appendix G

Linear Measure

12 inches	=	1 foot
3 feet	=	1 yard
36 inches	=	1 yard
5.5 yards	=	1 rod
16.5 feet	=	1 rod
40 rods	=	1 furlong
660 feet	=	1 furlong
8 furlongs	=	1 mile
320 rods	=	1 mile
1760 yards	=	1 mile
5280 feet	=	1 mile

Square Measure

144 square inches	=	1 square foot
9 square feet	=	1 square yard
30¼ square yards	=	1 square rod
272¼ square feet	=	1 square rod
160 square rods	=	1 acre
43,560 square feet	=	1 acre

Temperature Conversions

°F	=	$(9/5 \times °C) + 32$
°C	=	$5/9 \, (°F - 32)$

Cubic Measure

1728 cubic inches	=	1 cubic foot
27 cubic feet	=	1 cubic yard
128 cubic feet	=	1 cord
24¾ cubic fee	=	1 perch

Avoirdupois Weight

437.5 grains	=	1 ounce
16 ounces	=	1 pound
100 pounds	=	1 hundredweight
1000 pounds	=	1 kip
2 kips	=	1 tons

Time Measure

60 seconds	=	1 minute
60 minutes	=	1 hour
24 hours	=	1 day
7 days	=	1 week
52 weeks	=	1 year
365.26 days	=	1 year

Dry Measure

2 pints	=	1 quart
4 quarts	=	1 gallon
2 gallons	=	1 peck
8 quarts	=	1 peck
4 pecks	=	1 bushel

Troy Weight

24 grains	=	1 pennyweight
20 penny-weights	=	1 ounce
12 ounces	=	1 pounds

Liquid Measure

4 gills	=	1 pint
2 pints	=	1 quart
4 quarts	=	1 gallon
31½ gallons	=	1 barrel
2 barrels (63 gals)	=	1 hogshead

Surveyor's Measure

7.92 inches	=	1 link
16.5 feet	=	1 rod
25 links	=	1 rod
4 rods	=	1 chain
66 feet	=	1 chain
100 links	=	1 chain
80 chains	=	1 mile
625 square links	=	1 square rod
16 square rods	=	1 square chain
10 square chains	=	1 acre
640 acres	=	1 square mile
1 square mile	=	1 section
36 sections	=	1 township

Appendix H

Excavation of earth and rock involves the four basic operations:

- Loosening or digging

- Loading

- Hauling

- Dumping

Where permitted, explosives are used to loosen rock, hardpan, and frozen ground, although pneumatic spades also get the job done (although slower) in areas where explosives are not permitted.

In soft ground, loosening and loading becomes one operation; that is, a backhoe or dozer is used for both digging (scraping) and loading onto dump trucks.

Material removed from an excavation is measured by cubic yards "in place." That is, it is measured as solid ground and not as the loose material that is hauled away and dumped. The reason for this is that the latter occupies a volume about 25 percent greater than its original volume. Therefore, to measure the amount of material excavated requires the calculation of the resulting hole.

The volume in cubic yards of excavated material is found by finding the area in feet, multiplying the area by the depth in feet, and then dividing the resulting figure by 27 to convert the cubic feet to cubic yards.

When the ground is level and the shape of the hole is a square or rectangle, the calculations are quite simple. The same is true for a T-shaped hole, as the T can be divided into two rectangles. When other shapes are encountered, use equations for determining the area of geometric figures as described in Appendix H.

FOUNDATIONS

All foundation footings for conventional building construction are concrete. The walls, however, can be either poured concrete or else built up from either cinder or concrete block. A poured wall produces a better foundation than one built up from block, although block foundations are entirely adequate for most parts of the country for residential construction. The deciding factor is not personal preference; it's based on construction costs and the requirements of the local building code.

In areas where block is permitted, a poured foundation is always permitted, but the opposite may not be true. Where the local building codes permit concrete block foundations for residential and small commercial applications, these are the types most often found. Block foundations are almost always more economical that ones that utilize poured concrete. One factor is undisputable; poured concrete foundations are stronger and have a much better chance of being watertight than those made from concrete blocks.

Building codes require that footings be beneath the frost line. This dimension will vary for different parts of the country. For example, some northern states may require the footings to be a minimum of four feet; in more southerly parts of the country, it may be as little as 18 inches.

The depth or vertical dimension of the actual footing (not the distance beneath the frost line) is usually the same dimension as the thickness of the wall it supports (see Fig. AI-1). The width of the footing is twice the thickness of the wall it supports. These are minimum standard dimensions; they are adequate when soil is firm. Therefore, if the foundation wall is 8-inch thick poured concrete, the footing must be at least 8 inches deep and 16 inches wide, with the wall built in the center of the footing. This leaves 4 inches of footing on each side of the wall.

If the foundation wall is 10 inches thick, the footings must be increased to a depth of 10 inches and a width of 20 inches. In actual practice, however, the width of footings are usually oversized. The width of a back-hoe bucket is usually 24 inches. Therefore, almost all footings for residential construction are dug this width, regardless whether an 8-inch, 10-inch, or 12-inch foundation wall is used. The cost of the additional concrete is more than off-set by the savings in labor, if the footings would have to be dug by hand.

The type of concrete mix for most footing applications will be 1:2:4; that is, one part Portland cement, two parts sand, and four parts gravel. Just as important as the concrete mix is the type of soil the footings must rest on. It is absolutely necessary that the bottom of a footing must rest on firm soil. If the soil is not firm, the footings must be placed deeper into the ground until firm soil is

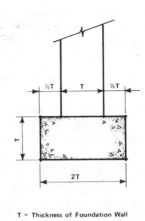

Figure AH-1: Typical foundation footing.

reached, or the footings must be strengthened by making them wider, deeper and adding reinforcement rods within the pour. In some areas of the country — especially in coastal areas — firm soil is not reached for great depths. In such areas, piling is normally used, driven deep into the ground by an apparatus appropriately called "pile-driver," which operates much the same as a non-rotary well-drilling apparatus. Several dozen such piles are driven into the ground from 10 to 30 feet, and then the house is constructed onto the piling — the piling acting as the house's foundation.

A footing must also rest on soil that never freezes. The reason for going below the *frost line* — the maximum depth to which soil freezes — is that frozen ground heaves in the spring as it thaws, with enough force to break concrete.

In most cases, it is best to determine the amount of concrete required for footings and foundation walls by using mathematical equations. To determine the volume of concrete, use the following equation:

$$volume = length \times width \times depth$$

If these dimensions are given in feet, they may be first calculated in feet, then divided by 27 to convert to cubic yards, or

$$volume\ in\ cu.\ yds. = \frac{length \times width \times depth\ (all\ in\ feet)}{27}$$

422

For example, to determine the number of cubic yards of concrete required to pour a 12′ × 24′ × 6″ concrete slab for, say, a residential patio, multiply 12 x 24 x 0.5 = 144 cu. ft., divided by 27 = 5.33 cu. yds. In most cases it is best to add about 10% to this calculated figure to compensate for spills and waste.

For large construction projects, contractors and their estimators almost always calculate the amount of concrete required by the appropriate equations. However, on residential and small commercial projects, many contractors and estimators use a "ball park" figure for their estimates. Time and experience has proven (to them) that their calculations are accurate enough for the type of projects they engage in. "Guessing at amounts and/or pricing," however, can be very risky for the novice or inexperienced person — the reason for many contracting firms to go out of business each year.

The following tables give useful information on excavation and foundation construction. For example, the following table will simplify calculating the amount of concrete mix required for poured concrete foundations walls. In using this table, choose the planned wall thickness and height and then multiply by the number of lineal feet of wall by the appropriate factor. For example, a 12-inch wall 7 feet 8 inches high and 128 feet long has a multiplication factor of .2840. So, multipy .2840 by 128 (lineal feet) and the answer is 36.35 cupic yards.

Concrete Calculation Table for Concrete Foundation Walls

Height of wall	12″ Thick	10″ Thick
7′-0″	.2592	.2160
7′-6″	.2777	.2314
7′-8″	.2840	.2366
8′-0″	.2962	.2468

To quickly find the amount of concrete required for footings, use the following table. Example: If the footing is to be 24 inches by 12 inches with a lineal measurement of 128 feet (centerline of keyway), multiply 128 by .074 to find the required amount of concrete — 9.47 cubic yards. Follow this example when calculating the aount of concrete requied for other applications. Merely substitute the value required.

Concrete Calculation Table for Footings

Footing Size	Note	Multiplication Factor
24″ × 12″	Measure length of keyway and multiply by	.0740
20″ × 10″	Measure length of keyway and multiply by	.0514
18″ × 9″	Measure length of keyway and multiply by	.0417

Quantities of Concrete Block and Mortar

Wall Thickness	For 100 sq. ft. of wall		For 100 Concrete Block
Inches	Number of block	Mortar Cu. Ft.	Mortar Cu. Ft.
8	112.5	2.6	2.3
12	112.5	2.6	2.3

Recommended Mortar Mixes

Type of Service	Cement	Hydrated Lime or Lime Putty	Mortar Sand (dampness, loose)
Ordinary service	1 — Portland cement	1 to 1¼	4 to 6
Subject to very heavy loads, hurricanes, earthquakes, severe frost action.	1 — Portland cement	0 to ¼	2 to 3

Recommended Mortar Mixes

Mortar Mixes			Quantities			
Cement Sack	Hydrated Lime or Lime Putty	Sand in Damp, Loose Condition Cu. Ft.	Massonry Cement Sack	Portland Cement Sack	Hydrated Lime or Lime Putty Cu. Ft.	Sand Cu. Ft.
1 Masonry cement	—	3	0.33	—	—	0.99
1 Portland cement	1	6	—	0.16	0.16	0.97

Recommended Thickness Of Concrete Slabs

Residential basement floors	4 inches
Private garage floors	5 inches
Porch floors	5 inches
Tile floor bases	2½ inches
Driveways	6 to 8 inches
Sidewalks	4 to 8 inches

Board Feet of Lumber Required for 1 Square Foot of Concrete Sidewalk Forms

Lumber Size'	Width of Sidewalk								
	2'	3'	4'	5'	6'	7'	8'	9'	10'
2" × 4"	.90	.625	.50	.375	.300	.25	.25	.20	.20
2" × 6"	1.25	.975	.625	.50	.40	.375	.33	.30	.25
2" × 8"	1.66	1.125	.875	.66	.60	.50	.40	.375	.33

Appendix I

Concrete

Concrete for Various Applications

Kind of Work	Cement (inches)	Sand (cu. ft.)	Gravel (cu. ft.)	Gallons Water per Sack			Maximum Aggregate Size
				Damp	Wet	Very Wet	
Footings, foundation walls (not watertight), columns, chimneys, retaining walls, garden walls	1	3	5	7	6	5	1½"
Same as above	1	2¾ - 3	4	6¼	5½	4¾	1½"
Watertight basement walls, swimming and wading pools, walls above grade, walks, driveways, terraces, tennis courts, steps, floors, septic tanks, storage tanks	1	2 -2¼	3	5½	5	4¼	1"
Same as above	1	2½	3½	6½	5	4½	1½"

Concrete for Various Applications (Cont.)

Kind of Work	Cement (inches)	Sand (cu. ft.)	Gravel (cu. ft.)	Gallons Water per Sack			Maximum Aggregate Size
				Damp	Wet	Very Wet	
Subject to severe wear, weather, or weak acid and alkali solutions	1	2	2¼	4½	4	3½	¾"
Topping for pavement steps, tennis courts, floors	1	1	1¾	4¾	4½	4¼	⅜"
Thin construction — 2 - 4 inches, fence and mailbox posts, garden furniture, tanks, flower boxes, bird baths	1	2	2	4½	3¾	3½	½"

Standard Sizes and Weights of Concrete Reinforcing Rods

Rod Reinforcement for Step Slabs

Slab Length (feet)	Slab Thickness (inches)	Rods, Lengthwise		Rods, Across	
		Diameter (inches)	Spacing (inches)	Diameter (inches)	Spacing (inches)
3	4	¼	10	¼	12 - 18
4	4	¼	5½	¼	12 - 18
5	5	¼	4½	¼	18 - 24
6	5	⅜	7	¼	18 - 24
7	6	⅜	6	¼	18 - 24
8	6	⅜	4	¼	18 - 24
9	7	½	7	¼	18 - 24

Standard Sizes and Weights of Concrete Reinforcing Rods

Rod Reinforcement for Porch or Other Slab

Length of span (ft.)	Slab Thickness (in.)	Rod Diameter (in.)	Rod Spacing (in.)
4	5	3/8	7½
6	5	3/8	6
8	5½	½	9½

Rod Reinforcement for Concrete Bridges

Length of span (ft.)	Slab Thickness (in.)	Rod Diameter (in.)	Rod Spacing (in.)
6	6½	5/8	8
9	6½	5/8	7
12	7	3/4	8

Welded-Wire Fabric Reinforcement for Typical Projects

Project	Style	Comment
Barbecue foundation slabs	6 x 6 - 8/8 to 4 x 4 - 6/6	Use heavier style for heavy, massive fireplaces and barbecues
Barn, dairy, poultry house floors	6 x 6 - 6/6	
Basement floors	6 x 6 - 8/8	Use 6/6 for larger areas or unstable soil
Driveway	6 x 6 - 6/6	Also used where cars cross sidewalks
Foundation slabs	6 x 6 - 10/10	Use heavier gauge where maximum dimension is over 15 ft. or subsoil is poorly drained
Foundation walls	6 x 6 - 6/6	
Garage floors	6 x 6 - 6/6	Position at midpoint of 5- or 6-inch slab
Porch floor	6 x 6 - 6/6	6" slab, 6' span, 1" from bottom
Sidewalks	6 x 6 - 10/10	Average use
Steps (free span)	6 x 6 - 6/6	Use heavier gauge if more than 5 risers
Steps on ground	6 x 6 - 6/6	Average
Terraces and patios	6 x 6 - 8/8	Average

When reinforced concrete is used for structural support, steel bars, called *reinforcing bars*, are used to give the required strength to the concrete. Deformed (not smooth) bars are used; that is, these bars have ridges to give a better hold to the concrete.

A size number appears on each bar and in general, this number can be translated into eights; that is, a bar with a number 4 on it would be ⅘″ or ½″ thick. The following table shows reinforcing bar sizes.

Bar designation No.	Unit weight lb/ft.	Diameter in.	Cross-sectional area, $in.^2$	Perimeter in.
3	.376	0.375	0.11	1.178
4	.668	0.5000	0.20	1.571
5	1.043	0.625	0.31	1.963
6	1.502	0.750	0.44	2.356
7	2.044	0.875	0.60	2.749
8	2.670	1.000	0.79	3.142
9	3.400	1.128	1.00	3.544
10	4.303	1.270	1.27	3.990
11	5.313	1.410	1.56	4.430
14	7.65	1.693	2.25	5.32
18	18.60	2.257	4.00	7.09

Appendix J

Masonry

Masonry construction of all types has proven its usefulness for centuries. Bricks, concrete blocks, tiles, and other variations have been used for all types of construction from gate posts to huge office buildings.

For economy in most types of commercial construction, the standard 8 x 8 x 16-inch and the 4 x 8 x 16-inch concrete blocks still dominate the field, but many new shapes and sizes are also available. Some, too, are colored, or they have polished or cut face to resemble stone. Large and small sizes and shapes are also used to vary the pattern in walls. Lintel and jamb blocks for doors and windows are prefabricated, ready for use.

Concrete blocks are used mostly for foundation work in residential construction, or to provide firewall construction in town houses or multifamily developments. For interior basement walls, furring strips are often used over the blocks, and then some type of wallboard is used for the interior finish.

Standard bricks are frequently used to face concrete block when the block is above ground. Wood structures are often brick veneered and usually provides better insulation than when brick is used on concrete block walls.

Regardless of the type of masonry used, all are bonded together with mortar. To be good for bonding together concrete blocks or bricks, mortar must have sufficient strength and be workable. Unless properly mixed and applied, the mortar will be the weakest part of any type of masonry. Both the strength and the waterproofing of masonry work depend largely on the strength of the mortar bond. Weak joints cause water leaks that will eventually weaken structures.

BRICKS AND BRICKWORK

Standard bricks manufactured in brick factories usually are 2¼ x 2¼ x 8 inches. Handmade bricks, along with English and Roman bricks, are sometimes of other dimensions.

Brick Veneer. In brick veneer construction the brick is used only as a facing mateial, without utilizing its load bearing properties. The brick is applied over wood framing and sheathing of both old and new houses.

Veneered walls resist fire exposure better than frame walls. The brick can be laid in any type bond or pattern formed by the use of half-bricks as headers. See Fig. AJ-1.

In new construction, the foundation walls should extend 5 inches outside the face of the wood sheathing to receive the brick veneer. On old buildings, the veneer sould be started on the projecting portion of the footing.

The wood walls should be covered with waterproof building paper. The brick veneer is then anchored in place with one noncorroding metal tie for each 2 square feet of brick area. The ties should be spaced not more than 24 inches apart horizontally and vertically.

Brick veneer can also be applied on concrete walls or more commonly, cement block walls. A shelf angle secured by an anchor bolt imbedded in the concrete supports the brick veneer or brick ties are embedded in the mortar joints of concrete block walls. Dovetail anchor slots are provided to hold the brick to the wall. The space between the brick and the concrete or concrete block wall is slushed with mortar or grout.

CONCRETE BLOCK WALLS

Because of their strength and fireproofing qualities, concrete blocks are one of the most widely used materials in masonry work. They are currently used almost exclusively for residential foundations and for commercial projects, especially when concrete block walls are brick veneered.

Blocks are available in colors that are durable and easily maintained. Concrete blocks can also be spray painted. A waterbase paint such as colored Portland cement or a paint with a latex-vinyl or acrylic-emulsion base may be used. Before painting the wall, a sealer coat is applied to close up all pores in the cement. An asphalt coating is normally applied to the outside of concrete block walls below grade.

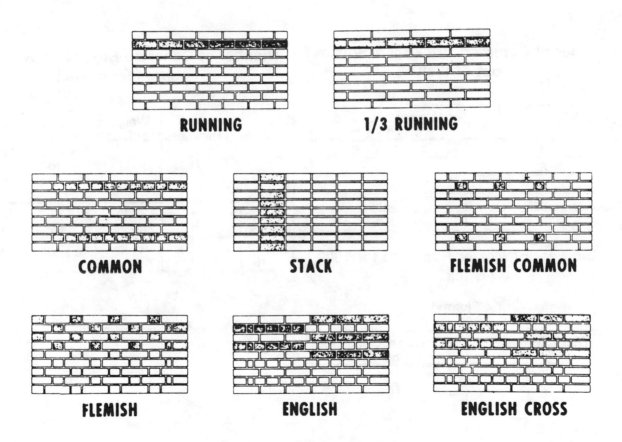

Figure AJ-1: Various brickwork patterns.

Flue Sizes for Fireplaces

Fireplace Dimensions			Flue Sizes			
			Rectangular		Circular	
Width of opening, inches	Approximate height, inches	Depth of opening, inches	Outside dimensions, inches	Effective area, square inches	Diameter, inches	Effective area, square inches
24	28	17 - 20	8½ × 8½	41	10	78
28	28	17 - 20	8½ × 13	70	10	78
30	30	17 - 21	8½ × 13	70	12	113
34	30	17 - 21	8½ × 13	70	12	113
36	30	21	8½ × 18	97	12	113
40	30	21 - 24	8½ × 18	97	15	177
42	30	21 - 25	8½ × 18	97	15	177

Number of Common Brick for 1 Square Foot of Wall

Wall Thickness	Number of brick to cover 1 sq. ft.
4	6.75
8	13.5
12	20.25
16	26.00
20	33.75

Cubic Feet of Mortar Required for 1000 Brick

(½″ joints)

Wall Thickness	Cubic Feet of Mortar Required
4	11.7
8	15.0
12	16.2
16	16.8
20	17.1

Number of Concrete Block to cover 100 square feet of wall

Size of Block	Wall Thickness	Weight per Block	Number Blocks Required for 100 Sq. Ft. of Wall Area	Cubic Feet of Mortar Required
4 x 8 x 16	4	28 lbs	110	3.25
8 x 8 x 16	8	50 lbs	110	3.25
8 x 12 x 16	12	85 lbs	110	3.25

Also see Appendix I for further information on masonry and mortar mixes.

Appendix K

LUMBER GRADING
Unfinished and Finished Sizes of Lumber

Original Lumber Size (Inches)	Actual Size (Inches)	Board Feet per Foot of Length
1 x 2	¾ x 1⅝	⅙
1 x 4	¾ x 3⅝	⅓
1 x 6	¾ x 5⅝	½
1 x 10	¾ x 9½	⅚
1 x 12	¾ x 11½	1
2 x 4	1⅝ x 3⅝	⅔
2 x 6	1⅝ x 5⅝	1
2 x 8	1⅝ x 7⅝	1⅓
2 x 10	1⅝ x 9½	1⅔
2 x 12	1⅝ x 11½	2
3 x 6	2⅝ x 5⅝	1½
4 x 4	3⅝ x 3⅝	1⅓
4 x 6	3⅝ x 5⅝	2

LUMBER GRADING

Lumber grading is done according to intended use. For residential framing (2 x 4, 2 x 6, etc.) it's graded mainly for strength not looks. For exposed use, however, as in trim, counters and cabinets, pieces are usually selected with suitable appearance. Select and finish grades are found in sizes and species used in cabinetwork and other applications where appearance is all important. The three top grades (also tops in price) are 1 & 2 clear (also B & Better), C Select, and D Select. In Idaho white pine, corresponding grades are Supreme, Choice, and Quality. Top grade in each group is nearly flawless. The other two grades may have very slight appearance flaws.

These grade systems apply to widely used species. Other species follow other systems suited to their characteristics. Common board grades are considerably lower in price than the top grades; these range from No. 1 Common to No. 5. A good choice for average woodworking is No. 2. It may have tight knots and a few other flaws, but has a generally good appearance and it's economical.

Actually, the difference between a grade of lumber and the grade both immediately above and below it is slight. The following grades from the Southern Pine Inspection Bureau illustrate this point.

- *Select*: Free from defects that reduce strength or stiffness. Used where high strength, stiffness, and good appearance are needed. The exposed deck in Fig. 2 is one example of its use.

- *No. 1*: Has small knots, good appearance, suitable for all types of construction. Not as good as select, but paint covers defects. Stair treads are often constructed from No. 1 lumber.

- *No. 2*: Has larger, tight knots. Suitable for all types of construction. No. 2 lumber is frequently used for rafters and roof trusses.

- *No. 3*: Provides high quality at low construction cost. No. 3 is almost as good as No. 2 except for slight added defects. No. 3 lumber may be used for floor joints, sills, and similar applications.

- *Stud*: For studs less than 10) long. Stiff, straight, and strong enough for bearing walls.

Other grades carried by some dealers are construction, standard, utility, and economy. Construction grade, used for light framing, has strength, similar to No. 2, but more visible defects. Each of the other grades down the line has less strength, poorer appearance, larger knots and more of them. Furthermore, you will see longer splits and checks, and more pitch pockets in these latter grades. Lumber of these last three grades is used for blocking, bracing, and forming where bending strength is not a requirement.

Treated lumber is being used more and more for outside applications. The treatment is usually chromated copper arsenate to protect the wood from insect attack and decay. Treated lumber should be used only where such protection is important. Although properly treated exposed wood may be used inside residences, it should never be used where the preservative may come into contact with food; that is, treated wood must not be used for cutting boards or countertops, nor for any type of food container.

Only treated wood that is visibly clean and free of surface residue should be used for patios, decks and walkways. Neither should it be used where it may come into direct or indirect contact with public drinking water, except for incidental contact such as docks and bridges.

LUMBER ABBREVIATIONS

Both softwood and hardwood are graded as to quality and structural soundness in accordance with American Softwood Lumber Standards. The following terms are traditionally abbeviated and sometimes appear as "stamps" on various types of wood.

Abbreviation	Meaning
AD	Air-dried.
ADF	After deducting freight
ALS	American Lumber Standard
AV	Average
B/L	Bill of Lading
B&B	B and better
B&S	Beams and Stringers
Bd. ft.	Broad foot (feet)
Bd	Board
Bdl	Bundle
Bev	Beveled
BM	Board Measure
Btr	Better
CB1S	Center bead one side
CB2S	Center bead two sides
CF	Cost and freight
CG2E	Center groove two edges
CIF	Cost, insurance, and freight
CIFE	Cost, insurance, freight, and exchange
Clg	Ceiling
Clr	Clear
Com	Center matched
CS	Caulking seam
Csg	Casing
Cu. Ft.	Cubic foot (feet)
CV1S	Center Vee one side
CV2S	Center Vee two sides
D/S	Drop Siding
D&H	Dressed and headed
D&M	Dressed and matched

Abbreviation	Meaning
DB. Clg.	Double-beaded ceiling
DB. Part	Double-headed partition
DET	Double end trimmed
Dim	Dimension
Dkg	Decking
E&CB1S	Edge and center bead one side
E&CBsS	Edge and center bead two sides
E&CV1S	Edge and cetner Vee one side
E&CV2S	Edge and center Vee two sides
EB1S	Edge bead one side
EB2S	Edge bead two sides
EE	Eased edges
EG	Edge grain
EM	End matched
EV1S	Edge Vee one side
EV2S	Edge Vee two sides
Fac	Factory
FAS	Free alongside (named vessel)
FBM	Foot or feet board measure
FG	Flat grain
Flg	Flooring
FOB	Free on board
FOHC	Free of heart center
FOK	Free of knots
Frt	Freight
Ft	Foot or feet
G/R	Grooved roofing
GM	Grade marked
H&M	Hit-and-miss
HB	Hollow back
Hrt CC	Heart cubical content
Hrt G	Heart girth
Hrt	Heart
IN	Inches
J&P	Joists and planks
KD	Kiln-dried

Abbreviation	Meaning
Lbr	Lumber
LCL	Less than carload
LFT	Linear foot or feet
Lgr	Longer
Lgth	Length
Lin	Linear
Lng	Lining
M	Thousand
MBM	Thousand board feet measure
MC	Moisture content
Merch	Merchantable
Mldg	Moulding
N1E	Nosed one edge
N2E	Nosed two edges
No.	Number
OG	Ogee
Ord	Order
P&T	Post and timber
Par	Paragraph
Pat	Pattern
Patrt	Partition
Pc	Piece
Pcs	Pieces
PE	Plain end
PO	Purchase order
R/L	Random lengths
R/W	Random widths
R/W&L	Random widths and lengths
Reg	Regular

Abbreviation	Meaning
Res	Resawed or resawn
Rfg	Roofing
Rgh	Rough
S/L	Shiplap
S1E	Surfaced one edge
S1S	Surfaced one side
S1S1E	Surfaced one side and one edge
S1S2E	Surfaced one side and two edges
S2E	Surfaced two edges
S2S	Surfaced two sides
S2S&CM	Surfaced two sides and center matched
S2S1E	Surfaced two sides and one edge
S4S	Surfaced four sides
S4S&CS	Surfaced four sides and cauking seam
S7E	Side and Edge
Sdg	Siding
SE & S	Square edge and sound
SE sdg	Square edge siding
Sel	Select
SL&C	Shipper's load and count
SM	Standard matched
Specs	Specifications
Std	Standard
Stpg	Stopping
Str.	Structural
T&G	Tongued and grooved
VG	Vertical grain
Wdr.	Wider
Wt	Weight

Board Feet of Lumber in Various Sizes

Board Measure 1½ Inch Thick

Width					Length	in	Feet					
(inches)	2	4	6	8	10	12	14	16	18	20	22	24
1	0 ft. 3 in.	0 ft. 6 in.	0 ft. 9 in.	1 ft. 0 in.	1 ft. 3 in.	1 fl. 6 in.	1 ft. 9 in.	2 ft. 0 in.	2 ft. 3 in.	2 ft. 6 in.	2 ft. 9 in.	3 ft. 0 in.
2	0 6	1 0	1 6	2 0	2 6	3 0	3 6	4 0	4 6	5 0	5 6	6 0
3	0 9	1 6	2 3	3 0	3 9	4 6	5 3	6 0	6 9	7 6	8 3	9 0
4	1 0	2 0	3 0	4 0	5 0	6 0	7 0	8 0	9 0	10 0	11 0	12 0
5	1 3	2 6	3 9	5 0	6 3	7 6	8 9	10 0	11 3	12 6	13 9	15 0
6	1 6	3 0	4 6	6 0	7 6	9 0	10 6	12 0	13 6	15 0	16 6	18 0
7	1 9	3 6	5 3	7 0	8 9	10 6	12 3	14 0	15 9	17 6	19 3	21 0
8	2 0	4 0	6 0	8 0	10 0	12 0	14 0	16 0	18 0	20 0	22 0	24 0
10	2 6	5 0	7 6	10 0	12 6	15 0	17 6	20 0	22 6	25 0	27 6	30 0
12	3 0	6 0	9 0	12 0	15 0	18 0	21 0	24 0	27 0	30 0	33 0	36 0
14	3 6	7 0	10 6	14 0	17 6	21 0	24 6	28 0	31 6	35 0	38 6	42 0
16	4 0	8 0	12 0	16 0	20 0	24 0	28 0	32 0	36 0	40 0	44 0	48 0
18	4 6	9 0	13 6	18 0	22 6	27 0	31 6	36 0	40 6	45 0	49 6	54 0

Board Feet of Lumber in Various Sizes

Board Measure 2 Inch Thick

Width (inches)	Length in Feet											
	2	4	6	8	10	12	14	16	18	20	22	24
1	0 ft. 4 in.	0 ft. 8 in.	1 ft. 0 in.	1 ft. 4 in.	1 ft. 8 in.	2 ft. 0 in.	2 ft. 4 in.	2 ft. 8 in.	3 ft. 0 in.	3 ft. 4 in.	3 ft. 8 in.	4 ft. 0 in.
2	0 8	1 4	2 0	2 8	3 4	4 0	4 8	5 4	6 0	6 8	7 4	8 0
3	1 0	2 0	3 0	4 0	5 0	6 0	7 0	8 0	9 0	10 0	11 0	12 0
4	1 4	2 8	4 0	5 4	6 8	8 0	9 4	10 8	12 0	13 4	14 8	16 0
5	1 8	3 4	5 0	6 8	8 4	10 0	11 8	13 4	15 0	16 8	18 4	20 0
6	2 0	4 0	6 0	8 0	10 0	12 0	14 0	16 0	18 0	20 0	22 0	24 0
7	2 4	4 8	7 0	9 4	11 8	14 0	16 4	18 8	21 0	23 4	25 8	28 0
8	2 8	5 4	8 0	10 8	13 4	16 0	18 8	21 4	24 0	26 8	29 4	32 0
10	3 4	6 8	10 0	13 4	16 8	20 0	23 4	26 8	30 0	33 4	36 8	40 0
12	4 0	8 0	12 0	16 0	20 0	24 0	28 0	32 0	36 0	40 0	44 0	48 0
14	4 8	9 4	14 0	18 8	23 4	28 0	32 8	37 4	42 0	46 8	51 4	56 0
16	5 4	10 8	16 0	21 4	26 8	32 0	37 4	42 8	48 0	53 4	58 8	64 0
18	6 0	12 0	18 0	24 0	30 0	36 0	42 0	48 0	54 0	60 0	66 0	72 0

Fir Plywood for Exterior Uses

Grade (Exterior)	Face	Back	Inner Plies	Uses
A-A	A	A	C	Outdoor, where appearance of both sides is important.
A-B	A	B	C	Alternate for A-A, where appearance of one side is less important.
A-C	A	C	C	Siding, soffits, fences. One "good" side grade.
B-C	B	C	C	For utility uses such as farm buildings, some kinds of fences.
C-C, Repaired	C	C	C	Excellent base for tile and linoleum, backing for wall covering.
C-C	C	C	C	Unsanded, for backing and rough construction exposed to weather.
B-B	B	B	B	Concrete forms. Re-use until wood literally wears out.

Fir Plywood for Interior Uses

Grade (Interior)	Face	Back	Inner Plies	Uses
A-A	A	A	D	Cabinet doors, built-ins, furniture where both sides will show.
A-B	A	B	D	Alternate of A-A — one side high standard, the other solid and smooth.
A-D	A	D	D	Good one side for paneling, built-ins, backing, underlay.
B-D	B	D	D	Utility grade. Has one good side. Backing, cabinet sides
C-D, Repaired	C	D	D	Underlay for tile, linoleum, and carpet and similar uses.
C-D	C	D	D	Sheathing and structural uses such as temporary enclosures, subfloor.
B-B	B	B	C	Concrete forms. Re-use until wood literally wears out.

Sizes of Common Wire Nails

Size of Nails	Number per Pound	Length (inches)	Number of Wire
3 penny	615	1¼	14½
4 penny	322	1½	13
6 penny	200	2	12
8 penny	106	2½	10½
10 penny	74	3	9½
16 penny	46	3½	9
20 penny	29	4	6
40 penny	17	5	4

Quanity of Nails Required

For	Nail Size	Amount
1000 laths	3 penny	6 lbs.
1000 shingles	4 penny	4 lbs.
1000 sq. ft. of beveled siding	6 penny	15 lbs.
1000 sq. ft. of sheathing	8 penny	16 lbs.
1000 sq. ft. of flooring	8 penny	24 lbs.
1000 sq. ft. of sheathing	10 penny	21 lbs.
1000 sq. ft. of studding	10 penny	12 lbs.
1000 sq. ft. of studding	20 penny	4 lbs.

GENERAL RULES OF NAILING

Do not use larger nails than are necessary for the job; choose the nail with the smallest diameter. Never feel that you are obligated to sink a nail with a single blow. The force required to drive a nail depends to some extent on the job; that is, a 16d nails requires more force than, say, a casing nail through a door trim, but none should be hid as hard as possible with the hammer.

Sharp nails often cause wood to split since they act like a chisel, separating the wood fibers. A blunt nail, driven with reasonable hammer blows, has less tendency to split. This is why many professionals blunt the nail with a hammer before driving it home.

Place nails as far in from the ends of boards as possible, and try not to drive through knots. If a board must be secured near a knot, drill a small pilot hole first. The same is true for wood that has a tendency to split or when working on very hard wood. A drilled pilot hole before driving the nail home can save much material. Lubricating the nail with soap before driving it into the wood will also help to avoid splitting the wood or bending the nail. Avoid driving a series of nails in a straight line. Stagger them to avoid splitting.

Appendix L

Written specifications that accompany some drawings are the descriptions of work involved, types and grades of materials, and other pertinent information to better help the project to be completed as the designer intended. Written specifications for a building construction project are the descriptions of work and duties required of the architect, engineer, or owner. Along with the working drawings of the project, these specifications form the basis of the contract requirements.

Specifications accompany nearly all drawings for government projects, as well as those for most manufacturers, to ensure that the completed object, product, or service will be exactly as the buyer wants.

For convenience in writing, speed in estimating work, and ease in reference, the most suitable organization of the specification is a series of sections dealing successively with the different trades, and in each section grouping all the work of the particular trade to which the section is devoted. all the work of each trade should be incorporated into the section devoted to that trade. Those people who use the specifications must be able to find all information needed without taking too much time in looking for it.

The Construction Specification Institute (CSI) developed the Uniform Construction Index some years ago that allowed all specifications, produce information, and cost data to be arranged into a uniform system. After Division 0 — Bidding and Contract Requirements — all remaining construction is divided into 16 divisions, and each division has several subsections. The following outline briefly describes the various sections normally included in a set of specifications for building construction.

Division 1 — General Requirements. This division summarizes the work, alternatives, project meetings, submissions, quality control, temporary facilities and controls, products, and the project closeout. Every responsible person involved with the project should become familiar with this division.

Division 2 — Site Work. This division outlines work involving such items as paving, sidewalks, outside utility lines (electrical, plumbing, gas, telephone, etc.), landscaping, grading, and other items pertaining to the outside of the building.

Division 3 — Concrete. This division covers work involving footings, concrete formwork, expansion and contraction; joints, cast-in-place concrete, specially finished concrete, precast concrete, concrete slabs, and the like.

Division 4 — Masonry. This division covers concrete mortar, stone, masonry accessories, and the like.

Division 5 — Metals. Metal roofs structural metal framing, metal joists, metal decking, ornamental metal, and expansion control normally fall under this division.

Division 6 — Carpentry. Items falling under this division include: rough carpentry, heavy timber construction, trestles, prefabricated structural wood, finish carpentry, wood treatment, architectural woodwork, and the like. Plastic fabrications may also be included in this division of the specifications.

Division 7 — Thermal and Moisture Protection. Waterproofing is the main topic discussed under this division. Other related items such as dampproofing, building insulation, shingles and roofing tiles, preformed roofing and siding, membrane roofing, sheet metal work, wall flashing, roof accessories, and sealants are also included.

Division 8 — Doors and Windows. All types of doors and frames are included under this division: metal, plastic, wood, etc. Windows and framing are also included along with hardware and other window and door accessories.

Division 9 — Finishes. Included in this division are the types, quality, and workmanship of lath and plaster, gypsum wallboard, tile, terrazzo, acoustical treatment, ceiling suspension systems, wood flooring, floor treatment, special coatings, painting, and wallcovering.

Division 10 — Specialties. Specialty items such as chalkboards and tackboards; compartments and cubicles, louvers and vents that are not connected with the heating, ventilating, and air conditioning system; wall and corner guards; access flooring; specialty modules; pest control; fireplaces; flagpoles; identifying devices; lockers; protective covers; postal specialties; partitions; scales; storage shelving; wardrobe specialties; and the like are covered in this division of the specifications.

Division 11 — Equipment. The equipment included in this division could include central vacuum cleaning systems, bank vaults, darkrooms, food service, vending machines, laundry equipment, and many similar items.

Division 12 — Furnishing. Items such as cabinets and storage, fabrics, furniture, rugs and mats, seating, and other similar furnishing accessories are included under this division.

Division 13 — Special Construction. Such items as air-supported structures, incinerators, and other special items will fall under this division.

Division 14 — Conveying Systems. This division covers conveying apparatus such as dumbwaiters, elevators, hoists and cranes, lifts, material-handling systems, turntables, moving stairs and walks, pneumatic tube systems, and powered scaffolding.

Division 15 — Mechanical. This division includes plumbing, heating, ventilating, and air conditioning and related work. Electric heat is sometimes covered under Division 16, if individual baseboard heating units are used in each room or area of the building.

Division 16 — Electrical. This division covers all electrical requirements for the building including lighting, power, alarm and communication systems, special electrical systems, and related electrical equipment.

CSI FORMAT

The following is a more detailed coverage of the CSI format, along with CSI reference numbers for various construction categories.

DIVISION 0 — BIDDING AND CONTRACT REQUIREMENTS

Reference
Number

00010	Pre-bid Information
00100	Instructions to Bidders
00200	Information Available to Bidders
00300	Bid/Tender Forms
00400	Supplements to Bid/Tender Forms
00500	Agreement Forms

05400   Cold-formed Metal Framing
05500   Metal Fabrications
05700   Ornamental Metal
05800   Expansion Control
05900   Metal Finishes

DIVISION 6 — WOOD AND PLASTICS

06050   Fasteners and Supports
06100   Rough Carpentry
06130   Heavy Timber Construction
06150   Wood-Metal Systems
06170   Prefabricated Structural Wood
06200   Finish Carpentry
06300   Wood Treatment
06400   Architectural Woodwork
06500   Prefabricated Structural Plastics
06600   Plastic Fabrications

DIVISION 7 — THERMAL AND MOISTURE PROTECTION

07100   Waterproofing
07150   Dampproofing
07200   Insulation
07250   Fireproofing
07300   Shingles and Roofing Tiles
07400   Preformed Roofing and Siding
07500   Membrane Roofing
07570   Traffic Topping
07600   Flashing and Sheet Metal
07800   Roof Accessories
07900   Sealants

DIVISION 8 — DOORS AND WINDOWS

08100   Metal Doors and Frames
08200   Wood and Plastic Doors
08250   Door Opening Assemblies
08300   Special Doors
08400   Entrances and Storefronts
08500   Metal Windows
08600   Wood and Plastic Windows

08650   Window Covering
08700   Hardware
08800   Glazing
08900   Glazed Curtain Walls

DIVISION 9 — FINISHES

09100   Metal Support Systems
09200   Lath and Plaster
09230   Aggregate Coatings
09300   Tile
09400   Terrazzo
09500   Acoustical Treatment
09550   Wood Flooring
09600   Stone and Brick Flooring
09650   Resilient Flooring
09680   Carpeting
09700   Special Flooring
09760   Floor Treatment
09800   Special Coatings
09900   Painting
09950   Wall Covering

DIVISION 10 — SPECIALTIES

10100   Chalkboards and Tackboards
10150   Compartments and Cubicles
10200   Louvers and Vents
10230   Grills and Screens
10240   Service Wall Systems
10260   Wall and Corner Guards
10270   Access Flooring
10280   Specialty Modules
10290   Pest Control
10300   Fireplaces and Stoves
10340   Prefabricated Steeples, Spires, and Cupolas
10350   Flagpoles
10400   Identifying Devices
10450   Pedestrian Control Devices
10500   Lockers
10520   Fire Extinguishers, Cabinets, and Accessories

10530	Protective Covers
10550	Postal Specialties
10600	Partitions
10650	Scales
10670	Storage Shelving
10700	Exterior Sun Control Devices
10750	Telephone Enclosures
10800	Toilet and Bath Accessories
10900	Wardrobe Specialties

DIVISION 11 — EQUIPMENT

11010	Maintenance Equipment
11020	Security and Vault Equipment
11030	Checkroom Equipment
11040	Ecclesiastical Equipment
11050	Library Equipment
11060	Theater and Stage Equipment
11070	Musical Equipment
11080	Registration Equipment
11090	Mercantile Equipment
11110	Commercial Laundry and Dry Cleaning Equipment
11120	Vending Equipment
11130	Audio-Visual Equipment
11140	Service Station Equipment
11150	Parking Equipment
11160	Loading Dock Equipment
11170	Waste Handling Equipment
11190	Detention Equipment
11200	Water Supply and Treatment Equipment
11300	Fluid Waste Disposal and Treatment Equipment
11400	Food Service Equipment
11450	Residential Equipment
11460	Unit Kitchens
11470	Darkroom Equipment
11480	Athletic, Recreational, and Therapeutic Equipment
11500	Industrial and Process Equipment
11600	Laboratory Equipment
11650	Planetarium and Observatory Equipment
11700	Medical Equipment
11780	Mortuary Equipment
11800	Telecommunication Equipment
11850	Navigation Equipment

DIVISION 12 — FURNISHINGS

12100	Artwork
12300	Manufactured Cabinets and Casework
12500	Window Treatment
12550	Fabrics
12600	Furniture and Accessories
12670	Rugs and Mats
12700	Multiple Seating
12800	Interior Plants and Plantings

DIVISION 13 — SPECIAL CONSTRUCTION

13010	Air-Supported Structures
13020	Integrated Assemblies
13030	Audiometric Rooms
13040	Clean Rooms
13050	Hyperbaric Rooms
13060	Insulated Rooms
13070	Integrated Ceilings
13080	Sound, Vibration, and Seismic Control
13090	Radiation Protection
13100	Nuclear Reactors
13110	Observatories
13120	Pre-engineered Structures
13130	Special-purpose Rooms and Buildings
13140	Vaults
13150	Pools
13160	Ice Rinks
13170	Kennels and Animal Shelters
13200	Seismographic Instrumentation
13210	Stress Recording Instrumentation
13220	Solar and Wind Instrumentation
13410	Liquid and Gas Storage Tanks
13510	Restoration of Underground Pipelines

13520   Filter Underdrains and Media
13530   Digestion Tank Covers and Appurtenances
13540   Oxygenation Systems
13550   Thermal Sludge Conditioning Systems
13560   Site Constructed Incinerators
13600   Utility Control Systems
13700   Industrial and Process Control Systems
13800   Oil and Gas Refining Installations and Control Systems
13900   Transportation Instrumentation
13940   Building Automation Systems
13970   Fire Suppression and Supervisory Systems
13980   Solar Energy Systems
13990   Wind Energy Systems

DIVISION 14 — CONVEYING SYSTEMS
14100   Dumbwaiters
14200   Elevators
14300   Hoists and Cranes
14400   Lifts
14500   Material Handling Systems
14600   Turntables
14700   Moving Stairs and Walks
14800   Powered Scaffolding
14900   Transportation Systems

DIVISION 15 — MECHANICAL
15050   HVAC and Piping Contractors
15200   Noise, Vibration, and Seismic Control
15250   Insulation
15300   Special Piping Systems
15400   Plumbing Systems
15450   Plumbing Fixtures and Trim
15500   Fire Protection
15600   Power or Heat Generation
15650   Refrigeration
15700   Liquid Heat Transfer
15800   Air Distribution
15900   Controls and Instrumentation

DIVISION 16 — ELECTRICAL
16050   Electrical Contractors
16200   Power Generation
16300   Power Transmission
16400   Service and Distribution
16500   Lighting
16600   Special Systems
16700   Communications
16850   Heating and Cooling
16900   Controls and Instrumentation

Appendix M

The ability of those involved in the construction and design of mechanical systems for buildings to quickly and accurately transmit technical information is vital to the development, construction, and use of such systems. The vital information is transmitted by working drawings and written mechanical specifications.

Ideally, mechanical contractors and their workers should be able to bid on, order material for, and construct a complete mechanical system solely from the information furnished. No further questions should arise. Working drawings and written specifications help to make this possible.

Division 15 of the specifications, in a format devised by Construction Specifications Institute, covers the mechanical work of a project, including grade of materials and the method of installation. The following is an outline of the sections usually included in Division 15 of the written specifications for building construction.

15010 General Provisions: The general provisions of the mechanical specifications consist of a selected group of considerations and regulations that apply to all sections of the division. Items covered my include the scope of the work in the mechanical contract, mechanical reference symbols, codes and fees, tests, a demonstration of the completed mechanical system, and identifications of the equipment and components used in the installation.

15050 Basic Materials and Methods: Definitive statements in this section of the specifications establish the means of identifying the type and quality of materials and equipment to be used. This section also establishes the accepted methods of installing various materials such as pipe and pipe fittings, traps, hangers and supports, valves, pumps, meters and gauges, and tanks.

15180 Insulation: All insulation for water piping, refrigerant piping, steam and condensate return piping, air ducts, breeching, and other mechanical equipment.

15200 Water Supply and Treatment: All water systems including pump and piping systems, water reservoirs and tanks, water treatment, metering, and related equipment.

15300 Wastewater Disposal and Treatment: All sewage, septic tank, and sewage treatment systems, including sewage ejectors, septic tanks, drainage fields, filtration equipment, aeration equipment, sludge digestion equipment, and the like.

15400 Plumbing: All plumbing systems, including water supply systems, waste piping systems, roof drainage systems, industrial waste drainage systems, drains, cleanouts, and fixtures.

15500 Fire Protection: All fire protection systems, including sprinkler, foam, CO_2, and standpipe equipment; fire hoses and fire hose connections, reels, cabinets, and accessories; portable fire extinguishers and cabinets and accessories, fire blankets, hood and duct fire protection equipment, and nonelectrical alarm equipment.

15600 Power or Heat Generation: All power and heat systems including furnaces, boilers, fuel storage tanks, exhaust and draft control equipment, and power and heat controls.

15650 Refrigeration: All refrigeration equipment including compressors, condensing units, chillers, cooling towers, ice banks, and evaporators.

15700 Liquid Heat Transfer: All heating and cooling piping systems, heat exchangers, humidifiers, and their controls.

15800 Air Distribution: All heating and cooling systems that do not use a liquid for heat transfer, including forced-air furnaces, fans, ventilators, air curtains, ductwork, dampers, grilles, filters, and similar devices.

15900 Controls and Instrumentation: All controls, meters, and instruments for the mechanical equipment, including electrical interlock identification, inspection, detection, and balancing; control piping, tubing and wiring; control air compressors and dryers; control panels; instrument panelboard; primary control devices; thermostats; humidistats; aquastats; relays and switches; timers; control dampers, valves, and motors; sequence controls, recording devices, alarm devices, and special process controls.

Most architects and consulting engineers specify the type and design of project items by listing manufacturers' names and catalog numbers, and only these particular items may be used. However, an item that is considered to be *an approved equal* will usually be accepted. Many architects, engineers, and owners are very reasonable when approving nonspecified components and equipment; others may require such close compliance tht no component will be accepted as an "equal." This is especially true when the architect/engineer allows a manufacturer to write a portion of the specifications because the manufacturer invariably writes an ironclad specification that only the manufacturer's equipment will meet.

CSI FORMAT

The first widely publicized document advocating a nationwide approach to uniform specification writing was published by the Construction Specifications Institute in March 1963 as *The CSI Format for Building Specifications.* It proposed the organized of technical specifications into sixteen basic groups designated as "Divisions," each of which was to be based on an interrelationship of place, trade, function, or material There has been widespread acceptance of this document within federal and state agencies throughout the United States and Canada and among private practitioners in the architectural and engineering profession. The format outline for Division 15 — Mechanical of these specifications are as follows:

DIVISION 15 — MECHANICAL

Reference
Number

15010 General Provisions
15015 Mechanical Drawing and Reference Symbols
15020 Work Included
15021 Work Not Included
15023 Codes, Fees, and Lateral Costs
15042 Tests
15043 Balancing of Air System
15046 Demonstration
15047 Identification

15050 Basic Materials and Methods
15060 Pipe and Pipe Fittings
15061 Steel Pipe
15062 Cast Iron Pipe

15063	Copper Pipe
15083	Strainers, Filters, and Dryers
15084	Vent Caps
15085	Traps
15087	Shock Absorbers
15090	Supports, Anchors, and Seals
15100	Valves, Cocks, and Faucets
15101	Gate Valves
15102	Blowdown Valves
15103	Butterfly Valves
15104	Ball Valves
15105	Globe Valves
15106	Refrigerant Valves
15110	Check Valves
15140	Pumps

15180 Insulation

15181	General
15182	Cold-Water Piping
15183	Chilled-Water Piping
15184	Refrigerant Piping
15185	Hot-Water Piping
15186	Steam and Condensate Return Piping
15187	Underground Piping
15188	Outside Piping
15190	Duct
15196	Equipment

15200 Water Supply and Treatment

15201	General
15220	Pump and Piping System
15230	Booster Pumping Equipment
15240	Water Reservoirs and Tanks
15270	Metering and Related Piping

15300 Wastewater Disposal and Treatment

15301	General
15310	Sewage Ejectors
15320	Grease Interceptors
15361	Septic Tanks
15362	Drainage Fields
15370	Manholes

15400 Plumbing

15401	General
15410	Roof Drainage System
15421	Floor and Shower Drains
15422	Roof Drains
15423	Cleanouts and Cleanout Access Covers
15424	Domestic Water Heaters
15450	Plumbing Fixtures and Trim

15500 Fire Protection

15501	General
15510	Sprinkler Equipment

15600 Power and Heat Generation

15601	General
15606	Oil Storage Tanks, Controls, and Piping
15615	Lined Breeching
15621	Boilers
15639	Boiler Accessories
15643	Boiler Tests

15650 Refrigeration

15651	General
15655	Refrigerant Compressors
15660	Condensing Units
15670	Chillers
15680	Cooling Tower

15700 Liquid Heat Transfer

15701	General
15710	Hot-Water Specialties
15715	Steam Specialties
15720	Condensate Pump and Receiver Set
15732	Converter
15750	Coils

Appendix N

Electrical Specifications

The General Conditions section of the written specifications for a building construction project consists of a selected group of regulations that apply to all subdivisions of the project. These conditions usually are the responsibility of the architect and for most projects, they are presented in the form of a standard document titled *General Conditions of the Contract for Construction*, which is available from the American Institute of Architects (AIA). All trades should become familiar with this document.

Division 16 of written specifications for building construction covers the electrical and related work on a given project, including the grade of materials to be used and the manner of installation for the electrical system. The following is an outline of the various sections normally included in Division 16 — Electrical.

16010 General Provisions: The general provisions of the electrical specifications normally consist of a selected group of considerations and regulations that apply to all sections of the division. Items covered may include the scope of the work (work included and not included) in the electrical contract, electrical reference symbols, codes and fees, tests, demonstration of the completed electrical system, and identification of the equipment and components used in the installation.

16100 Basic Materials and Methods: Definitive statements in this portion of the specifications should establish the means of identifying the type and quality of materials and equipment selected for use. This section should further establish the accepted methods of installing various materials such as raceways, conduits, bus ducts, underfloor ducts, cable trays, wires and cables, wire connections and devices, outlet boxes, floor boxes, cabinets, panelboards, switches and receptacles, motors, motor starters, disconnects, overcurrent protective devices, supporting devices, and electronic devices.

16200 Power Generation: This section normally covers items of equipment used for emergency or standby power facilities, the type used to take over essential electrical service during a normal power source outage. This section usually cites requirements for a complete installation of all emergency circuits on a given project, including emergency service or standby power in the form of a generator set or storage batteries, automatic

449

control facilities, feeders, panelboards disconnects, branch circuits, and outlets.

Items to be fully described in this section include the generator and its engine (reciprocating or turbine), cooling equipment, exhaust equipment, starting equipment, and automatic or manual transfer equipment.

16300 Power Transmission: Unlike the feeders, branch circuits, and the like that carry electrical power inside a building, high voltage (over 600 volts) power transmission is the subject of this section. Normally the specifications that require this section are for those projects constructed on government reservations and large industrial sites.

Cable and equipment specified in this section almost always is over 2.4 kV and includes such items as substations, switchgear, transformers, vaults, manholes, rectifiers, converters, and capacitors. Size and type of enclosures, as well as instrumentation, may also be included here.

16400 Service and Distribution: Power distribution facilities (under 600 volts) for the project's service entrance, metering, distribution switchboards, branch circuit panelboards, feeder circuits, and the like are described in this section by paragraphs or clauses covering selected related equipment items.

In addition to the electric service characteristics — voltage, frequency, phase, etc., the quality and capacity levels of all items involved in the service entrance and power distribution system should be clearly defined. Typical items include the size and number of conductors, installation and supporting methods, location, rating, type and circuit protection features of all main circuit breakers, and other disconnecting means. The interrupting capacity of fuses is especially important and should be a major consideration in this section of the electrical specifications.

Other items for consideration are grounding, transformers (usually dry type), underground or overhead service, primary load interrupters, converters, and rectifiers.

16500 Lighting: This section covers general conditions relating to selected lighting equipment to ensure that all such equipment is furnished and installed exactly as designated by the architect or engineer. Further clauses establish the quality and type of interior lighting

fixtures, luminous ceilings, signal lighting, exterior lighting fixtures, stadium lighting, roadway lighting, accessories, lamps, ballasts and related accessories, poles, and standards. Methods of installation also are included in most sets of specifications.

Where special lighting equipment is specified, the specifications normally call for large, detailed shop drawings to be submitted to the architect or engineer for approval prior to installation.

16600 Special System: Items that may be covered here include a wide variety of special systems unusual to conventional electrical installations. Examples of such items are lighting protection systems, special emergency light and power systems, storage batteries, battery charging equipment, and perhaps cathodic protection. However, this section is by no means limited to these few items. Many other special systems can be described in this section of the electrical specifications.

16700 Communications: Equipment items that are interconnected to permit audio or visual contact between two or more stations or to monitor activity and operation at remote points are covered here. Most clauses deal with a particular manufacturer's equipment and state what items will be furnished and what is expected of the system once it is in operation.

Items covered under this section include radio, shortwave and microwave transmission; alarm and detection systems; smoke detectors; clock and program equipment; telephone and telegraph equipment; intercommunication and public address equipment; television systems; master TV antenna equipment, and learning laboratories.

16850 Heating and Cooling: Because of working agreements among labor unions, most heating and cooling equipment is installed by workers other than electricians, and the requirements are usually covered in Division 15 — Mechanical of the written specifications. In some cases, however, the electrical contractor is responsible for installing certain pieces of heating and cooling equipment, especially on residential and apartment projects.

The main point that this section of the specifications should make is that the system installation meet with the design requirements. To do this, electric heat specifications should include pertinent data about the factors of the building insulation on which the design is based, as well as installation instructions for the selected equipment. To further aid in ensuring a proper installation, exact descriptions (manufacturer, catalog number, wattage rating, etc.) of the units normally are specified.

Items in this category of the electrical specifications include snow melting cables and mats, heating cable, electric heating coil, electric baseboard heaters, packaged room air conditioners, radiant heaters, duct heaters, and fan-type floor, ceiling, and wall heaters.

16900 Controls and Instrumentation: As the name implies, this section covers all types of controls and instrumentation used on a given project. Examples include recording and indicating devices, motor control centers, lighting control equipment, electrical interlocking devices and applications, control of electric heating and cooling, limit switches, and numerous other such devices and systems.

Other divisions of the specifications besides Division 16, especially Division 15 — Mechanical, may involve a certain amount of electrical work. The responsibility for such work should be clearly defined. Without great care by all concerned, confusion about responsibility may result in the electrical contractor paying for work he had assumed was not under his division (Division 16).

The most common sources of confusion are in control wiring for boilers, heating, ventilating and air conditioning systems; control, signal and power wiring beyond the machine room; disconnect switches for elevator construction; automatic machinery controls; wiring on machine tools; wiring on overhead cranes and hoists; mounting and connecting of motors; connecting hospital, laundry and restaurant equipment; connecting of motors; connecting hospital, laundry and restaurant equipment; connecting electric signs; connecting motion picture projection and sound equipment; installing electric lighting fixtures furnished by someone other than the electrical contractor; connecting unit heaters, unit ventilators, electric fans, electric water heaters, electric water coolers, electric ranges, and other appliances when they are not furnished by the electrical contractor; connecting transformers, and similar items.

The details of discovering and resolving conflicting statements in written specifications should be a major concern to all who use them. More important, however, is avoiding the conflicts altogether.

CSI FORMAT

In order to standardize specification formats, the Construction Specifications Institute, Inc. (CSI) in Washington, DC developed a very practical format that most architects and engineers have now adopted. The format outline of Division 16 — Electrical is as follows:

DIVISION 16 — ELECTRICAL

16010 General Provisions
16015 Electrical Reference Symbols
16020 Work Included
16021 Work Not Included
16025 Codes and Fees
16030 Test
16031 Demonstration of Complete Electrical System
16031 Identification

16100 Basic Materials and Methods
16101 General
16110 Raceways
16111 Conduits
16112 Bus Ducts
16113 Underfloor Ducts
16114 Cable Trays
16120 Wires and Cables
16121 Wire Connections and Devices
16125 Pulling Cables
16130 Outlet Boxes
16131 Pull and Junction Boxes
16132 Floor Boxes
16133 Cabinets
16134 Panelboards
16140 Switches and Receptacles
16150 Motors
16160 Motor Starters
16170 Disconnects (motor and circuit)

16180   Overcurrent Protective Devices
16190   Supporting Devices
16199   Electronic Devices
16200   Power Generation
16201   General
16210   Generator
16220   Engine
16221   Reciprocating Engine
16471   Branch Circuit Panelboard
16480   Feeder Circuit
16490   Converters
16491   Rectifiers

16500   Lighting
16501   General
16510   Interior Lighting Fixtures
16511   Luminous Ceiling
16515   Signal Lighting
16530   Exterior Lighting Fixtures
16531   Stadium Lighting
16532   Roadway Lighting
16550   Accessories
16551   Lamps
16552   Ballasts and Accessories
16570   Poles and Standards

16600   Special Systems
16601   General
16610   Lightning Protection
16620   Emergency Light and Power
16621   Storage Batteries
16622   Battery Charging Equipment
16640   Cathodic Protection

16700   Communications
16701   General
16710   Radio Transmission
16711   Shortwave Transmission
16712   Microwave Transmission
16720   Alarm and Detection
16721   Fire Alarm and Detection
16725   Smoke Detection
16727   Burglar Alarm
16730   Clock and Program Equipment
16740   Telephone
16750   Telegraph
16760   Intercommunication Equipment
16770   Public Address Equipment
16780   Television System
16781   Master TV Antenna Equipment
16790   Learning Laboratories

16850 Heating and Cooling
16851   General
16858   Snow Melting Cable and Mat
16859   Heating Cable
16860   Electric Heating Coil
16865   Electric Baseboard
16870   Packaged Room Air Conditioners
16880   Radiant Heaters
168   Electric Heaters (Prop Fan Type)

16900 Controls and Instrumentation
16901   General
16910   Recording and Indicating Devices
16920   Motor Control Centers
16930   Lighting Control Equipment
16940   Electrical Interlock
16950   Control of Electric Heating
16960   Limit Switches

Appendix O

Trade Organizations

ASA
Acoustical Society of America
500 Sunnyside Blvd.
Woodberry, NY 11797
(516) 349-7800

ASC
Adhesive and Sealant Council, Inc.
627 K. St. NW
Washington, DC 20001
(202) 452-1500

ARI
Air-Conditioning and Refrigeration Institute
1501 Wilson Blvd.
Arlington, VA 22209
(703) 524-8800

ACCA
Air Conditioning Contractors of America
1513 16th St.
Washington, DC 20036
(202) 583-9370

ACEA
Allied Construction Employers Association
180 N. Executive Drive
Brookfield, WI 53008
(414) 785-1430

AA
Aluminum Association
900 19th ST., NW
Washington, DC 20006
(202) 862-5100

AAA
American Arbitration Association
140 W. 51st St.
New York, NY 10020
(212) 484-4000

American Association of State Highway and Transportation Officials
444 N. Capitol St., NW, Suite 225
Washington, DC 20001
(202) 624-5800

ABCA
American Building Contrators Association
11100 Valley Blvd., Suite 120
El Monte, CA 91731
(818) 401-0071

ACI
American Concrete Institute
22400 W. Seven Mile Rd.
Detroit, MI 48219
(313) 532-2600

ACPA
American Concrete Pavement Association
3800 N. Wilke Rd., Suite 490
Arlington Heights, IL 60004
(708) 394-5577

ACPA
American Concrete Pipe Association
8300 Boone Blvd.
Vienna, VA 22180
(703) 821-1990

ACEC
American Consulting Engineers Council
1015 15th St., NW, Suite 802
Washington, DC 20005
(202) 347-7474

AGA
American Gas Association, Inc.
1515 Wilson Blvd.
Arlington, VA 22209
(703) 841-8400

AHA
American Hardboard Association
520 N. Hicks Rd.
Palatine, IL 60067
(708) 934-8800

AHMA
American Hardware Manufacturers Association
931 N. Plum Grove Rd.
Schaumburg, IL 60173
(708) 605-1025

AIA
American Institute of Architects
1735 New York Ave., NW
Washington, DC 20006
(202) 626-7300

ASID
American Society of Interior Designers
200 Lexington Ave.
New York, NY 10016
(212) 685-3480

AISC
American Institute of Steel Construction, Inc.
400 N. Michigan Ave.
Chicago, IL 60611
(312) 670-2400

AITC
Amerian Institute of Timber Construction
11818 S.E. Mill Plain Blvd.
Vancouver, WA 98684
(206) 254-9132

AISI
American Iron and Steel Institute
1133 15th St. NW, Suite 300
Washington, DC 20005
(202) 452-7100

ALSC
American Lumber Standards Committee
P.O. Box 210
Germantown, MD 20874
(301) 972-1700

ANSI
American National Standards Institute
1430 Broadway
New York, NY 10018
(212) 642-4900

APFA
American Pipe Fitting Association
6203 Old Keene Mill Ct.
Springfield, VA 22152
(703) 644-0001

APA
American Plywood Association
PO Box 11700
Tacoma, WA 98411
(509) 565-6600

ARTBA
American Road and Transportation Builders Association
525 School St. SW
Washington, DC 20024
(202) 488-2722

ASTM
American Society for Testing and Materials
1916 Race St.
Philadelphia, PA 19103
(215) 299-5400

ASCE
American Society of Civil Engineers
345 E. 47th St.
New York, NY 10017
(212) 705-7496

ASCC
American Society of Concrete Construction
426 S. Westgate
Addison, IL 60101
(708) 543-0870

ASHRAE
Americn Society of Heating, Refrigerating, and Air—Conditioning Engineers, Inc.
1791 Tullie Circle, NE
Atlanta, GA 30329
(404) 636-8400

ASHI
American Society of Home Inspectors
1735 N. Lynn St., Suite 950
Arlington, VA 22209-2022
(703) 524-2008

ASID
American Society of Interior Designers
1430 Broadway
New York, NY 10018
(212) 685-3480

ASME
American Society of Mechanical Engineers
345 E. 47th St.
New York, NY 10017
(212) 705-7800

ASSE
American Society of Sanitary Engineers
PO Box 40362
Bay Village, OH 44140
(216) 835-3040

ASA
American Subcontractors Association
1004 Duke St.
Alexandria, VA 22314
(703) 684-3450

AWS
Amerian Welding Society, Inc.
550 N.W. LeJeune Rd.
Miami, FL 33126
(305) 443-9353

APA
Architectural Precast Association
825 E. 64th St.
Indianapolis, IN 46220
(317) 251-1214

AWI
Architectural Woodwork Institute
2310 S. Walter Reed Dr.
Arlington, VA 22206
(703) 671-9100

AIA/NA
Asbestos Information Association/North America
1745 Jefferson Davis HWY., Suite 509
Arlington, VA 22202
(703) 979-1150

AI
Asphalt Institute
Asphalt Institute Building
College Park, MD 20740
(301) 779-9354

ABC
Associated Builders and Contractors, Inc.
729 15th Street, N.W.
Washington, DC 20005
(202) 637-8800

AGC
Associated General Contractors of America
1957 East St., NW
Washington, DC 20005
(202) 393-2040

SMACNA
Associated Sheet Metal Contractors, Inc.
3121 W. Hallandale Beach Blvd., Suite 114
Hallandale, FL 33009
(305) 961-0440

ABC
Association of Bituminous Contractors
2020 K St. NW, Suite 800
Washington, DC 20006

AWCI
Association of the Wall and Ceiling Industries International
1600 Cameron St.
Alexandria, VA 22314
(703) 684-2924

BIA
Brick Institute of America
11490 Commerce Park Dr., Suite 300
Reston, VA 22091
(703) 620-0010

BHMA
Builder's Hardware Manufacturers Association, Inc.
60 E. 42nd St., Rm. 511
New York, NY 10165
(212) 661-4261

BRB
Building Research Board
2101 Constitution Ave., NW
Washington, DC 20418

BSC
Building Systems Contractors
6710 Persimmon Tree Rd.
Bethesda, MD 20817
(303) 320-2505

CRA
California Redwood Association
405 Enfrente Dr., Suite 200
Novalto, CA 94949
(415) 382-0662

CRI
Carpet and Rug Institute
PO Box 2048
Dalton, GA 30722-2048
(404) 278-3176

CISCA
Ceilings and Interior Systems Construction Association
104 Wilmot, Suite 201
Deerfield, IL 60015
(708) 940-8800

CTI
Ceramic Tile Institute
700 N. Virgil Ave.
Los Angeles, CA 90029
(213) 660-1911

CRSI
Concrete Reinforcing Steel Institute
933 N. Plum Grove Rd.
Schaumburg, IL 60195
(708) 517-1200

CIEA
Construction Industry Employers Association
625 Ensminger Rd.
Tonawanda, NY 14150
(716) 875-4744

CCIC
Construction Consultants International Corp.
8133 Leesburg Pike
Vienna, VA 22180
(703) 734-2393

CCC
Construction Contractors Council
6120 Brandon Ave.
Springfield, VA 22150
(703) 644-2215

CC&M
Construction Costs and Management
6575 Edsall Rd.
Sterling, VA
(703) 354-7991

CEMC
Construction Education Management Corp.
8133 Leesburg Pike
Vienna, VA 22180
(703) 734-2399

CEI
Construction Environment Inc.
5655-D General Washington Dr.
Alexandria, VA 22312
(703) 750-0525

CIMA
Construction Industry Manufacturers Association
111 E. Wisconsin Ave., Suite 940
Milwaukee, WI 53202-4879
(414) 272-0943

CMC
Construction Management Collaborative
901 N. Pitt St.
Alexandria, VA 22314
(703) 836-3344

CM&C
Constsruction Managers and Consultants, Inc.
12107 Stallion Ct.
Woodbridge, VA 22192
(703) 690-1635

CSI
Construction Specifications Institute
601 Madison St.
Alexandria, VA 22314
(703) 684-0300

Corps of Engineers/U.S. Department of the Army
20 Massachuesetts Ave., NW
Washington, DC 20314
(202) 272-0660

CABO
Council of American Building Officials
5205 Leesburg Pike, Suite 708
Falls Church, VA 22041
(703) 931-4533

DHI
Door and Hardware Institute
7711 Old Springhouse Rd.
McLean, VA 22102-3474
(703) 556-3990

DIPRA
Ductile Iron Pipe Research Association
245 Riverchase Parkway E., Suite 0
Birmingham, AL 35244
(205) 988-9870

EPA
Environmental Protection Agency
401 M St., SW
Washington, DC 20460
(202) 382-2090

FPRS
Forest Products Research Society
2801 Marshall Ct.
Madison, WI 53705
(608) 231-1361

GBCA
General Building Contractors Association
36 S. 18th St.
PO Box 15959
Philadelphia, PA 19103
(215) 568-7015

HPMA
Hardwood Plywood Manufacturers Association
PO Box 2789
Reston, VA 22090
(703) 435-2900

IESNA
Illuminating Engineering Society of North America
345 E. 47th St.
New York, NY 10017
(212) 705-7926

ILIA
Indiana Limestone Institute of America
Stone City Bank Building, Suite 400
Bedford, IN 47421
(812) 275-4426

IDSA
Industrial Designers Society of America
1142 E. Walker St.
Great Falls, VA 22066
(703) 759-0100

IFI
Industrial Fasteners Institute
1505 E. Ohio Building
Cleveland, OH 44114
(216) 241-1482

IHEA
Industrial Heating Equipment Association
1901 N. Moore St.
Arlington, VA 22209
(703) 525-2513

ISEA
Industrial Safety Equipment Association
1901 N. Moore St.
Arlington, VA 22209
(703) 525-1695

IEEE
Institute of Electrical and Electronics Engineers
345 E. 47th St.
New York, NY 10017
(212) 705-7900

ICAA
Insulation Contractors Association of America
15819 Crabbs Branch Way
Rockville, MD 20855
(301) 590-0030

International Association of Bridge, Structural and Ornamental Iron Workers
1750 New York Ave., NW, Suite 400
Washington, DC 20006
(202) 383-4800

IALD
International Association of Lighting Designers
18 E. 16th St., Suite 208
New York, NY 10003
(212) 206-1281

IAPMO
International Association of Plumbing and Mechanical Officials
20001 Walnut Dr., S
Walnut, CA 91789
(714) 595-8449

IBB
International Brotherhood of Boilermakers, Iron Ship Builders, Blacksmiths, Forgers and Helpers
753 State Avenue, Suite 565
Kansas City, KS 66101
(913) 371-2640

IBEW
International Brotherhood of Electrical Workers
1125 15th St. NW
Washington, DC 20005
(202) 833-7000

IBPAT
International Brotherhood of Painters and Allied Trades
1750 New York Avenue, NW
Washington, DC 20005
(202) 637-0700

IMI
International Masonry Institute
823 15th St. NW, Suite 1001
Washington, DC 20005
(202) 783-3908

IRF
International Road Federation
525 School St. SW
Washington, DC 20024
(202) 554-2106

IUBAC
International Union of Bricklayers and Allied Craftsmen
Bowen Building
815 15th St. NW
Washington, DC 20005
(202) 783-3788

IUOE
International Union of Operating Engineers
1125 17th St., NW
Washington, DC 20036
(202) 429-9100

LIUNA
Laborers' International Union of North America
905 16th St., NW
Washington, DC 20006-1765
(202) 737-8320

Manufacturers Standardization Society of the Valve and Fittings Industry
127 Park St., NE
Vienna, VA 22180
(703) 281-6613

MFMA
Maple Flooring Manufacturers Association
60 Revere Dr., Suite 500
Northbrook, IL 60062
(708) 480-9080

MIA
Marble Institute of America
33505 State St.
Farmington, MI 48024
(313) 476-5558

MCAA
Mason Contractors Association of America
17W 601 14th St.
Oakbrook Terrace, IL 60181
(708) 620-6767

MCAA
Mechanical Contractors Association of America
1385 Piccard Dr.
Rockville, MD 20850
(301) 869-5800

MBMA
Metal Building Manufacturers Association
1230 Keith Building
Cleveland, OH 44115
(216) 241-7333

MLSFA
Metal Lath/Steel Framing Association
600 S. Federal, Suite 400
Chicago, IL 60605
(312) 922-6222

NAPA
National Asphalt Pavement Association
6811 Kenilworth Ave., Suite 620
PO Box 517
Riverdale, MD 20737
(301) 779-4880

NABD
National Association of Brick Distributors
212 S. Henry St.
Alexandria, VA 22314
(703) 549-2555

NADC
National Association of Demolition Contractors
4415 W. Harrison St.
Hillside, IL 60162
(708) 449-5959

NADCO
National Association of Development Companies
1730 Rhode Island Ave., NW
Washington, DC 20036
(202) 785-8484

NADO
National Association of Development Organizations
400 N. Capitol St.
Washington, DC 20001
(202) 624-7806

NADC
National Association of Dredging Contractors
1733 King St.
Alexandria, VA 22314
(703) 548-8300

NAEC
National Association of Elevator Contractors
4053 LaVista Rd., Suite 120
Tucker, GA 30084
(404) 496-1270

NAHB
National Association of Home Builders
15th and M St., NW
Washington, DC 20005
(202) 822-0200

NAPHCC
National Association of Plumbing, Heating, and Cooling Contractors
PO Box 6808
Falls Church, VA 22046
(703) 237-8100

NARSC
National Association of Reinforcing Steel Contractors
10382 Main St.
PO Box 225
Fairfax, VA 22030
(703) 591-1870

NAWIC
National Association of Women in Construction
327 S. Adams St.
Fort Worth, TX 76104
(817) 877-5551

NBMDA
National Building Material Distributors Association
1417 Lake Cook Rd.
Deerfield, IL 60015
(708) 945-7201

NCMA
National Concrete Masonry Association
PO Box 781
Herndon, VA 22070
(703) 435-4900

NCSBCS
National Conference of States on Building Codes and Standards
505 Huntmar Park Dr.
Herndon, VA 22070
(703) 437-0100

NCA
National Constructors Association
1730 M St. NW
Washington, DC 20036

NCIC
National Construction Industry Council
1919 Pennsylvania Ave. NW
Washington, DC 20006
(202) 887-1494

NCRP
National Council on Radiation Protection and Measurement
7910 Woodmont Ave., Suite 800
Bethesda, MD 20814
(301) 657-2625

NECA
National Electrical Contractors Association
7315 Wisconsin Ave.
13th Floor, West Building
Bethesda, MD 20814
(301) 657-3110

NEMA
National Electrical Manufacturers Association
2101 L St., NW, Suite 300
Washington, DC 20037
(202) 457-8400

NFPA
National Fire Protection Association
Batterymarch Park
Quincy, MA 02269
(617) 770-3000

NFPA
National Forest Products Association
1250 Connecticut Ave., NW, Suite 200
Washington, DC 20036
(202) 463-2700

NGA
National Glass Association
8200 Greensboro Dr., Suite 302
McLean, VA 22102
(703) 442-4890

NHRA
National Housing Rehabilitation Association
1726 18th St. NW
Washington, DC 20009
(202) 328-9171

NKCA
National Kitchen Cabinet Association
6711 Lee Highway
Arlington, VA 22205
(703) 237-7580

NLA
National Lime Association
3601 N. Fairfax Dr.
Arlington, VA 22201
(703) 243-5463

NLBMDA
National Lumber and Building Material Dealers Association
40 Ivy St., SE
Washington, DC 20003
(202) 547-2230

NOFMA
National Oak Flooring Manufacturers Association
PO Box 3009
Memphis, TN 38173-0009
(901) 526-5016

NPCA
National Paint and Coatings Association
1500 Rhode Island Ave., NW
Washington, DC 20005
(202) 462-6272

NPA
National Particleboard Association
18928 Premiere Ct.
Gaithersburg, MD 20879
(301) 690-0604

NPCA
National Precast Concrete Association
825 E. 64th St.
Indianapolis, IN 46220
(317) 253-0486

NRMCA
National Ready Mixed Concrete Association
900 Spring St.
Silver Spring, MD 20910
(301) 587-1400

NRCA
National Roofing Contractors Association
1 O'Hare Center
6250 River Rd.
Rosemont, IL 60018
(708) 299-9070

NSPE
National Society of Professional Engineers
1420 King St.
Alexandria, VA 22314
(703) 684-2800

NSA
National Stone Association
1415 Elliot Pl., NW
Washington, DC 20007
(202) 342-1100

NTMA
National Terrazzo and Mosaic Association
3166 Des Plaines Ave., Suite 132
Des Plaines, IL 60018
(708) 635-7744

NWWDA
National Wood Window and Door Association
1400 E. Touhey Ave.
Des Plaines, IL 60018
708) 299-5200

OPCMIA
Operative Plasterers' and Cement Masons' International Association of the United States and Canada
1125 17th St., NW, 6th Floor
Washington, DC 20036
(202) 393-6569

PDCA
Painting and Decorating Contractors of America
3913 Old Lee Hwy.
Fairfax, VA 22030
(703) 359-0826

PPI
Plastics Pipe Institute
355 Lexington Ave.
New York, NY 10017
(212) 351-5420

PHCIB
Plumbing-Heating-Cooling Information Bureau
303 E. Wacker Dr., Suite 711
Chicago, IL 60601
(312) 372-7331

PMI
Plumbing Manufacturers Institute
800 Roosevelt Rd., Building C, Suite 20
Glen Ellyn, IL 60137
(708) 858-9172

PTI
Post-Tensioning Institute
1717 W. Northern Ave., Suite 218
Phoenix, AZ 85021
(602) 870-7540

PCA
Portland Cement Association
5420 Old Orchard Rd.
Skokie, IL 60077
(708) 966-6200

PCI
Prestressed Concrete Institute
175 W. Jackson Blvd., Suite 1859
Chicago, IL 60604
(312) 786-0300

RFCI
Resilient Floor Covering Institute
966 Hungerford Dr., Suite 12B
Rockville, MD 20850
(301) 340-8580

SSFI
Scaffolding, Shoring, and Forming Institute, Inc.
1230 Keith Building
Cleveland, OH 44115
(216) 241-7333

SMA
Screen Manufacturers Association
655 Irving Park, Suite 201
Chicago, IL 60613-3198
(312) 525-2644

SWI
Sealant and Waterproofers Institute
3101 Broadway, Suite 300
Kansas City, MO 64111
(816) 561-8230

SIGMA
Sealed Insulating Glass Manufacturers Association
111 E. Wacker Dr., Suite 600
Chicago, IL 60601
(312) 644-6610

SMACNA
Sheet Metal and Air Conditioning Contractors National Association, Inc.
4201 Lafayette Center Dr.
Chantilly, VA 22021
(703) 803-2980

SMWIA
Sheet Metal Workers International Association
1750 New York Ave., NW
Washington, DC 20006
(202) 783-5880

SBCCI
Southern Building Code Congress International, Inc.
900 Montclair Rd.
Birmingham, AL 35213
(205) 591-1853

SDI
Steel Door Institute
712 Lakewood Center N
14600 Detroit Ave.
Cleveland, OH 44107
(216) 899-0010

SJI
Steel Joist Institute
1205 48th Ave., N, Suite A
Myrtle Beach, SC 29577
(803) 449-0487

SSPC
Steel Structures Painting Council
4400 5th Ave.
Pittsburgh, PA 15213
(412) 268-3327

SWI
Steel Window Institute
1230 Keith Building
Cleveland, OH 44115
(216) 241-7333

SBA
Systems Builders Association
PO Box 117
West Milton, OH 45383
(513) 698-4127

TCAA
Tile Contractors Association of America, Inc.
112 N. Alfred St.
Alexandria, VA 22314
(703) 836-5995

TCA
Tile Council of America
PO Box 2222
Princeton, NJ 08542
(609) 921-7050

UL
Underwriters' Laboratories, Inc.
333 Pfingsten Rd.
Northbrook, IL 60062
(708) 272-8800

USJ&P
United Association of Journeymen and Apprentices of the Plumbing and Pipe Fitting Industry of the United States and Canada
901 Massachusetts Ave., NW
Washington, DC 20001
(202) 628-5823

UBC
United Brotherhood of Carpenters and Joiners of America
101 Constitution Ave., NW
Washington, DC 20001
(202) 546-6206

US/OSHA
U.S. Department of Labor/Occupational Safety and Health Administration
200 Constitution Ave., NW
Washington, DC 20210
(202) 523-8148

U.S. Forest Products Laboratory
One Gifford Pinchot Dr.
Madison, WI 53705-2398
(608) 231-9200

UURWAW
United Union of Roofers, Waterproofers and Allied Workers
1125 17th St. NW, 5th Floor
Washington, DC 20036
(202) 638-3228

VMA
Valve Manufacturers Association of America
1050 17th St., NW Suite 701
Washington, DC 20036
(202) 331-8105

WMA
Wallcovering Manufacturers Association
355 Lexington Ave.
New York, NY 10017
(212) 661-4261

WWPA
Western Wood Products Association
Yeon Building
522 S.W. 5th Ave.
Portland, OR 97204
(503) 224-3930

WRI
Wire Reinforcement Institute
1760 Reston Pkwy.
Reston, VA 22090
(703) 709-9207